On Central Critical Values of the Degree Four L-functions for GSp(4): The Fundamental Lemma

of the
American Mathematical Society

Number 782

On Central Critical Values of the Degree Four L-functions for GSp(4): The Fundamental Lemma

Masaaki Furusawa
Joseph A. Shalika

July 2003 • Volume 164 • Number 782 (fourth of 5 numbers) • ISSN 0065-9266

American Mathematical Society
Providence, Rhode Island

2000 *Mathematics Subject Classification.*
Primary 11F46, 11F70, 11L05; Secondary 11F67, 11F72, 22E50, 22E55.

Library of Congress Cataloging-in-Publication Data

Furusawa, Masaaki.
 On central critical values of the degree four L-functions for GSp(4) : the fundamental lemma / Masaaki Furusawa, Joseph A. Shalika.
  p. cm. — (Memoirs of the American Mathematical Society, ISSN 0065-9266 ; no. 782)
 "Volume 164, number 782 (fourth of 5 numbers)."
 Includes bibliographical references.
 ISBN 0-8218-3328-6 (alk. paper)
 1. Siegel domains. 2. Modular groups. 3. Automorphic forms. I. Shalika, Joseph A., 1941– II. Title. III. Series.

QA3.A57 no. 782
[QA331]
510 s–dc21
[512′.7]
                                                                                        2003048027

Memoirs of the American Mathematical Society

 This journal is devoted entirely to research in pure and applied mathematics.

 Subscription information. The 2003 subscription begins with volume 161 and consists of six mailings, each containing one or more numbers. Subscription prices for 2003 are $555 list, $444 institutional member. A late charge of 10% of the subscription price will be imposed on orders received from nonmembers after January 1 of the subscription year. Subscribers outside the United States and India must pay a postage surcharge of $31; subscribers in India must pay a postage surcharge of $43. Expedited delivery to destinations in North America $35; elsewhere $130. Each number may be ordered separately; *please specify number* when ordering an individual number. For prices and titles of recently released numbers, see the New Publications sections of the *Notices of the American Mathematical Society*.
 Back number information. For back issues see the *AMS Catalog of Publications*.
 Subscriptions and orders should be addressed to the American Mathematical Society, P. O. Box 845904, Boston, MA 02284-5904, USA. *All orders must be accompanied by payment.* Other correspondence should be addressed to 201 Charles Street, Providence, RI 02904-2294, USA.
 Copying and reprinting. Individual readers of this publication, and nonprofit libraries acting for them, are permitted to make fair use of the material, such as to copy a chapter for use in teaching or research. Permission is granted to quote brief passages from this publication in reviews, provided the customary acknowledgment of the source is given.
 Republication, systematic copying, or multiple reproduction of any material in this publication is permitted only under license from the American Mathematical Society. Requests for such permission should be addressed to the Acquisitions Department, American Mathematical Society, 201 Charles Street, Providence, Rhode Island 02904-2294, USA. Requests can also be made by e-mail to reprint-permission@ams.org.

 Memoirs of the American Mathematical Society is published bimonthly (each volume consisting usually of more than one number) by the American Mathematical Society at 201 Charles Street, Providence, RI 02904-2294, USA. Periodicals postage paid at Providence, RI. Postmaster: Send address changes to Memoirs, American Mathematical Society, 201 Charles Street, Providence, RI 02904-2294, USA.

 © 2003 by the American Mathematical Society. All rights reserved.
 This publication is indexed in *Science Citation Index*®, *SciSearch*®, *Research Alert*®, *CompuMath Citation Index*®, *Current Contents*®/*Physical, Chemical & Earth Sciences*.
 Printed in the United States of America.
 ∞ The paper used in this book is acid-free and falls within the guidelines established to ensure permanence and durability.
 Visit the AMS home page at http://www.ams.org/

10 9 8 7 6 5 4 3 2 1   08 07 06 05 04 03

*Dedicated to Hervé Jacquet
on the occasion of his sixtieth birthday*

Contents

Abstract	viii
Introduction	ix
Chapter 1. Statement of Results	1
1.1. The Global Relative Trace Formulas	1
1.2. Notation	9
1.3. Main Results	10
Chapter 2. Gauss Sum, Kloosterman Sum and Salié Sum	14
2.1. Gauss sum	14
2.2. Kloosterman sum and Salié sum	17
2.3. The Davenport-Hasse relation	22
Chapter 3. Matrix Argument Kloosterman Sums	26
3.1. Kloosterman sum over n by n symmetric matrices	26
3.2. Evaluation of the split Kloosterman sum	28
3.3. Evaluation of the anisotropic Kloosterman sum	38
3.4. Evaluation of the quadratic Kloosterman sum	57
Chapter 4. Evaluation of the Novodvorsky Orbital Integral	81
4.1. Double coset decomposition	81
4.2. Relevant double cosets	85
4.3. Evaluation of the Novodvorsky orbital integral	91
Chapter 5. Evaluation of the Bessel Orbital Integral	104
5.1. Relevant double cosets	104
5.2. Evaluation of the Bessel orbital integral	110
Chapter 6. Evaluation of the Quadratic Orbital Integral	116
6.1. Double coset decomposition	116
6.2. Relevant double cosets	124
6.3. Evaluation of the quadratic orbital integral	129
Bibliography	138

Abstract

In this paper we prove two equalities of local Kloosterman integrals on $\mathrm{GSp}(4)$, the group of 4 by 4 symplectic similitude matrices. One is an equality between the Novodvorsky orbital integral and the Bessel orbital integral and the other one is an equality between the Bessel orbital integral and the quadratic orbital integral. We conjecture that both of Jacquet's relative trace formulas for the central critical values of the L-functions for $\mathrm{GL}(2)$ in [**J1**] and [**J2**], where Jacquet has given another proof of Waldspurger's result [**W2**], generalize to the ones for the central critical values of the degree four spinor L-functions for $\mathrm{GSp}(4)$. We believe that our approach will lead us to a proof and also a precise formulation of a conjecture of Böcherer [**B**] and its generalization. Support for this conjecture may be found in the important paper of Böcherer and Schulze-Pillot [**BSP**]. Also a numerical evidence has been recently given by Kohnen and Kuss [**KK**]. Our results serve as the fundamental lemmas for our conjectural relative trace formulas for the main relevant double cosets.

2000 *Mathematics Subject Classification.* Primary 11F46, 11F67, 11F70, 11L05; Secondary 111F72, 22E50, 22E55.

Key words and phrases. Relative trace formula, fundamental lemma, Kloosterman integral, relevant double coset, Bessel subgroup, Novodvorsky subgroup.

The first author was partially supported by Stipendium at the Max-Planck-Institut für Mathematik in Bonn, the Bell Fund at the Institute for Advanced Study, the Sumitomo Foundation and Grant-in-Aid for Scientific Research No. 10640028, Ministry of Education, Science, Sports and Culture of Japan.

Received by the editor August 1, 2000.

Introduction

It is of considerable significance to investigate the central values of automorphic L-functions as the numerous arithmetic applications of Waldspurger's results [**W1, W2**] on $\mathrm{GL}(2)$ have demonstrated over the years. One of the natural candidates for the group to be studied after $\mathrm{GL}(2)$ is $\mathrm{GSp}(4)$, the group of 4 by 4 symplectic similitude matrices.

Concerning the group $\mathrm{GSp}(4)$, Böcherer [**B**] has proclaimed the following remarkable conjecture.

BÖCHERER'S CONJECTURE . *Let Φ be a holomorphic Siegel cusp form of degree two and weight k with respect to $\mathrm{Sp}(4,\mathbb{Z})$, which is a Hecke eigenform. Let*

$$\Phi(Z) = \sum_{T>0} a(T,\Phi)\exp\left(2\pi\sqrt{-1}\operatorname{tr}(TZ)\right)$$

be its Fourier expansion. For a fundamental discriminant $-D$, i.e. a discriminant of an imaginary quadratic field $\mathbb{Q}(\sqrt{-D})$, let us define $B_D(\Phi)$ by

$$B_D(\Phi) = \sum_{\{T|\det T = D/4\}/\sim} \frac{a(T,\Phi)}{\epsilon(T)},$$

where $\sim$ denotes the equivalence relation defined by

$$T_1 \sim T_2 \iff \text{ there exists } \gamma \in \mathrm{SL}_2(\mathbb{Z}) \text{ such that } {}^t\gamma T_1 \gamma = T_2,$$

and $\epsilon(T) = \#\{\gamma \in \mathrm{SL}_2(\mathbb{Z}) \mid {}^t\gamma T\gamma = T\}$.

Then there exists a constant C_Φ which depends on Φ but not on D such that

$$L\left(\frac{1}{2}, \pi_\Phi \otimes \chi_D\right) = C_\Phi \cdot D^{-k+1} \cdot |B_D(\Phi)|^2,$$

where π_Φ denotes the automorphic representation of $\mathrm{GSp}_4(\mathbb{A}_\mathbb{Q})$ associated with Φ, χ_D denotes the quadratic character of $\mathbb{A}_\mathbb{Q}^\times$ associated with the quadratic extension $\mathbb{Q}(\sqrt{-D})$ in the sense of the classfield theory and $L(s, \pi_\Phi \otimes \chi_D)$ denotes the degree four spinor L-function twisted by χ_D, which is normalized so that its functional equation is with respect to $s \mapsto 1 - s$.

Böcherer [**B**] proved this assertion in the case of Eisenstein series and the Saito-Kurokawa lifting. Later in [**BSP**] he and Schulze-Pillot gave further support for this conjecture by proving the assertion in the case of the Yoshida lifting. We mention also that Kohnen and Kuss [**KK**] recently gave a numerical evidence for this conjecture for a cusp form which does not belong to the image of the Saito-Kurokawa lifting.

In this paper we prove two equalities of local Kloosterman integrals on $\mathrm{GSp}(4)$. The first main theorem (Theorem 1.13) is an equality between the Novodvorsky orbital integral and the Bessel orbital integral and the second main theorem (Theorem 1.14) is an equality between the Bessel orbital integral and the quadratic orbital integral, which is a different type of Novodvorsky orbital integral on $\mathrm{GSp}(4)$ over a quadratic extension.

Let us explain the relevance of these equalities with the central values of the degree four spinor L-functions for $\mathrm{GSp}(4)$. Jacquet has given another proof of Waldspurger's result [**W2**] by establishing two distinct relative trace formulas for $\mathrm{GL}(2)$ in [**J1**] and [**J2**] respectively. We conjecture that both of Jacquet's relative trace formulas generalize to $\mathrm{GSp}(4)$. Our conjectural trace formulas are described in Section 1.1 as (1.12) and (1.13). In order to establish a global relative trace formula, the first but essential step is to prove the "fundamental lemma." Our results serve as the "fundamental lemmas" for our conjectural global relative trace formulas for the main relevant double cosets. We believe that this approach will lead us to a proof and also a precise formulation of Böcherer's conjecture and its generalization.

This paper is organized as follows. In Chapter 1 we introduce the necessary notation and state the main results of the paper. In Chapter 2 we review some basic facts on the one variable classical Kloosterman sum. In Chapter 3 we introduce certain Kloosterman sums over n by n symmetric matrices and then we evaluate them explicitly in terms of the one variable Kloosterman sum when $n = 2$ in the cases which will be needed later for the evaluation of the orbital integrals. In particular, later in Chapter 3 we generalize the Davenport-Hasse relation to a matrix argument Kloosterman sum. This chapter is the heart of the matter of this paper. Also the results in this chapter may be of some independent interest. In Chapters 4, 5 and 6, we evaluate the Novodvorsky orbital integral, the Bessel orbital integral and the quadratic orbital integral respectively using the results obtained in Chapter 3.

The first author would like to express his gratitude to his colleagues at the Osaka City University for supporting him to take a research leave for the academic year 1999–2000 to work on this project. He also would like to thank the hospitality of the Max-Planck-Institut für Mathematik in Bonn and the Institute for Advanced Study where the part of this work was done.

Masaaki Furusawa
Joseph A. Shalika

CHAPTER 1

Statement of Results

1.1. The Global Relative Trace Formulas

In this section first we introduce some notation and then we describe our conjectural relative trace formulas.

Let F be a number field and let G be the group $\mathrm{GSp}(4)$, an algebraic group defined over F by

$$(1.1) \quad G = \left\{ g \in \mathrm{GL}(4) \mid {}^t g \begin{pmatrix} 0 & 1_2 \\ -1_2 & 0 \end{pmatrix} g = \lambda(g) \begin{pmatrix} 0 & 1_2 \\ -1_2 & 0 \end{pmatrix}, \lambda(g) \in \mathrm{GL}(1) \right\},$$

where ${}^t g$ denotes the transpose of g.

DEFINITION 1.1. Let D be a central simple quaternion algebra over F and let us denote the canonical involution of D by $x \mapsto \bar{x}$. Then we define G_D, the *quaternion similitude unitary group of degree* 2 *over* D, by

$$(1.2) \quad G_D = \left\{ g \in \mathrm{GL}(2, D) \mid g^* \begin{pmatrix} 0 & 1 \\ 1 & 0 \end{pmatrix} = \mu(g) \begin{pmatrix} 0 & 1 \\ 1 & 0 \end{pmatrix}, \mu(g) \in \mathrm{GL}(1) \right\},$$

where

$$(1.3) \quad g^* = \begin{pmatrix} \bar{a} & \bar{c} \\ \bar{b} & \bar{d} \end{pmatrix} \quad \text{for} \quad g = \begin{pmatrix} a & b \\ c & d \end{pmatrix}.$$

We regard G_D as an algebraic group over F. We remark that G_D is an inner form of $G = \mathrm{GSp}(4)$ and we have the following lemma.

LEMMA 1.2. *Suppose that* $D = \mathrm{Mat}_{2 \times 2}(F)$. *Then, in* $\mathrm{GL}_4(F)$, *we have*

$$\xi G_D \xi^{-1} = G \quad \text{where} \quad \xi = \begin{pmatrix} 1 & 0 & 0 & 0 \\ 0 & 1 & 0 & 0 \\ 0 & 0 & 0 & 1 \\ 0 & 0 & -1 & 0 \end{pmatrix}.$$

PROOF. First we note that for $a = \begin{pmatrix} \alpha & \beta \\ \gamma & \delta \end{pmatrix} \in D = \mathrm{Mat}_{2 \times 2}(F)$, we have

$$\bar{a} = \begin{pmatrix} \delta & -\beta \\ -\gamma & \alpha \end{pmatrix} = w_0^{-1}\, {}^t a w_0 \quad \text{where} \quad w_0 = \begin{pmatrix} 0 & 1 \\ -1 & 0 \end{pmatrix}.$$

Hence, for $g = \begin{pmatrix} a & b \\ c & d \end{pmatrix} \in \mathrm{Mat}_{2 \times 2}(D)$, we have

$$g^* = \begin{pmatrix} \bar{a} & \bar{c} \\ \bar{b} & \bar{d} \end{pmatrix} = \begin{pmatrix} w_0 & 0 \\ 0 & w_0 \end{pmatrix}^{-1} {}^t g \begin{pmatrix} w_0 & 0 \\ 0 & w_0 \end{pmatrix}.$$

Therefore
$$G_D = \left\{ g \in \mathrm{GL}(2, D) \mid g^* \begin{pmatrix} 0 & 1 \\ 1 & 0 \end{pmatrix} g = \mu(g) \begin{pmatrix} 0 & 1 \\ 1 & 0 \end{pmatrix}, \mu(g) \in \mathrm{GL}(1, F) \right\}$$
$$= \left\{ g \in \mathrm{GL}(4, F) \mid {}^t g \begin{pmatrix} 0 & w_0 \\ w_0 & 0 \end{pmatrix} g = \mu(g) \begin{pmatrix} 0 & w_0 \\ w_0 & 0 \end{pmatrix}, \mu(g) \in \mathrm{GL}(1, F) \right\}.$$

Our assertion follows since
$$\xi \begin{pmatrix} 0 & w_0 \\ w_0 & 0 \end{pmatrix} \xi^{-1} = \begin{pmatrix} 0 & 1_2 \\ -1_2 & 0 \end{pmatrix} \quad \text{and} \quad {}^t(\xi g \xi^{-1}) = \xi \, {}^t g \, \xi^{-1}.$$

$\square$

Let $d \in F^\times$ such that $d \notin (F^\times)^2$. Let us take $\eta = \sqrt{d}$ and let E be the corresponding quadratic extension of F, $E = F(\eta)$.

Then for each $\epsilon \in F^\times$, let

(1.4)
$$D_\epsilon = \left\{ \begin{pmatrix} a & b\epsilon \\ b^\sigma & a^\sigma \end{pmatrix} \mid a, b \in E \right\} \subset \mathrm{Mat}_{2 \times 2}(E),$$

where $\alpha \mapsto \alpha^\sigma$ denotes the unique non-trivial element in $\mathrm{Gal}(E/F)$. For the sake of completeness, here we give a proof of the following well known statement.

LEMMA 1.3. *Let $X(E : F)$ denote the isomorphism classes of central simple quaternion algebras over F containing a field isomorphic to E. Then the mapping*
$$F^\times \ni \epsilon \mapsto D_\epsilon$$
induces a bijection between $F^\times / N_{E/F}(E^\times)$ and $X(E : F)$.

Under this correspondence, we have in particular
$$D_1 \simeq \mathrm{Mat}_{2 \times 2}(F).$$

PROOF. First we note that since
$$\begin{pmatrix} a & b\epsilon \cdot N_{E/F}(x) \\ b^\sigma & a^\sigma \end{pmatrix} = \begin{pmatrix} x & 0 \\ 0 & 1 \end{pmatrix} \begin{pmatrix} a & bx^\sigma \epsilon \\ b^\sigma x & a^\sigma \end{pmatrix} \begin{pmatrix} x & 0 \\ 0 & 1 \end{pmatrix}^{-1}$$

for $x \in E^\times$, we have $D_\epsilon \simeq D_{\epsilon \cdot N_{E/F}(x)}$ as F-algebras. We also note that since
$$\begin{pmatrix} 1 & 1 \\ \eta & -\eta \end{pmatrix} \begin{pmatrix} a & b \\ b^\sigma & a^\sigma \end{pmatrix} \begin{pmatrix} 1 & 1 \\ \eta & -\eta \end{pmatrix}^{-1} = \begin{pmatrix} \frac{1}{2}(a + a^\sigma + b + b^\sigma) & \frac{1}{2\eta}(a - a^\sigma + b^\sigma - b) \\ \frac{\eta}{2}(a - a^\sigma + b - b^\sigma) & \frac{1}{2}(a + a^\sigma - b - b^\sigma) \end{pmatrix},$$

we have $D_1 \simeq \mathrm{Mat}_{2 \times 2}(F)$ as F-algebras.

Conversely let D be a central simple quaternion algebra over F containing E. Then we remark the following fact. Let $\iota : E \hookrightarrow D$ be another embedding of E into D over F. Then ι can be extended to an inner automorphism of D (see, e.g. Jacobson [**Ja**, Theorem 4.9]). In other words, the embedding of E into D over F is unique up to an inner automorphism of D.

Let t denote the reduced trace mapping of D over F. Then we recall that $t(xy) = t(yx)$ for $x, y \in D$. Let us define $E^\perp \subset D$ by
$$E^\perp = \{ x \in D \mid t(ay) = 0 \text{ for any } a \in E \}.$$

Since $D \times D \ni (x, y) \mapsto t(xy) \in F$ is a non-degenerate symmetric F-bilinear form, $E^\perp$ is a two dimensional vector space over F and $E^\perp$ is closed under both left and right multiplications by elements of E. Let $a \in E \cap E^\perp$. Then
$$0 = t(a^\sigma a) = t(N_{E/F}(a)) = 2 \cdot N_{E/F}(a).$$

Hence $a = 0$ and $E \cap E^\perp = \{0\}$. Let us take $\alpha \neq 0$ in $E^\perp$. Then $E^\perp = \alpha E$. Let us define an F-algebra homomorphism $\nu : D \ni x \mapsto \begin{pmatrix} a & b \\ c & d \end{pmatrix} \in \mathrm{Mat}_{2 \times 2}(E)$ by

$$(xe_1, xe_2) = (e_1, e_2) \begin{pmatrix} a & b \\ c & d \end{pmatrix},$$

where $e_1 = 1$ and $e_2 = \alpha$. Since D is simple, ν is injective.

For $a \in E$, we have $a\alpha = \alpha a'$ for some $a' \in E$. Then $a \mapsto a'$ is an F-algebra homomorphism of E. If this is trivial, then E is in the center of D, This is a contradiction since D is an F-central algebra. Hence $a' = a^\sigma$ where σ is the unique non-trivial element of $\mathrm{Gal}(E/F)$.

Since $\mathrm{tr}(\nu(\alpha)) = t(\alpha) = 0$ and $\alpha e_1 = e_2$, we may write $\nu(\alpha) = \begin{pmatrix} 0 & \epsilon \\ 1 & 0 \end{pmatrix}$ for some $\epsilon \in E^\times$.

Let us write $\alpha^2 = a + b\alpha$ where $a, b \in E$. Then by applying ν to both sides, we have

$$\begin{pmatrix} \epsilon & 0 \\ 0 & \epsilon \end{pmatrix} = \begin{pmatrix} a & b\epsilon \\ b^\sigma & a^\sigma \end{pmatrix}.$$

Hence $b = 0$ and $\epsilon = a = a^\sigma$. Thus $\epsilon \in F^\times$. Therefore we have shown that ν is an F-algebra isomorphism between D and D_ϵ.

Here we note that for $c \in E^\times$, $(\alpha c)^2 = \alpha c \alpha c = \alpha \alpha c^\sigma c = \epsilon \cdot N_{E/F}(c)$. Thus

$$(1.5) \qquad \epsilon \cdot N_{E/F}(E^\times) = \{y^2 \mid y \in E^\perp, y \neq 0\}.$$

Since the right hand side of (1.5) is intrinsically defined, ϵ is uniquely determined up to a multiplication by an element of $N_{E/F}(E^\times)$. □

We note that for $x \in \begin{pmatrix} a & b\epsilon \\ b^\sigma & a^\sigma \end{pmatrix} \in D_\epsilon$, the canonical involution $x \mapsto \bar{x}$ of D_ϵ is given by $\bar{x} = \begin{pmatrix} a^\sigma & -b\epsilon \\ -b^\sigma & a \end{pmatrix}$ and we have

$$n(x) = \begin{pmatrix} aa^\sigma - \epsilon bb^\sigma & 0 \\ 0 & aa^\sigma - \epsilon bb^\sigma \end{pmatrix}, \quad t(x) = \begin{pmatrix} a + a^\sigma & 0 \\ 0 & a + a^\sigma \end{pmatrix},$$

where $n(x)$ (resp. $t(x)$) denotes the reduced norm (resp. trace) of x.

From now on let us simply write G_ϵ for G_{D_ϵ} where $\epsilon \in F^\times$. We remark that, in particular, G_1 is another realization of $G = \mathrm{GSp}(4)$ and we have

$$(1.6) \qquad \alpha G_1 \alpha^{-1} = G \quad \text{where} \quad \alpha = \begin{pmatrix} 1 & 0 & 0 & 0 \\ 0 & 1 & 0 & 0 \\ 0 & 0 & 0 & 1 \\ 0 & 0 & -1 & 0 \end{pmatrix} \begin{pmatrix} 1 & 1 & 0 & 0 \\ \eta & -\eta & 0 & 0 \\ 0 & 0 & 1 & 1 \\ 0 & 0 & \eta & -\eta \end{pmatrix}.$$

DEFINITION 1.4. Let us define the *upper Novodvorsky subgroup* H of G by

$$H = \left\{ \begin{pmatrix} a & 0 & 0 & 0 \\ 0 & b & 0 & 0 \\ 0 & 0 & b & 0 \\ 0 & 0 & 0 & a \end{pmatrix} \begin{pmatrix} 1_2 & X \\ 0 & 1_2 \end{pmatrix} \;\middle|\; a, b \in \mathrm{GL}(1), X \in \mathrm{Sym}^2 \right\}$$

and the *lower Novodvorsky subgroup* $\bar{H}$ of G by $\bar{H} = \{{}^t h \mid h \in H\}$, i.e.

$$\bar{H} = \left\{ \begin{pmatrix} a & 0 & 0 & 0 \\ 0 & b & 0 & 0 \\ 0 & 0 & b & 0 \\ 0 & 0 & 0 & a \end{pmatrix} \begin{pmatrix} 1_2 & 0 \\ Y & 1_2 \end{pmatrix} \mid a, b \in \mathrm{GL}(1), Y \in \mathrm{Sym}^2 \right\}$$

where Sym^2 denotes the set of 2 by 2 symmetric matrices.

DEFINITION 1.5. For $\epsilon \in F^\times$, let us define the *upper Bessel subgroup* R_ϵ of G_ϵ by

$$R_\epsilon = \left\{ \begin{pmatrix} a & 0 & 0 & 0 \\ 0 & a^\sigma & 0 & 0 \\ 0 & 0 & a & 0 \\ 0 & 0 & 0 & a^\sigma \end{pmatrix} \begin{pmatrix} 1_2 & X \\ 0 & 1_2 \end{pmatrix} \mid a \in E^\times, X \in D_\epsilon^- \right\},$$

where $D_\epsilon^- = \{\alpha \in D_\epsilon \mid t(\alpha) = 0\}$. The *lower Bessel subgroup* $\bar{R}_\epsilon$ of G_ϵ is defined by $\bar{R}_\epsilon = \{{}^t r \mid r \in R_\epsilon\}$, i.e.

$$\bar{R}_\epsilon = \left\{ \begin{pmatrix} a & 0 & 0 & 0 \\ 0 & a^\sigma & 0 & 0 \\ 0 & 0 & a & 0 \\ 0 & 0 & 0 & a^\sigma \end{pmatrix} \begin{pmatrix} 1_2 & 0 \\ Y & 1_2 \end{pmatrix} \mid a \in E^\times, Y \in D_\epsilon^- \right\}.$$

Now let us describe our conjectural relative trace formulas. Let $\mathbb{A}_F$ denote the adele ring of F and let ψ be a non-trivial additive character of $\mathbb{A}_F/F$. We denote by χ_E the quadratic character of $\mathbb{A}_F^\times$ corresponding to the quadratic extension E over F in the sense of the classfield theory. We define an additive character ψ_E of $\mathbb{A}_E/E$ by

$$\psi_E(x) = \psi\left[\mathrm{tr}_{E/F}(x)\right].$$

Let Ω be a character of $\mathbb{A}_E^\times/E^\times$ and let ω denote its restriction to $\mathbb{A}_F^\times/F^\times$. Let ω' be the lift of ω to $\mathbb{A}_E^\times/E^\times$, i.e.

$$\omega'(x) = \omega(xx^\sigma) = (\omega \circ N_{E/F})(x) \quad \text{for} \quad x \in \mathbb{A}_E^\times$$

and let $\Omega' = \Omega^{-1} \omega'$, i.e.

$$\Omega'(x) = \Omega(x^\sigma) \quad \text{for} \quad x \in \mathbb{A}_E^\times.$$

DEFINITION 1.6. We define a character θ of $H(\mathbb{A}_F)$ by

$$(1.7) \quad \theta\left[\begin{pmatrix} a & 0 & 0 & 0 \\ 0 & b & 0 & 0 \\ 0 & 0 & b & 0 \\ 0 & 0 & 0 & a \end{pmatrix} \begin{pmatrix} 1_2 & X \\ 0 & 1_2 \end{pmatrix} \right] = \chi_E(ab) \cdot \psi\left[\mathrm{tr}\left(\begin{pmatrix} 0 & 1 \\ 1 & 0 \end{pmatrix} X\right)\right].$$

By abuse of notation, we denote by ψ, a character of $\bar{H}(\mathbb{A}_F)$ defined by

$$(1.8) \quad \psi\left[\begin{pmatrix} a & 0 & 0 & 0 \\ 0 & b & 0 & 0 \\ 0 & 0 & b & 0 \\ 0 & 0 & 0 & a \end{pmatrix} \begin{pmatrix} 1_2 & 0 \\ Y & 1_2 \end{pmatrix} \right] = \psi\left[\mathrm{tr}\left(\begin{pmatrix} 0 & 1 \\ 1 & 0 \end{pmatrix} Y\right)\right].$$

Again by abuse of notation we denote by Ω, a character of $H(\mathbb{A}_E)$ defined by

$$(1.9) \quad \Omega\left[\begin{pmatrix} \alpha & 0 & 0 & 0 \\ 0 & \beta & 0 & 0 \\ 0 & 0 & \beta & 0 \\ 0 & 0 & 0 & \alpha \end{pmatrix}\begin{pmatrix} 1_2 & Z \\ 0 & 1_2 \end{pmatrix}\right] = \Omega(\alpha\beta) \cdot \psi_E\left[\operatorname{tr}\left(\begin{pmatrix} 0 & 1 \\ 1 & 0 \end{pmatrix} Z\right)\right].$$

DEFINITION 1.7. We define a character τ of $R_\epsilon(\mathbb{A}_F)$ by

$$(1.10) \quad \tau\left[\begin{pmatrix} a & 0 & 0 & 0 \\ 0 & a^\sigma & 0 & 0 \\ 0 & 0 & a & 0 \\ 0 & 0 & 0 & a^\sigma \end{pmatrix}\begin{pmatrix} 1_2 & x \\ 0 & 1_2 \end{pmatrix}\right] = \Omega(a) \cdot \psi\left[\operatorname{tr}\left(\begin{pmatrix} -\eta & 0 \\ 0 & \eta \end{pmatrix} x\right)\right].$$

We also define a character ξ of $\bar{R}_\epsilon(\mathbb{A}_F)$ by

$$(1.11) \quad \xi\left[\begin{pmatrix} a & 0 & 0 & 0 \\ 0 & a^\sigma & 0 & 0 \\ 0 & 0 & a & 0 \\ 0 & 0 & 0 & a^\sigma \end{pmatrix}\begin{pmatrix} 1_2 & 0 \\ y & 1_2 \end{pmatrix}\right] = \Omega'(a) \cdot \psi\left[\operatorname{tr}\left(\begin{pmatrix} -\eta^{-1} & 0 \\ 0 & \eta^{-1} \end{pmatrix} y\right)\right].$$

Then our conjectural relative trace formulas are as follows.

CONJECTURE 1.8. *Suppose that the character Ω of $\mathbb{A}_E^\times$ is trivial. Let Φ be a smooth function with compact support on $G(\mathbb{A}_F)/Z(\mathbb{A}_F)$, where Z denotes the center of G. Define as usual a kernel function K_Φ by*

$$K_\Phi(g_1, g_2) = \sum_{\gamma \in Z(F)\backslash G(F)} \Phi\left(g_1^{-1} \gamma g_2\right).$$

Similarly for each ϵ in the set of representatives of $F^\times/N_{E/F}(E^\times)$, let Φ_ϵ be a smooth function with compact support on $G_\epsilon(\mathbb{A}_F)/Z_\epsilon(\mathbb{A}_F)$, where Z_ϵ denotes the center of G_ϵ. Then a kernel function K_{Φ_ϵ} is defined accordingly. We assume that $\Phi_\epsilon \equiv 0$ for almost all ϵ.

Then we conjecture that the relative trace formula

$$(1.12) \quad \int_{\bar{H}(\mathbb{A}_F)/\bar{H}(F)Z(\mathbb{A}_F)} \int_{H(\mathbb{A}_F)/H(F)Z(\mathbb{A}_F)} K_\Phi\left(\bar{h}^{-1}, h\right) \psi(\bar{h}) \theta(h) d\bar{h} \, dh$$
$$= \sum_\epsilon \int_{\bar{R}_\epsilon(\mathbb{A}_F)/\bar{R}_\epsilon(F)Z_\epsilon(\mathbb{A}_F)} \int_{R_\epsilon(\mathbb{A}_F)/R_\epsilon(F)Z_\epsilon(\mathbb{A}_F)} K_{\Phi_\epsilon}\left(\bar{r}^{-1}, r\right) \xi(\bar{r}) \tau(r) d\bar{r} \, dr$$

holds, where ϵ runs over the set of representatives of $F^\times/N_{E/F}(E^\times)$, and, Φ and Φ_ϵ are matching functions. (Actually one needs to interpret both sides of the above equality in a suitably regularized sense.)

CONJECTURE 1.9. *Let Ψ be a smooth function on $G(\mathbb{A}_E)$ such that*

$$\Psi(zg) = \omega'(z)^{-1} \Psi(g) \quad \text{for} \quad z \in Z(\mathbb{A}_E)$$

and the support of Ψ is compact modulo $Z(\mathbb{A}_E)$. Let K_Ψ denote the kernel function defined by Ψ.

On the other hand for each ϵ in the set of representatives of $F^\times/N_{E/F}(E^\times)$, let Ψ_ϵ be a smooth function on $G_\epsilon(\mathbb{A}_F)$ such that

$$\Psi_\epsilon(zg) = \omega(z)^{-1} \Psi_\epsilon(g) \quad \text{for} \quad z \in Z_\epsilon(\mathbb{A}_E)$$

and the support of Ψ_ϵ is compact modulo $Z_\epsilon(\mathbb{A}_F)$. Let K_{Ψ_ϵ} denote the corresponding kernel function and we assume that $\Psi_\epsilon \equiv 0$ for almost all ϵ.

Then we conjecture that the relative trace formula

$$(1.13) \quad \int_{G(\mathbb{A}_F)/G(F)Z(\mathbb{A}_F)} \int_{H(\mathbb{A}_E)/H(E)Z(\mathbb{A}_E)} K_\Psi\left(g^{-1}, h\right) \Omega\left(h\right) \left(\omega \chi_E\right) \left(\lambda\left(g\right)\right) dg\, dh$$

$$= \sum_\epsilon \int_{\bar{R}_\epsilon(\mathbb{A}_F)/\bar{R}_\epsilon(F)Z_\epsilon(\mathbb{A}_F)} \int_{R_\epsilon(\mathbb{A}_F)/R_\epsilon(F)Z_\epsilon(\mathbb{A}_F)} K_{\Psi_\epsilon}\left(\bar{r}^{-1}, r\right) \xi\left(\bar{r}\right) \tau\left(r\right) d\bar{r}\, dr$$

holds, where ϵ runs over the set of representatives of $F^\times/N_{E/F}\left(E^\times\right)$, and, Ψ and Ψ_ϵ are matching functions.

The relative trace formula (1.12) (resp. (1.13)) is exactly a generalization to $\mathrm{GSp}\left(4\right)$ of Jacquet's relative trace formula for $\mathrm{GL}\left(2\right)$ in [**J1**] (resp. [**J2**]). Jacquet has given another proof of Waldspurger's result in [**W2**] based on these relative trace formulas. By extracting the cuspidal parts of the both sides of the above equalities, we should be able to draw conclusions similar to his for $\mathrm{GSp}\left(4\right)$.

Let us discuss the conclusions expected. First we recall Novodvorsky's integral representation [**N**] for the degree four spinor L-function for $\mathrm{GSp}\left(4\right)$ (see also Bump [**Bu**] for the global unfolding). Let π be an irreducible cuspidal automorphic representation of $G\left(\mathbb{A}_F\right) = \mathrm{GSp}_4\left(\mathbb{A}_F\right)$ and let V_π be its space of cusp forms. For $\phi \in V_\pi$ and a character ρ of $\mathbb{A}_F^\times/F^\times$, we consider the integral

$$(1.14) \quad Z(s, \phi, \rho) = \int_{\mathbb{A}_F^\times/F^\times} \int_{(\mathbb{A}_F/F)^3} \phi \left[\begin{pmatrix} 1 & x_2 & x_4 & 0 \\ 0 & 1 & 0 & 0 \\ 0 & 0 & 1 & 0 \\ 0 & z & -x_2 & 1 \end{pmatrix} \begin{pmatrix} y & 0 & 0 & 0 \\ 0 & y & 0 & 0 \\ 0 & 0 & 1 & 0 \\ 0 & 0 & 0 & 1 \end{pmatrix}\right]$$
$$\cdot \psi\left(-2x_2\right) \rho\left(y\right) |y|^{s-\frac{1}{2}} dx\, d^\times y.$$

First we note that this integral converges absolutely for *any* $s \in \mathbb{C}$. Then we note that $Z(s, \phi, \rho)$ unfolds to

$$Z(s, \phi, \rho) = \int_{\mathbb{A}_F^\times} \int_{\mathbb{A}_F} W_\phi \left[\begin{pmatrix} y & 0 & 0 & 0 \\ 0 & y & 0 & 0 \\ 0 & 0 & 1 & 0 \\ 0 & xy & 0 & 1 \end{pmatrix}\right] \rho\left(y\right) |y|^{s-\frac{1}{2}} dx\, d^\times y$$

where

$$W_\phi(g) = \int_{(\mathbb{A}_F/F)^4} \phi \left[\begin{pmatrix} 1 & x & 0 & 0 \\ 0 & 1 & 0 & 0 \\ 0 & 0 & 1 & 0 \\ 0 & 0 & -x & 1 \end{pmatrix} \begin{pmatrix} 1 & 0 & a & b \\ 0 & 1 & b & c \\ 0 & 0 & 1 & 0 \\ 0 & 0 & 0 & 1 \end{pmatrix} g\right]$$
$$\cdot \psi\left[-2\left(x+c\right)\right] dx\, da\, db\, dc,$$

i.e. the Whittaker Fourier coefficient of ϕ. Suppose that π is *globally generic*, i.e. the Whittaker Fourier coefficient does not vanish identically on V_π. Then we may choose $\phi \in V_\pi$ so that

$$Z(s, \phi, \rho) = L(s, \pi \otimes \rho)$$

where the right hand side denotes the degree four spinor L-function for π twisted by ρ. We refer the reader to Takloo-Bighash [**T**] for the determination of the non-archimedean local factors for $\mathrm{GSp}\left(4\right)$ and to Soudry [**S1, S2**] for the general local theory for $\mathrm{SO}_{2\ell+1} \times \mathrm{GL}_n$.

Here we remark that when we define ϕ_1 on $G(\mathbb{A}_F)$ by

$$\phi_1(g) = \phi\left(w_2^{-1} g w_2\right) \quad \text{where} \quad w_2 = \begin{pmatrix} 1 & 0 & 0 & 0 \\ 0 & 0 & 0 & 1 \\ 0 & 0 & 1 & 0 \\ 0 & -1 & 0 & 0 \end{pmatrix},$$

we have

$$(1.15) \quad Z(s,\phi,\rho) = \int_{\mathbb{A}_F^\times/F^\times} \int_{(\mathbb{A}_F/F)^3} \phi_1 \left[\begin{pmatrix} 1 & 0 & a & b \\ 0 & 1 & b & c \\ 0 & 0 & 1 & 0 \\ 0 & 0 & 0 & 1 \end{pmatrix} \begin{pmatrix} y & 0 & 0 & 0 \\ 0 & 1 & 0 & 0 \\ 0 & 0 & 1 & 0 \\ 0 & 0 & 0 & y \end{pmatrix}\right]$$
$$\cdot \psi\left[\mathrm{tr}\left(\begin{pmatrix} 0 & 1 \\ 1 & 0 \end{pmatrix} \begin{pmatrix} a & b \\ b & c \end{pmatrix}\right)\right] \rho(y) |y|^{s-\frac{1}{2}} da\, db\, dc\, d^\times y.$$

Thus the equality (1.15) reveals the relationship between the Novodvorsky period and the central critical value $L\left(\frac{1}{2}, \pi \otimes \rho\right)$.

In light of these observations, first we conjecture that the following conclusion should come out from the relative trace formula (1.12).

CONJECTURE 1.10. *Let π be a globally generic irreducible cuspidal automorphic representation of $G(\mathbb{A}_F)$ with the trivial central character.*

Then we have

$$L\left(\frac{1}{2}, \pi\right) L\left(\frac{1}{2}, \pi \otimes \chi_E\right) \neq 0$$

if and only if there exists a triple $(\epsilon, \pi_\epsilon, \varphi_\epsilon)$, where $\epsilon \in F^\times$, π_ϵ an irreducible cuspidal automorphic representation of $G_\epsilon(\mathbb{A}_F)$, corresponding to π in the functorial sense, and φ_ϵ a cusp form in the space of π_ϵ such that

$$(1.16) \quad \int_{Z_\epsilon(\mathbb{A}_F) R_\epsilon(F) \backslash R_\epsilon(\mathbb{A}_F)} \varphi_\epsilon(r) \tau(r) dr \neq 0$$

where we take Ω to be trivial in (1.10).

We first remark that when $F = \mathbb{Q}$, $E = \mathbb{Q}\left(\sqrt{-D}\right)$, $\epsilon = 1$ and $\pi_\epsilon = \pi_\Phi$, the irreducible cuspidal representation associated with the holomorphic Siegel cuspidal eigenform Φ, the period integral (1.16) is equal to the sum of Fourier coefficients $B_D(\Phi)$ in Böcherer's conjecture, up to some elementary factors. We also remark on the genericity condition in the statement of Conjecture 1.10. Let σ be an irreducible cuspidal automorphic representation of $G(\mathbb{A}_F)$ and let $L\{\sigma\}$ denote the global L-packet containing σ. Then the global L-packet conjecture asserts that $L\{\sigma\}$ contains a unique globally generic element when σ is tempered. In other words when σ is tempered, we may utilize the globally generic cusp form belonging to the space of the generic element in $L\{\sigma\}$ to express the L-value for σ, *regardless* whether σ itself is generic or not. It is known that when $\sigma = \pi_\Phi$, the irreducible cuspidal representation associated with a holomorphic Siegel eigenform Φ, σ is *not* generic. On the other hand Piatetski-Shapiro and Soudry [**PS**] have shown that the non-vanishing of the period integral (1.16) indeed implies the existence of the near equivalent generic cuspidal representation when σ is not the Saito-Kurokawa lifting.

Now let us state the conjecture which should follow from the relative trace formula (1.13).

CONJECTURE 1.11. *Let π be an irreducible cuspidal automorphic representation of $G(\mathbb{A}_F)$. Let Ω be a character of $\mathbb{A}_E^\times/E^\times$ and let us denote by $\pi(\Omega)$, the automorphic representation of $\mathrm{GL}(2, \mathbb{A}_F)$ associated with Ω. Assume that the central character of π is equal to the inverse of $\omega = \Omega\mid_{\mathbb{A}_F^\times}$.*

Then we have

$$L\left(\frac{1}{2}, \pi \otimes \pi(\Omega)\right) \neq 0$$

if and only if there exists a triple $(\epsilon, \pi_\epsilon, \varphi_\epsilon)$, where $\epsilon \in F^\times$, π_ϵ an irreducible cuspidal automorphic representation of $G_\epsilon(\mathbb{A}_F)$, corresponding to π in the functorial sense, and φ_ϵ a cusp form in the space of π_ϵ such that

$$\int_{Z_\epsilon(\mathbb{A}_F) R_\epsilon(F) \backslash R_\epsilon(\mathbb{A}_F)} \varphi_\epsilon(r) \tau(r) \, dr \neq 0.$$

In particular when π denotes the cuspidal representation of $G(\mathbb{A}_\mathbb{Q})$ corresponding to a holomorphic Siegel eigen cusp form Φ, by looking at the Fourier expansion of Φ, it should follow as a corollary of Conjecture 1.11 that there exist an imaginary quadratic field E and a *finite order* Hecke character Ω on $\mathbb{A}_E^\times$ such that

$$L\left(\frac{1}{2}, \pi \otimes \pi(\Omega)\right) \neq 0.$$

We also remark that when we denote the base change of π to $G(\mathbb{A}_E)$ by Π, we have $L(s, \pi \otimes \pi(\Omega)) = L(s, \Pi \otimes \Omega)$. Thus we are lead to speculate that the $G(F)$ distinguishedness characterizes the generic cuspidal representations of $G(\mathbb{A}_E)$ which are in the image of the quadratic base change lifting from F. Here we note that when G' denotes the intersection of $G(E)$ and the similitude unitary group with respect to the Hermitian matrix $\begin{pmatrix} 0 & \eta 1_2 \\ -\eta 1_2 & 0 \end{pmatrix}$, i.e.

$$G' = \left\{ g \in G(E) \mid {}^t g^\sigma \begin{pmatrix} 0 & \eta 1_2 \\ -\eta 1_2 & 0 \end{pmatrix} g = \lambda'(g) \begin{pmatrix} 0 & \eta 1_2 \\ -\eta 1_2 & 0 \end{pmatrix}, \lambda'(g) \in \mathrm{GL}_1(F) \right\},$$

we have

$$G' = Z(F) G(F).$$

We refer the interested reader to Gurevich [**Gur**] and Jacquet, Lapid and Rogawski [**JLR**, Introduction].

The reader should compare these assertions of our conjectures with the paper of Gross and Prasad [**GP1**] where a general conjecture of this type has been formulated.

Moreover, the detailed analysis should yield an identity that expresses the central critical value $L\left(\frac{1}{2}, \pi\right) L\left(\frac{1}{2}, \pi \otimes \chi_E\right)$ (resp. $L\left(\frac{1}{2}, \pi \otimes \pi(\Omega)\right)$) as the square norm of one of these period integrals multiplied by a constant C_π (resp. C'_π), where C_π depends only on π and *not* on the quadratic extension E, and, C'_π depends only on π, E and *not* on the character Ω of $\mathbb{A}_E^\times$. One expects also a simple explicit formula for the constants C_π and C'_π. We note that Böcherer did not conjecture what the constant of proportionality C_Φ in his conjecture was in [**B**]. Further elaboration of these ideas would depend heavily on the (local) conjectures of Gross and Prasad in [**GP2**].

We mention that Guo [**Gu1**] developed Jacquet's central idea in [**Gu1**] where he proved in the case of GL(2) corresponding to Jacquet's relative trace formula in

[**J1**], the positivity of the central critical value. Recently Chen [**C**], and, Chen and Jacquet [**CJ**] proved the positivity for GL(2) in the case corresponding to Jacquet's relative trace formula in [**J2**]. Guo also proved in [**Gu2**] the analogue for GL(2n) of Jacquet's fundamental lemma for GL(2). With more work it is likely that both Guo's approach and ours will lead to a proof of the positivity of the central values for GSp(4).

1.2. Notation

For the rest of the paper, we move to the local situation. Throughout the paper we fix the notation as follows.

Let F denote a non-archimedean local field whose residual characteristic is not equal to two, $\mathcal{O}_F$ be its ring of integers, and ϖ be a prime element of F. Let q denote the cardinality of the residue field $\mathcal{O}_F/\varpi\mathcal{O}_F$ and $|\cdot|$ denote the normalized absolute value on F, so that $|\varpi| = q^{-1}$. Let us fix an additive character ψ of F of exponent zero, i.e. the largest fractional ideal on which ψ is trivial is precisely $\mathcal{O}_F$.

Let E be the unique unramified quadratic extension of F and let $\mathcal{O}_E$ be its ring of integers. Let us take $\eta \in \mathcal{O}_E^\times$ such that $E = F(\eta)$ and $\eta^2 = d \in F^\times$. Let us denote by σ the unique non-trivial element in $\mathrm{Gal}(E/F)$, the Galois group of E/F. Then we have $\eta^\sigma = -\eta$. When there is no fear of confusion, we denote the normalized absolute value on E also by $|\cdot|$, and in case we need to be careful, we employ $|\cdot|_E$ instead. We note that ϖ is also a prime element of E and we have $|\varpi|_E = q^{-2}$. For $x \in E^\times$, we denote by $\mathrm{ord}(x)$ the ordinal of x. Hence we have $x \in \varpi^{\mathrm{ord}(x)} \mathcal{O}_E^\times$ and $|x|_E = q^{-2\,\mathrm{ord}(x)}$. We define an additive character ψ_E of E by

$$\psi_E(x) = \psi(\mathrm{tr}_{E/F}(x)) \quad \text{for} \quad x \in E.$$

Let n be a positive integer. Then let us denote $\mathcal{O}_F/\varpi^n \mathcal{O}_F$ (resp. $\mathcal{O}_E/\varpi^n \mathcal{O}_E$) by $\mathcal{O}_{F,n}$ (resp. $\mathcal{O}_{E,n}$). We denote by $\psi^{(n)}$ the additive character of $\mathcal{O}_{F,n}$ defined by

$$\psi^{(n)}(\bar{x}) = \psi(\varpi^{-n} x)$$

where $\bar{x}$ denotes the image of $x \in \mathcal{O}_F$ by the natural homomorphism $\mathcal{O}_F \to \mathcal{O}_{F,n}$. When there is no fear of confusion, we often identify $a \in \mathcal{O}_E$ with its image in $\mathcal{O}_{E,n}$ by the natural homomorphism.

By abuse of notation we denote by σ the automorphism of $\mathcal{O}_{E,n}$ induced by $\sigma : \mathcal{O}_E \to \mathcal{O}_E$. Then we define $\mathcal{O}_{E,n}^1$ by

$$\mathcal{O}_{E,n}^1 = \left\{ u \in \mathcal{O}_{E,n}^\times \mid uu^\sigma \equiv 1 \pmod{\varpi^n} \right\}.$$

Also we define $\mathcal{O}_E^1$ by

$$\mathcal{O}_E^1 = \left\{ x \in E^\times \mid xx^\sigma = 1 \right\}.$$

For our later use we recall the following lemma here.

LEMMA 1.12. *1. Let n be a positive integer. Then for $x \in \mathcal{O}_{E,n}^\times$,*

$$x \in \mathcal{O}_{E,n}^1 \iff x \equiv z^\sigma z^{-1} \pmod{\varpi^n} \text{ for some } z \in \mathcal{O}_E^\times.$$

In other words the natural mapping $\mathcal{O}_E^1 \to \mathcal{O}_{E,n}^1$ is surjective.

Here we have

$$\#\mathcal{O}_{E,n}^1 = q^n (1 + q^{-1}).$$

2. For $m \geq n > 0$, let $\phi_{m,n}$ denote the restriction to $\mathcal{O}^1_{E,m}$ of the natural homomorphism $\mathcal{O}_{E,m} \to \mathcal{O}_{E,n}$.

Then $\phi_{m,n} : \mathcal{O}^1_{E,m} \to \mathcal{O}^1_{E,n}$ is surjective and the number of elements in the kernel of $\phi_{m,n}$, $\#\ker \phi_{m,n}$, is given by

$$\#\ker \phi_{m,n} = q^{m-n}.$$

PROOF. 1. Suppose that $xx^\sigma \equiv 1 \pmod{\varpi^n}$. Then $xx^\sigma \in 1 + \varpi^n \mathcal{O}_F$. Since $y \mapsto y^2$ is an automorphism of $1 + \varpi^n \mathcal{O}_F$, there exists $y \in 1 + \varpi^n \mathcal{O}_F$ such that $xx^\sigma = y^2$. Thus $N_{E/F}\left(xy^{-1}\right) = 1$ and there exists $z \in \mathcal{O}_E^\times$ such that $xy^{-1} = z^\sigma z^{-1}$. Hence $x \equiv xy^{-1} \equiv z^\sigma z^{-1} \pmod{\varpi^n}$. The other direction is clear. Thus

$$\#\mathcal{O}^1_{E,n} = \frac{\#\mathcal{O}^\times_{E,n}}{\#\mathcal{O}^\times_{F,n}} = \frac{q^{2n}\left(1-q^{-2}\right)}{q^n\left(1-q^{-1}\right)} = q^n\left(1+q^{-1}\right).$$

2. The surjectivity follows from the previous assertion. Then the rest is clear. □

Let us denote the character of $F^\times$ associated with the quadratic extension E/F, in the sense of the local classfield theory, by χ_E. We note that χ_E is uniquely determined by

$$\chi_E(\varpi) = -1, \quad \chi_E\mid_{\mathcal{O}_F^\times} = 1.$$

Let Ω be an unramified character of $E^\times$. We denote by ω the restriction of Ω to $F^\times$ and by ω' the lift of ω to $E^\times$, i.e.

$$\omega'(x) = \omega(xx^\sigma) \text{ for } x \in E^\times.$$

Let us denote by Ω' the conjugate of Ω, i.e.

$$\Omega'(x) = \left(\Omega^{-1}\omega'\right)(x) = \Omega(x^\sigma) \text{ for } x \in E^\times.$$

Here we note that since $E^\times = F^\times \mathcal{O}_E^\times$ and Ω is unramified, we have

$$\Omega' = \Omega.$$

1.3. Main Results

Let Ξ be the characteristic function of $G(\mathcal{O}_F) = \mathrm{GSp}(\mathcal{O}_F)$. For $x \in F \setminus \{0,1\}$ and $\lambda \in F^\times$, we define the Novodvorsky orbital integral $\mathcal{N}(x,\lambda)$ by

(1.17)
$$\mathcal{N}(x,\lambda) = \int_{\bar{H}} \int_{Z \backslash H} \Xi\left[\bar{h}\left(\begin{pmatrix} 1 & x \\ 1 & 1 \end{pmatrix} \quad 0 \\ 0 \quad \frac{\lambda}{1-x}\begin{pmatrix} 1 & -1 \\ -x & 1 \end{pmatrix}\right) h\right] \psi(\bar{h}) \theta(h)\, dh\, d\bar{h}.$$

Let Ξ_1 be the characteristic function of $G_1 \cap \mathrm{GL}_4(\mathcal{O}_E)$. For $u \in E^\times$ such that $uu^\sigma \neq 1$ and $\mu \in F^\times$, we define the Bessel orbital integral $\mathcal{B}(u,\mu)$ by

(1.18) $\mathcal{B}(u,\mu) = \Omega(u) \int_{\bar{R}_1} \int_{Z_1 \backslash R_1}$

$$\Xi_1\left[\bar{r}_1\left(\begin{pmatrix} 1 & u \\ u^\sigma & 1 \end{pmatrix} \quad 0 \\ 0 \quad \frac{\mu}{1-uu^\sigma}\begin{pmatrix} 1 & u \\ u^\sigma & 1 \end{pmatrix}\right) r_1\right] \xi(\bar{r}_1) \tau(r_1)\, d\bar{r}_1\, dr_1.$$

1.3. MAIN RESULTS

We recall that the characters ψ, θ, ξ and τ are defined as (1.7),(1.8), (1.10) and (1.11), respectively.

Then our first main theorem is as follows.

THEOREM 1.13. *Assume that Ω is trivial. Then for $x \in F \setminus \{0,1\}$ and $\lambda \in F^\times$, we have*

$$(1.19) \qquad \mathcal{N}(x,\lambda) = \begin{cases} \mathcal{B}(u,\lambda), & \text{when } x = uu^\sigma \text{ for } u \in E^\times \\ 0, & \text{when } x \notin N_{E/F}(E^\times). \end{cases}$$

This equality serves as the fundamental lemma for the relative trace formula (1.12) for the main regular relevant double cosets.

We explicitly evaluate the Novodvorsky orbital integral $\mathcal{N}(x,\lambda)$ in Chapter 4 and the Bessel orbital integral $\mathcal{B}(u,\mu)$ in Chapter 5. We obtain the equality (1.19) by comparing Theorem 4.13 with Theorem 5.11.

The announcement of Theorem 1.13 appeared as [**FS1**].

Now let us state our second main theorem. For $\epsilon \in \{1, \varpi\}$, let us denote by $\mathcal{C}_\epsilon$ the matrix in $G_E = \mathrm{GSp}_4(E)$ defined by

$$(1.20) \qquad \mathcal{C}_\epsilon = \begin{pmatrix} 1 & 0 & -\epsilon & 0 \\ 0 & 1 & 0 & 1 \\ -\epsilon^{-1}\eta & 0 & -\eta & 0 \\ 0 & \eta & 0 & -\eta \end{pmatrix}.$$

Let Ξ_E be the characteristic function of $G(\mathcal{O}_E) = \mathrm{GSp}_4(\mathcal{O}_E)$. Let $\lambda \in E^\times$ and $y \in E^\times$ such that $\epsilon^{-1} yy^\sigma \neq 1$. Then we define the quadratic orbital integral $\mathcal{Q}(\epsilon, y, \lambda)$ by

$$(1.21) \quad \mathcal{Q}(\epsilon, y, \lambda) = \Omega\left(\epsilon^{-1} y\right) \int_{G_F} \int_{Z_F \setminus H_E} \Xi_E \left[g\, \mathcal{C}_\epsilon \begin{pmatrix} \begin{pmatrix} 1 & y \\ 0 & 1 \end{pmatrix} & 0 \\ 0 & \lambda \begin{pmatrix} 1 & 0 \\ -y & 1 \end{pmatrix} \end{pmatrix} h \right] (\omega\chi_E)(\lambda(g))\, \Omega(h)\, dg\, dh.$$

Here Ω denotes the character on H_E defined as (1.9).

THEOREM 1.14. *For $\epsilon \in \{1, \varpi\}$, $\lambda \in E^\times$ and $y \in E^\times$ such that $\epsilon^{-1}yy^\sigma \neq 1$, let*

$$s_\epsilon(y) = \frac{-\epsilon^{-1} yy^\sigma}{1 - \epsilon^{-1} yy^\sigma} \quad \text{and} \quad t_\epsilon(y, \lambda) = \frac{\epsilon \lambda \lambda^\sigma}{1 - \epsilon^{-1} yy^\sigma}.$$

Then we have

$$(1.22) \qquad \mathcal{Q}(\epsilon, y, \lambda) = \begin{cases} \mathcal{B}(u, \mu), & \text{when } uu^\sigma = s_\epsilon(y) \text{ and } \mu = t_\epsilon(y, \lambda) \\ 0, & \text{when } s_\epsilon(y) \notin N_{E/F}(E^\times). \end{cases}$$

This equality serves as the fundamental lemma for the relative trace formula (1.13) for the main regular relevant double cosets.

We explicitly evaluate the quadratic orbital integral $\mathcal{Q}(\epsilon, y, \lambda)$ in Chapter 6. We obtain the equality (1.22) by comparing Theorem 5.11 with Theorem 6.17.

The announcement of Theorem 1.14 appeared as [**FS2**].

The matching of the double cosets in Theorem 1.13 and Theorem 1.14 may be interpreted as follows. First we have a mapping $\Upsilon : \bar{H}_F \backslash G_F / H_F \to \bar{H}_E \backslash G_E / H_E$

defined by $\Upsilon\left(\bar{H}_F\, g\, H_F\right) = \bar{H}_E\, g\, H_E$. Then by Proposition 4.1 and 4.3, the mapping Υ is injective on the double cosets of the form

$$\bar{H}_F \begin{pmatrix} \begin{pmatrix} 1 & x \\ 1 & 1 \end{pmatrix} & 0 \\ 0 & \lambda\, {}^t\begin{pmatrix} 1 & x \\ 1 & 1 \end{pmatrix}^{-1} \end{pmatrix} H_F \quad \text{where} \quad x \in F \setminus \{0, 1\},\ \lambda \in F^\times.$$

Now we note that by Lemma 1.2, we have $\xi G_\epsilon \xi^{-1} \subset G_E$. We also note that $\xi R_\epsilon \xi^{-1} \subset H_E$ and $\xi \bar{R}_\epsilon \xi^{-1} \subset \bar{H}_E$. Thus we may define a mapping

$$\Theta_\epsilon : \bar{R}_\epsilon \backslash G_\epsilon / R_\epsilon \to \bar{H}_E \backslash G_E / H_E$$

by $\Theta_\epsilon\left(\bar{R}_\epsilon\, g\, R_\epsilon\right) = \bar{H}_E\, \xi g \xi^{-1}\, H_E$. Here we note that for $u \in E^\times$ such that $uu^\sigma \neq 1$ and $\mu \in F^\times$, we have

(1.23)
$$\Theta_1\left[\bar{R}_1 \begin{pmatrix} \begin{pmatrix} 1 & u \\ u^\sigma & 1 \end{pmatrix} & 0 \\ 0 & \mu \begin{pmatrix} 1 & u \\ u^\sigma & 1 \end{pmatrix}^{-1} \end{pmatrix} R_1\right] = \bar{H}_E \begin{pmatrix} \begin{pmatrix} 1 & uu^\sigma \\ 1 & 1 \end{pmatrix} & 0 \\ 0 & \mu\, {}^t\begin{pmatrix} 1 & uu^\sigma \\ 1 & 1 \end{pmatrix}^{-1} \end{pmatrix} H_E$$

since

$$\begin{pmatrix} \begin{pmatrix} 1 & u \\ u^\sigma & 1 \end{pmatrix} & 0 \\ 0 & \frac{\mu}{1-uu^\sigma}\begin{pmatrix} 1 & -u^\sigma \\ -u & 1 \end{pmatrix} \end{pmatrix}$$
$$= \begin{pmatrix} u^\sigma & 0 & 0 & 0 \\ 0 & 1 & 0 & 0 \\ 0 & 0 & 1 & 0 \\ 0 & 0 & 0 & u^\sigma \end{pmatrix}^{-1} \begin{pmatrix} \begin{pmatrix} 1 & uu^\sigma \\ 1 & 1 \end{pmatrix} & 0 \\ 0 & \mu\, {}^t\begin{pmatrix} 1 & uu^\sigma \\ 1 & 1 \end{pmatrix}^{-1} \end{pmatrix} \begin{pmatrix} u^\sigma & 0 & 0 & 0 \\ 0 & 1 & 0 & 0 \\ 0 & 0 & 1 & 0 \\ 0 & 0 & 0 & u^\sigma \end{pmatrix}^{-1}.$$

Thus by Proposition 4.1, 4.3 and 5.8, the mapping Θ_1 is injective on such double cosets. Then the matching condition for the double cosets in Theorem 1.13 is equivalent to

(1.24)
$$\Upsilon\left[\bar{H}_F \begin{pmatrix} \begin{pmatrix} 1 & x \\ 1 & 1 \end{pmatrix} & 0 \\ 0 & \lambda\, {}^t\begin{pmatrix} 1 & x \\ 1 & 1 \end{pmatrix}^{-1} \end{pmatrix} H_F\right] = \Theta_1\left[\bar{R}_1 \begin{pmatrix} \begin{pmatrix} 1 & u \\ u^\sigma & 1 \end{pmatrix} & 0 \\ 0 & \mu \begin{pmatrix} 1 & u \\ u^\sigma & 1 \end{pmatrix}^{-1} \end{pmatrix} R_1\right].$$

As for the matching in Theorem 1.14, we define a mapping

$$\Theta : G_F \backslash G_E / H_E \to \bar{H}_E \backslash G_E / H_E$$

by

$$\Theta\left(G_F\, g\, H_E\right) = \bar{H}_E\, J\theta(g)\, H_E \quad \text{where} \quad \theta(g) = g^{-\sigma} g,\ J = \begin{pmatrix} 0 & 1_2 \\ -1_2 & 0 \end{pmatrix}.$$

Then for $\lambda \in E^\times$ and $y \in E^\times$ such that $\epsilon^{-1} yy^\sigma \neq 1$, we have

(1.25)
$$\Theta\left[G_F C_\epsilon \begin{pmatrix} \begin{pmatrix} 1 & y \\ 0 & 1 \end{pmatrix} & 0 \\ 0 & \lambda \begin{pmatrix} 1 & 0 \\ -y & 1 \end{pmatrix} \end{pmatrix} H_E\right] = \bar{H}_E \begin{pmatrix} \begin{pmatrix} 1 & s_\epsilon(y) \\ 1 & 1 \end{pmatrix} & 0 \\ 0 & t_\epsilon(y, \lambda)\, {}^t\begin{pmatrix} 1 & s_\epsilon(y) \\ 1 & 1 \end{pmatrix}^{-1} \end{pmatrix} H_E$$

since

$$J\theta\left[C_\epsilon \begin{pmatrix} \begin{pmatrix} 1 & y \\ 0 & 1 \end{pmatrix} & 0 \\ 0 & \lambda\begin{pmatrix} 1 & 0 \\ -y & 1 \end{pmatrix} \end{pmatrix}\right] = \lambda^{-\sigma} \begin{pmatrix} y^{-\sigma} & 0 & 0 & 0 \\ 0 & 1 & 0 & 0 \\ 0 & 0 & 1 & 0 \\ 0 & 0 & 0 & y^{-\sigma} \end{pmatrix}$$
$$\begin{pmatrix} \begin{pmatrix} 1 & s_\epsilon(y) \\ 1 & 1 \end{pmatrix} & 0 \\ 0 & t_\epsilon(y, \lambda)\, {}^t\begin{pmatrix} 1 & s_\epsilon(y) \\ 1 & 1 \end{pmatrix}^{-1} \end{pmatrix} \begin{pmatrix} -\epsilon^{-1} y^\sigma & 0 & 0 & 0 \\ 0 & 1-\epsilon^{-1} yy^\sigma & 0 & 0 \\ 0 & 0 & 1-\epsilon^{-1} yy^\sigma & 0 \\ 0 & 0 & 0 & -\epsilon^{-1} y^\sigma \end{pmatrix}.$$

Thus by Proposition 4.1, 4.3 and 6.9, the mapping Θ is injective on such double cosets. Then the matching condition for the double cosets in Theorem 1.14 is equivalent to

(1.26)
$$\Theta_1 \left[\bar{R}_1 \begin{pmatrix} \begin{pmatrix} 1 & u \\ u^\sigma & 1 \end{pmatrix} & 0 \\ 0 & \mu \overline{\begin{pmatrix} 1 & u \\ u^\sigma & 1 \end{pmatrix}}^{-1} \end{pmatrix} R_1 \right] = \Theta \left[G_F \mathcal{C}_\epsilon \begin{pmatrix} \begin{pmatrix} 1 & y \\ 0 & 1 \end{pmatrix} & 0 \\ 0 & \lambda \begin{pmatrix} 1 & 0 \\ -y & 1 \end{pmatrix} \end{pmatrix} H_E \right].$$

We would like to thank Hervé Jacquet for suggesting us to include an explanation about the matching of the double cosets.

CHAPTER 2

Gauss Sum, Kloosterman Sum and Salié Sum

The material in this section are all well known. Here we give the proofs for the sake of completeness. We refer the reader to two excellent books, Berndt, Evans and Williams [**BEW**] and Lidl and Niederreiter [**LN**, Chapter 5], for encyclopedic treatment on exponential sums.

We recall the notation. We denote the natural homomorphism $\mathcal{O}_F \to \mathcal{O}_F/\varpi\mathcal{O}_F$ by $x \mapsto \bar{x}$. Then $\psi^{(1)}$ denotes the character of $\mathbb{F}_q = \mathcal{O}_F/\varpi\mathcal{O}_F$ defined by

$$\psi^{(1)}(\bar{x}) = \psi(\varpi^{-1}x).$$

DEFINITION 2.1. We define a character $\mathrm{sgn} : \mathcal{O}_F^\times \to \mathbb{C}^\times$ by

$$\mathrm{sgn}(x) = \begin{cases} 1, & \text{when } x \in (\mathcal{O}_F^\times)^2 \\ -1, & \text{when } x \notin (\mathcal{O}_F^\times)^2. \end{cases}$$

We also denote by sgn the character of $\mathbb{F}_q^\times$ defined by $\mathrm{sgn}(\bar{x}) = \mathrm{sgn}(x)$ for $x \in \mathcal{O}_F^\times$.

We normalize the Haar measure dx on F so that $\mathcal{O}_F$ has measure 1.

2.1. Gauss sum

DEFINITION 2.2. Let $\rho : \mathbb{F}_q^\times \to \mathbb{C}^\times$ be a character. Then we define the *Gauss sum* $\mathfrak{G}(\rho)$ by

$$(2.1) \qquad \mathfrak{G}(\rho) = \sum_{y \in \mathbb{F}_q} \rho(y)\,\psi^{(1)}(y)$$

where we extend ρ to $\mathbb{F}_q$ by $\rho(0) = 0$.

DEFINITION 2.3. For $a \in F$ such that $|a| > 1$, we define $C(a) \in \mathbb{C}^\times$ by

$$(2.2) \qquad C(a) = \begin{cases} q^{-\frac{n}{2}}, & \text{when } \mathrm{ord}(a) = -n \text{ is even} \\ q^{-\frac{n+1}{2}}\,\mathrm{sgn}(\varpi^n a)\,\mathfrak{G}(\mathrm{sgn}), & \text{when } \mathrm{ord}(a) = -n \text{ is odd.} \end{cases}$$

For a positive integer $n > 0$ we simply write C_n for $C(\varpi^{-n})$.

We note that for positive integers m, n such that $m \equiv n \pmod{2}$, we have

$$(2.3) \qquad C_m C_n = q^{-\frac{m+n}{2}}\,\mathrm{sgn}(-1)^m$$

since $\mathfrak{G}(\mathrm{sgn})^2 = q \cdot \mathrm{sgn}(-1)$.

DEFINITION 2.4. For $a, b \in F$, let us define the *Gaussian integral* $\mathcal{G}(a, b)$ by

$$(2.4) \qquad \mathcal{G}(a, b) = \int_{\mathcal{O}_F} \psi\left(ax^2 + 2bx\right) dx.$$

Then we may evaluate the Gaussian integral $\mathcal{G}(a, b)$ *explicitly* as follows.

PROPOSITION 2.5. *1. When $|a| \leq 1$, we have*
$$\mathcal{G}(a,b) = \begin{cases} 1, & \text{when } |b| \leq 1 \\ 0, & \text{when } |b| > 1. \end{cases}$$

2. When $1 < |a| < |b|$, we have
$$\mathcal{G}(a,b) = 0.$$

3. When $|a| > 1$ and $|a| \geq |b|$, we have
$$\mathcal{G}(a,b) = C(a) \cdot \psi\left(-a^{-1}b^2\right)$$

where $C(a)$ is as defined in (2.2).

PROOF. 1. This is clear from the definition.
2. Let $k = -\mathrm{ord}(a) > 0$. Then we have
$$\mathcal{G}(a,b) = \int_{\mathcal{O}_F} \psi\left(ax^2 + 2bx\right) dx$$
$$= \int_{\mathcal{O}_F} \int_{\mathcal{O}_F} \psi\left[a\left(x + \varpi^k y\right)^2 + 2b\left(x + \varpi^k y\right)\right] dx\, dy$$
$$= \int_{\mathcal{O}_F} \psi\left(ax^2 + 2bx\right) \left(\int_{\mathcal{O}_F} \psi\left[2b\varpi^k\left(1 + ab^{-1}x\right)y\right] dy\right) dx,$$

since $a\varpi^{2k} \in \mathcal{O}_F$. Here the inner integral vanishes because
$$|2b\varpi^k(1 + ab^{-1}x)| = |\varpi^k b| > 1.$$

3. By completing the square, we have
$$\mathcal{G}(a,b) = \psi\left(-a^{-1}b^2\right) \int_{\mathcal{O}_F} \psi\left[a\left(x + a^{-1}b\right)^2\right] dx$$
$$= \psi\left(-a^{-1}b^2\right) \int_{\mathcal{O}_F} \psi\left(ax^2\right) dx,$$

since $a^{-1}b \in \mathcal{O}_F$. Hence it is enough for us to prove that
$$\int_{\mathcal{O}_F} \psi\left(ax^2\right) dx = C(a). \tag{2.5}$$

First we review some basic facts about the Weil constant. Let us recall the formula
$$\int_F \hat{\Phi}(x) \psi\left(\frac{1}{2}\alpha x^2\right) dx = |\alpha|^{-\frac{1}{2}} \gamma(\alpha, \psi) \int_F \Phi(x) \psi\left(-\frac{1}{2}\alpha^{-1} x^2\right) dx \tag{2.6}$$

which serves to define the Weil constant $\gamma(\alpha, \psi)$. Here Φ is a Schwartz-Bruhat function on F and $\hat{\Phi}$ is its Fourier transform defined with respect to ψ as
$$\hat{\Phi}(x) = \int_F \Phi(y) \psi(xy) dy.$$

Since ψ is of exponent zero, for ϕ the characteristic function of $\mathcal{O}_F$, we have $\hat{\phi} = \phi$. Hence for $\alpha \in F^\times$ such that $|\alpha| \leq 1$, we have
$$\int_{\mathcal{O}_F} \psi\left(-\frac{1}{2}\alpha^{-1} x^2\right) dx = |\alpha|^{\frac{1}{2}} \gamma(\alpha, \psi)^{-1}.$$

It is also easily seen that $\gamma(\alpha\beta^2, \psi) = \gamma(\alpha, \psi)$ for $\alpha, \beta \in F^\times$ and $\gamma(\varepsilon, \psi) = 1$ for $\varepsilon \in \mathcal{O}_F^\times$.

Thus when $\mathrm{ord}(a) = -2r$ where r is a positive integer, we have

$$\int_{\mathcal{O}_F} \psi\left(ax^2\right) dx = \left|-2^{-1}a^{-1}\right|^{\frac{1}{2}} \gamma\left(-2^{-1}a^{-1}, \psi\right)^{-1} = |a|^{-\frac{1}{2}} = C(a).$$

Suppose that $\mathrm{ord}(a) = -2r - 1$ where r is an integer such that $r \geq 0$. Then we have

$$\gamma\left(-2^{-1}a^{-1}, \psi\right) = \gamma\left(-2^{-1}\varpi^{-2r}a^{-1}, \psi\right).$$

Further we have

$$\gamma\left(-2^{-1}\varpi^{-2r}a^{-1}, \psi\right)^{-1} = \left|-2^{-1}\varpi^{-2r}a^{-1}\right|^{-\frac{1}{2}} \int_{\mathcal{O}_F} \psi\left(\varpi^{2r}ax^2\right) dx$$

$$= q^{\frac{1}{2}} \int_{\mathcal{O}_F} \psi\left(\varpi^{2r}ax^2\right) dx.$$

Now for $\varepsilon \in \mathcal{O}_F^\times$, let us define $\mathfrak{H}(\varepsilon)$ by

$$\mathfrak{H}(\varepsilon) = \int_{\mathcal{O}_F} \psi\left(\varpi^{-1}\varepsilon t^2\right) dt = q^{-1} \sum_{x \in \mathbb{F}_q} \psi^{(1)}\left(\bar{\varepsilon}x^2\right).$$

It is clear that $\mathfrak{H}(\varepsilon) = \mathfrak{H}(1)$ for $\varepsilon \in \left(\mathcal{O}_F^\times\right)^2$. Suppose that $\varepsilon \notin \left(\mathcal{O}_F^\times\right)^2$. Then we have

$$\mathfrak{H}(1) + \mathfrak{H}(\varepsilon) = q^{-1}\left\{\sum_{x \in \mathbb{F}_q} \psi^{(1)}\left(x^2\right) + \sum_{y \in \mathbb{F}_q} \psi^{(1)}\left(\bar{\varepsilon}y^2\right)\right\} = 2q^{-1}\sum_{z \in \mathbb{F}_q} \psi^{(1)}(z) = 0.$$

Hence we have $\mathfrak{H}(\varepsilon) = \mathrm{sgn}(\varepsilon)\,\mathfrak{H}(1)$ for $\varepsilon \in \mathcal{O}_F^\times$.

Again suppose that $\varepsilon \notin \left(\mathcal{O}_F^\times\right)^2$. Then we have

$$\mathfrak{G}(\mathrm{sgn}) = \sum_{x \in \left(\mathbb{F}_q^\times\right)^2} \psi^{(1)}(x) - \sum_{x \in \mathbb{F}_q^\times \setminus \left(\mathbb{F}_q^\times\right)^2} \psi^{(1)}(x)$$

$$= \frac{1}{2}\left\{\left(\sum_{y \in \mathbb{F}_q} \psi^{(1)}\left(y^2\right) - 1\right) - \left(\sum_{y \in \mathbb{F}_q} \psi^{(1)}\left(\bar{\varepsilon}y^2\right) - 1\right)\right\}$$

$$= \frac{q}{2}\{\mathfrak{H}(1) - \mathrm{sgn}(\varepsilon)\,\mathfrak{H}(1)\} = q\,\mathfrak{H}(1).$$

Thus for $\varepsilon \in \mathcal{O}_F^\times$, we have

$$\mathfrak{H}(\varepsilon) = q^{-1}\mathrm{sgn}(\varepsilon)\,\mathfrak{G}(\mathrm{sgn}).$$

Hence we have

$$\int_{\mathcal{O}_F} \psi\left(ax^2\right) dx = q^{-\frac{1}{2}}|a|^{-\frac{1}{2}}\mathrm{sgn}\left(\varpi^{2r+1}a\right)\mathfrak{G}(\mathrm{sgn}) = C(a).$$

□

2.2. Kloosterman sum and Salié sum

DEFINITION 2.6. For $r, s \in F^\times$, we define the *Kloosterman sum* $\mathcal{K}\ell(r, s)$ by

$$\mathcal{K}\ell(r, s) = \int_{\mathcal{O}_F^\times} \psi\left(r\varepsilon + s\varepsilon^{-1}\right) d\varepsilon. \tag{2.7}$$

Here $d\varepsilon$ denotes the restriction to $\mathcal{O}_F^\times$ of the Haar measure on the *additive* group F. Thus $d\varepsilon$ is also a Haar measure on the multiplicative group $\mathcal{O}_F^\times$ and we note that we have

$$\int_{\mathcal{O}_F^\times} d\varepsilon = 1 - q^{-1}.$$

It is clear from the definition that we have

$$\mathcal{K}\ell(r, s) = \mathcal{K}\ell(s, r) \tag{2.8}$$

and

$$\mathcal{K}\ell(r\varepsilon', s) = \mathcal{K}\ell(r, s\varepsilon') \tag{2.9}$$

for $\varepsilon' \in \mathcal{O}_F^\times$.

The following proposition allows one to evaluate the Kloosterman sum $\mathcal{K}\ell(r, s)$ *explicitly* when $|r| \neq |s|$.

PROPOSITION 2.7. *Suppose that $|r| > |s|$. Then the Kloosterman sum $\mathcal{K}\ell(r, s)$ satisfies the following properties.*
1. *If $|r| \leq 1$, then $\mathcal{K}\ell(r, s) = 1 - q^{-1}$.*
2. *If $|r| > q$ and $|r| > |s|$, then $\mathcal{K}\ell(r, s) = 0$.*
3. *If $|r| = q$ and $|s| \leq 1$, then $\mathcal{K}\ell(r, s) = -q^{-1}$.*

Only the second assertion when $|r| > |s| > 1$ is non-trivial. We prove the following lemma which clearly implies it.

LEMMA 2.8. *Suppose that m, n and ℓ are integers such that $m > n \geq \ell > 0$. Then for $\zeta_1, \zeta_2, \rho \in \mathcal{O}_F^\times$, we have*

$$\int_{\rho(1+\varpi^\ell \mathcal{O}_F)} \psi\left(\varpi^{-m}\zeta_1 \varepsilon + \varpi^{-n}\zeta_2 \varepsilon^{-1}\right) d\varepsilon = 0.$$

PROOF. Since $1 + \varpi^{m-1}\mathcal{O}_F$ is a subgroup of $1 + \varpi^\ell \mathcal{O}_F$, we have

$$\int_{\rho(1+\varpi^\ell \mathcal{O}_F)} \psi\left(\varpi^{-m}\zeta_1 \varepsilon + \varpi^{-n}\zeta_2 \varepsilon^{-1}\right) d\varepsilon$$

$$= \int_{\mathcal{O}_F} \int_{\rho(1+\varpi^\ell \mathcal{O}_F)} \psi\left[\varpi^{-m}\zeta_1 \varepsilon \left(1+\varpi^{m-1}y\right)^{-1} + \varpi^{-n}\zeta_2 \varepsilon^{-1}\left(1+\varpi^{m-1}y\right)\right] d\varepsilon\, dy.$$

Here we note that

$$\varpi^{-m}\zeta_1 \varepsilon \left(1+\varpi^{m-1}y\right)^{-1} + \varpi^{-n}\zeta_2 \varepsilon^{-1}\left(1+\varpi^{m-1}y\right)$$

$$= \varpi^{-m}\zeta_1 \varepsilon \left\{1 + \sum_{j=1}^{\infty} \left(-\varpi^{m-1}y\right)^j\right\} + \varpi^{-n}\zeta_2 \varepsilon^{-1}\left(1+\varpi^{m-1}y\right)$$

$$= \left(\varpi^{-m}\zeta_1 \varepsilon + \varpi^{-n}\zeta_2 \varepsilon^{-1}\right) + \varpi^{-1}\left(\varpi^{m-n}\zeta_2 \varepsilon^{-1} - \zeta_1 \varepsilon\right) y$$

$$+ \sum_{j=2}^{\infty} (-1)^j \varpi^{j(m-1)-m} \zeta_1 \varepsilon y^j$$

and the last term belongs to $\mathcal{O}_F$, since
$$j(m-1) - m \geq 2(m-1) - m = m - 2 \geq 0$$
for $j \geq 2$. Thus we have
$$\int_{\rho(1+\varpi^\ell \mathcal{O}_F)} \psi\left(\varpi^{-m}\zeta_1\varepsilon + \varpi^{-n}\zeta_2\varepsilon^{-1}\right) d\varepsilon$$
$$= \int_{\rho(1+\varpi^\ell \mathcal{O}_F)} \psi\left(\varpi^{-m}\zeta_1\varepsilon + \varpi^{-n}\zeta_2\varepsilon^{-1}\right)$$
$$\left(\int_{\mathcal{O}_F} \psi\left[\varpi^{-1}\left(\varpi^{m-n}\zeta_2\varepsilon^{-1} - \zeta_1\varepsilon\right)y\right] dy\right) d\varepsilon.$$

Here we note that we have
$$\varpi^{m-n}\zeta_2\varepsilon^{-1} - \zeta_1\varepsilon \in \mathcal{O}_F^\times,$$
since $m > n$. Therefore the inner integral vanishes. □

We note the following useful statement separately as a corollary.

COROLLARY 2.9. *When $|r| > q$ and $|r| \neq |s|$, we have $\mathcal{K}\ell(r,s) = 0$.*

Now let us consider the case when $|r| = |s|$. It is clear that $\mathcal{K}\ell(r,s) = 1$ when $|r| = |s| \leq 1$.

Suppose that $|r| = |s| = q$. Let us write $r = \varpi^{-1}a$ and $s = \varpi^{-1}b$. Then we have

(2.10) $\quad \mathcal{K}\ell(r,s) = \displaystyle\int_{\mathcal{O}_F^\times} \psi\left(\varpi^{-1}a\varepsilon + \varpi^{-1}b\varepsilon^{-1}\right) d\varepsilon = q^{-1} \sum_{y \in \mathbb{F}_q^\times} \psi^{(1)}\left(\bar{a}y + \bar{b}y^{-1}\right).$

In other words, we obtain the *honest Kloosterman sum* over the finite field $\mathbb{F}_q$ in this case.

Let us consider the case when $|r| = |s| > q$. Then we may evaluate the Kloosterman sum $\mathcal{K}\ell(r,s)$ *explicitly*. In order to evaluate it, first we prove the following lemma.

LEMMA 2.10. *Let m, n be integers such that $0 < 2m \leq n$. For $r, s \in \varpi^{-n}\mathcal{O}_F^\times$ and $\rho \in \mathcal{O}_F^\times$, let us define the integral $\mathcal{I}(r,s,\rho)$ by*
$$\mathcal{I}(r,s,\rho) = \int_{\rho(1+\varpi^m \mathcal{O}_F)} \psi\left(r\varepsilon + s\varepsilon^{-1}\right) d\varepsilon.$$

1. The integral $\mathcal{I}(r,s,\rho)$ vanishes unless
$$r^{-1}s \in \left(\mathcal{O}_F^\times\right)^2 \quad \text{and} \quad \rho^2 \equiv r^{-1}s \pmod{\varpi^m}.$$

2. Suppose that $\rho^2 \equiv r^{-1}s \pmod{\varpi^m}$. Let $\zeta \in \mathcal{O}_F^\times$ be the unique square root of $r^{-1}s$ such that $\rho \equiv \zeta \pmod{\varpi^m}$. Then we have
$$\mathcal{I}(r,s,\rho) = C(r\zeta)\,\psi(2r\zeta).$$

PROOF. Let us define an integer ℓ by
$$\ell = \begin{cases} \frac{n}{2}, & \text{when } n \text{ is even} \\ \frac{n+1}{2}, & \text{when } n \text{ is odd}. \end{cases}$$

Then we have
$$\mathcal{I}(r,s,\rho) = \int_{\mathcal{O}_F} \int_{\rho(1+\varpi^m \mathcal{O}_F)} \psi\left[r\varepsilon\left(1+\varpi^\ell y\right) + s\varepsilon^{-1}\left(1+\varpi^\ell y\right)^{-1}\right] d\varepsilon\, dy$$
$$= \int_{\rho(1+\varpi^m \mathcal{O}_F)} \psi\left(r\varepsilon + s\varepsilon^{-1}\right) \left(\int_{\mathcal{O}_F} \psi\left[\varpi^\ell r\varepsilon^{-1}\left(\varepsilon^2 - r^{-1}s\right) y\right] dy\right) d\varepsilon,$$
since $\ell \geq m$ and
$$\left(1+\varpi^\ell y\right)^{-1} = 1 - \varpi^\ell y + \sum_{j=2}^\infty \left(-\varpi^\ell y\right)^j \in \left(1 - \varpi^\ell y\right) + \varpi^n \mathcal{O}_F.$$

Since $|\varpi^\ell r| > 1$, the inner integral vanishes unless there exists $\varepsilon \in \rho\left(1+\varpi^m \mathcal{O}_F\right)$ such that $\varepsilon^2 \equiv r^{-1}s \pmod{\varpi^{n-\ell}}$. Suppose that there exists such ε. Then we have $\rho^2 \equiv r^{-1}s \pmod{\varpi^m}$, since $\varepsilon \equiv \rho \pmod{\varpi^m \mathcal{O}_F}$ and $n - \ell \geq m$.

Suppose that $\rho^2 \equiv r^{-1}s \pmod{\varpi^m}$. Let ζ be the unique square root of $r^{-1}s$ such that $\rho \equiv \zeta \pmod{\varpi^m}$. Then we have
$$\mathcal{I}(r,s,\rho) = \mathcal{I}(r,s,\zeta) = \int_{\zeta(1+\varpi^{n-\ell}\mathcal{O}_F)} \psi\left(r\varepsilon + s\varepsilon^{-1}\right) d\varepsilon$$
$$= q^{-n+\ell} \int_{\mathcal{O}_F} \psi\left[r\zeta\left\{\left(1+\varpi^{n-\ell}y\right) + \left(1+\varpi^{n-\ell}y\right)^{-1}\right\}\right] dy.$$

Here we have
$$\left(1+\varpi^{n-\ell}y\right) + \left(1+\varpi^{n-\ell}y\right)^{-1} = \left(1+\varpi^{n-\ell}y\right) + \left(1 + \sum_{j=1}^\infty \left(-\varpi^{n-\ell}y\right)^j\right)$$
$$\equiv \begin{cases} 2 \pmod{\varpi^n}, & \text{when } n \text{ is even} \\ 2 + \varpi^{n-1}y^2 \pmod{\varpi^n}, & \text{when } n \text{ is odd.} \end{cases}$$

Thus by Proposition 2.5, we have
$$\mathcal{I}(r,s,\rho) = C(r\zeta)\, \psi(2r\zeta).$$
$\square$

PROPOSITION 2.11. *Suppose that $|r| = |s| = q^n$ where $n \geq 2$.*
1. *Then $\mathcal{K}\ell(r,s)$ vanishes unless $rs \in (F^\times)^2$.*
2. *When $rs \in (F^\times)^2$, we have*
$$\mathcal{K}\ell(r,s) = C\left(\sqrt{rs}\right) \psi\left(2\sqrt{rs}\right) + C\left(-\sqrt{rs}\right) \psi\left(-2\sqrt{rs}\right).$$

PROOF. Let us define a positive integer m by
$$m = \begin{cases} \frac{n}{2}, & \text{when } n \text{ is even} \\ \frac{n-1}{2}, & \text{when } n \text{ is odd.} \end{cases}$$
Then we have
$$\mathcal{K}\ell(r,s) = \sum_{\rho \in \mathcal{O}_F^\times/(1+\varpi^m \mathcal{O}_F)} \int_{\rho(1+\varpi^m \mathcal{O}_F)} \psi\left(r\varepsilon + s\varepsilon^{-1}\right) d\varepsilon$$
$$= \sum_{\rho \in \mathcal{O}_F^\times/(1+\varpi^m \mathcal{O}_F)} \mathcal{I}(r,s,\rho).$$
The rest is clear from Lemma 2.10. $\square$

Now we consider a variant of the Kloosterman sum.

DEFINITION 2.12. For $r, s \in F^\times$, let us define the *Salié sum* $\mathcal{S}(r, s)$ by

$$\mathcal{S}(r, s) = \int_{\mathcal{O}_F^\times} \operatorname{sgn}(\varepsilon) \cdot \psi\left(r\varepsilon + s\varepsilon^{-1}\right) d\varepsilon. \tag{2.11}$$

We note that since $\operatorname{sgn}(\varepsilon) = \operatorname{sgn}(\varepsilon^{-1})$, we have

$$\mathcal{S}(r, s) = \mathcal{S}(s, r).$$

The following two propositions allow one to evaluate the Salié sum explicitly in *every* case.

PROPOSITION 2.13. *The Salié sum $\mathcal{S}(r, s)$ satisfies the following properties.*
1. *If $|r| \leq 1$ and $|s| \leq 1$, then $\mathcal{S}(r, s) = 0$.*
2. *If $|r| > q$ and $|r| > |s|$, then $\mathcal{S}(r, s) = 0$.*
3. *If $|r| = q$ and $|s| \leq 1$, then*

$$\mathcal{S}(r, s) = C(r).$$

PROOF. 1. Since sgn is a non-trivial character of $\mathcal{O}_F^\times$, we have

$$\mathcal{S}(r, s) = \int_{\mathcal{O}_F^\times} \operatorname{sgn}(\varepsilon) d\varepsilon = 0.$$

2. Let us write $r = \varpi^{-m}\zeta_1$ and $s = \varpi^{-n}\zeta_2$ where $\zeta_1, \zeta_2 \in \mathcal{O}_F^\times$. Then by the assumption, we have $m > 1$ and $m > n$. Since $1 + \varpi^{m-1}\mathcal{O}_F \subset (\mathcal{O}_F^\times)^2$, by the same argument as in the proof of Proposition 2.7, we have

$$\mathcal{S}(r, s) = \int_{\mathcal{O}_F^\times} \operatorname{sgn}(\varepsilon) \psi\left(r\varepsilon + s\varepsilon^{-1}\right)$$

$$\left(\int_{\mathcal{O}_F} \psi\left[\varpi^{-1}\left(\varpi^{m-n}\zeta_2\varepsilon^{-1} - \zeta_1\varepsilon\right)y\right] dy\right) d\varepsilon$$

and the inner integral vanishes.

3. Let us write $r = \varpi^{-1}a$. Then we have

$$\mathcal{S}(r, s) = \int_{\mathcal{O}_F^\times} \operatorname{sgn}(\varepsilon) \psi(r\varepsilon) = q^{-1} \mathfrak{G}(\operatorname{sgn}) \operatorname{sgn}(a) = C(r).$$

□

PROPOSITION 2.14. *Suppose that $|r| = |s| = q^n$ where $n \geq 1$.*
Then $\mathcal{S}(r, s)$ vanishes unless $rs \in (F^\times)^2$.
Suppose that $rs \in (F^\times)^2$.
1. *When n is even, we have*

$$\mathcal{S}(r, s) = C(r) \left\{ \operatorname{sgn}\left(\frac{\sqrt{rs}}{r}\right) \psi\left(2\sqrt{rs}\right) + \operatorname{sgn}\left(-\frac{\sqrt{rs}}{r}\right) \psi\left(-2\sqrt{rs}\right) \right\}.$$

2. *When n is odd, we have*

$$\mathcal{S}(r, s) = C(r) \left\{ \psi\left(2\sqrt{rs}\right) + \psi\left(-2\sqrt{rs}\right) \right\}.$$

2.2. KLOOSTERMAN SUM AND SALIÉ SUM

PROOF. Let $\varepsilon' = r^{-1}s \in \mathcal{O}_F^\times$. Then

$$\mathcal{S}(r,s) = \int_{\mathcal{O}_F^\times} \operatorname{sgn}(\varepsilon\varepsilon') \psi\left(r\varepsilon\varepsilon' + s\varepsilon^{-1}\varepsilon'^{-1}\right) d\varepsilon$$

$$= \operatorname{sgn}(\varepsilon') \int_{\mathcal{O}_F^\times} \operatorname{sgn}(\varepsilon) \psi\left(s\varepsilon + r\varepsilon^{-1}\right) d\varepsilon$$

$$= \operatorname{sgn}(\varepsilon') \mathcal{S}(s,r) = \operatorname{sgn}(\varepsilon') \mathcal{S}(r,s).$$

Hence $\mathcal{S}(r,s)$ vanishes unless $\operatorname{sgn}(\varepsilon') = \operatorname{sgn}(r^{-1}s) = 1$, i.e. $rs \in (F^\times)^2$.

Suppose that $rs \in (F^\times)^2$. Let us consider the case when $n \geq 2$ first. Then we define a positive integer m by

$$m = \begin{cases} \frac{n}{2}, & \text{when } n \text{ is even} \\ \frac{n-1}{2}, & \text{when } n \text{ is odd.} \end{cases}$$

Then we have

$$\mathcal{S}(r,s) = \sum_{\rho \in \mathcal{O}_F^\times/(1+\varpi^m\mathcal{O}_F)} \int_{\rho(1+\varpi^m\mathcal{O}_F)} \operatorname{sgn}(\varepsilon) \psi\left(r\varepsilon + s\varepsilon^{-1}\right) d\varepsilon$$

$$= \sum_{\rho \in \mathcal{O}_F^\times/(1+\varpi^m\mathcal{O}_F)} \operatorname{sgn}(\rho) \mathcal{I}(r,s,\rho)$$

since $1 + \varpi^m \mathcal{O}_F \subset (\mathcal{O}_F^\times)^2$. Then the assertion follows from Lemma 2.10.

Now let us consider the case when $n = 1$. For this computation, we follow Iwaniec [**I**, Lemma 4] (who attributes this computation to Williams [**Wi**]). See also Jacquet [**J3**, (124)] (who explicitly refers to Iwaniec [**I**]). Let $\zeta \in \mathcal{O}_F^\times$ be one of the square roots of $r^{-1}s$. Then we have

$$\mathcal{S}(r,s) = \int_{\mathcal{O}_F^\times} \operatorname{sgn}(\varepsilon) \psi\left(r\varepsilon + r\zeta^2 \varepsilon^{-1}\right) d\varepsilon = \operatorname{sgn}(\zeta) \int_{\mathcal{O}_F^\times} \operatorname{sgn}(\varepsilon) \psi\left(r\zeta\varepsilon + r\zeta\varepsilon^{-1}\right) d\varepsilon$$

by a change of variable $\varepsilon \mapsto \zeta\varepsilon$. Let $\alpha = \varpi r\zeta \pmod{\varpi} \in \mathbb{F}_q^\times$. Then we have

$$\mathcal{S}(r,s) = q^{-1}\operatorname{sgn}(\zeta) \sum_{t \in \mathbb{F}_q^\times} \operatorname{sgn}(t) \psi^{(1)}\left[\alpha\left(t + t^{-1}\right)\right].$$

We note that by changing the order of summation, we have

$$\sum_{t \in \mathbb{F}_q^\times} \operatorname{sgn}(t) \psi^{(1)}\left[\alpha\left(t + t^{-1}\right)\right] = \sum_{u \in \mathbb{F}_q} \psi^{(1)}(\alpha u) \sum_{t \in \mathbb{F}_q^\times, t+t^{-1}=u} \operatorname{sgn}(t).$$

We recall that we extend sgn to $\mathbb{F}_q$ by $\operatorname{sgn}(0) = 0$. Now we claim that

$$\sum_{t \in \mathbb{F}_q^\times, t+t^{-1}=u} \operatorname{sgn}(t) = \operatorname{sgn}(u-2) + \operatorname{sgn}(u+2).$$

This is clear when $u = \pm 2$ since $t = \pm 1$ is the only solution of $t + t^{-1} = \pm 2$. Thus suppose that $u \neq \pm 2$. Then

$$\operatorname{sgn}(u-2) \sum_{t \in \mathbb{F}_q^\times, t+t^{-1}=u} \operatorname{sgn}(t) = \sum_{t \in \mathbb{F}_q^\times, t+t^{-1}=u} \operatorname{sgn}\left[(t+t^{-1}-2)t\right]$$

$$= \sum_{t \in \mathbb{F}_q^\times, t+t^{-1}=u} \operatorname{sgn}\left[(t-1)^2\right] = \sum_{t \in \mathbb{F}_q^\times, t+t^{-1}=u} 1 = 1 + \operatorname{sgn}[(u-2)(u+2)],$$

since $t + t^{-1} = u$ has a solution in $\mathbb{F}_q^\times$ if and only if the quadratic equation
$$t^2 - ut + 1 = 0$$
has a solution in $\mathbb{F}_q$, i.e. $u^2 - 4 \in (\mathbb{F}_q)^2$, and then the number of solutions is equal to two since $u \neq \pm 2$.

Thus
$$\sum_{t \in \mathbb{F}_q^\times} \operatorname{sgn}(t)\, \psi^{(1)}\left[\alpha\left(t + t^{-1}\right)\right]$$
$$= \sum_{u \in \mathbb{F}_q} \psi^{(1)}(\alpha u)\{\operatorname{sgn}(u-2) + \operatorname{sgn}(u+2)\}$$
$$= \left\{\psi^{(1)}(2\alpha) + \psi^{(1)}(-2\alpha)\right\} \sum_{u \in \mathbb{F}_q} \psi^{(1)}(\alpha u)\, \operatorname{sgn}(u)$$
$$= \left\{\psi^{(1)}(2\alpha) + \psi^{(1)}(-2\alpha)\right\} \mathfrak{G}(\operatorname{sgn})\, \operatorname{sgn}(\alpha).$$

Since
$$\operatorname{sgn}(\zeta)\operatorname{sgn}(\alpha) = \operatorname{sgn}(\zeta)\operatorname{sgn}(\varpi r \zeta) = \operatorname{sgn}(\varpi r),$$
we have
$$\mathcal{S}(r,s) = q^{-1} \mathfrak{G}(\operatorname{sgn})\, \operatorname{sgn}(\varpi r)\{\psi(2\varpi r\zeta) + \psi(-2\varpi r\zeta)\}$$
$$= C(r)\left\{\psi\left(2\sqrt{rs}\right) + \psi\left(-2\sqrt{rs}\right)\right\}. \qquad \square$$

2.3. The Davenport-Hasse relation

DEFINITION 2.15. Let n be a positive integer and let a be an element of $\mathcal{O}_F^\times$. Then we define the integral $\mathcal{H}_n(a)$ by

(2.12) $$\mathcal{H}_n(a) = \int_{\{\xi \in \mathcal{O}_E^\times \mid N_{E/F}(\xi) \in a(1+\varpi^n \mathcal{O}_F)\}} \psi_E\left(2\varpi^{-n}\xi\right) d\xi.$$

Here $d\xi$ denotes the Haar measure on the additive group E normalized so that
$$\int_{\mathcal{O}_E} d\xi = 1.$$

Our goal in this section is to prove the following proposition, which is a p-adic version of Zagier [**Z**, Proposition,p.24]. Though Ye treats in full generality in [**Y1, Y2**], we give here a proof for the convenience of the reader. This may be regarded as the Davenport-Hasse relation for the Kloosterman sums.

PROPOSITION 2.16. We have

(2.13) $$\mathcal{H}_n(a) = (-1)^n\, q^{-n} \cdot \mathcal{K}\ell\left(2\varpi^{-n}, 2\varpi^{-n}a\right).$$

Here we note that (2.13) is also expressed as

(2.14) $$\sum_{\{x \in \mathcal{O}_{E,n}^\times \mid xx^\sigma \equiv a \pmod{\varpi^n}\}} \psi^{(n)}\left[2(x + x^\sigma)\right]$$
$$= (-1)^n \sum_{y \in \mathcal{O}_{F,n}^\times} \psi^{(n)}\left[2\left(y + ay^{-1}\right)\right].$$

As for the proof, we consider the case when $n \geq 2$ and the case when $n = 1$ separately.

Let us consider the case when $n \geq 2$ first. Then by Proposition 2.11, it is enough to prove the following lemma.

LEMMA 2.17. *Suppose that $n \geq 2$. Then $\mathcal{H}_n(a)$ vanishes unless $a \in \left(\mathcal{O}_F^\times\right)^2$. Suppose that $a \in \left(\mathcal{O}_F^\times\right)^2$. Then $\mathcal{H}_n(a)$ is equal to*

$$(-1)^n q^{-n} \left\{ C\left(2\varpi^{-n}\sqrt{a}\right) \psi\left(4\varpi^{-n}\sqrt{a}\right) + C\left(-2\varpi^{-n}\sqrt{a}\right) \psi\left(-4\varpi^{-n}\sqrt{a}\right) \right\}.$$

PROOF. Let us define a positive integer m by

$$m = \begin{cases} \frac{n}{2}, & \text{when } n \text{ is even} \\ \frac{n+1}{2}, & \text{when } n \text{ is odd.} \end{cases}$$

Then for $\varepsilon \in 1 + \varpi^m \mathcal{O}_E$ such that $N_{E/F}(\varepsilon) \in 1 + \varpi^n \mathcal{O}_F$, we have

$$\mathcal{H}_n(a) = \int_{\left\{\xi \in \mathcal{O}_E^\times \mid N_{E/F}(\xi) \in a(1+\varpi^n \mathcal{O}_F)\right\}} \psi_E\left(2\varpi^{-n}\xi\varepsilon\right) d\xi.$$

Thus

$$\mathcal{H}_n(a) = \int_{x \in \mathcal{O}_F} \int_{y \in \mathcal{O}_F} \int_{\left\{\xi \in \mathcal{O}_E^\times \mid N_{E/F}(\xi) \in a(1+\varpi^n \mathcal{O}_F)\right\}}$$
$$\cdot \psi_E\left[2\varpi^{-n}\xi\left\{1 + \varpi^m\left(\varpi^{n-m}x + \eta y\right)\right\}\right] dx\, dy\, d\xi.$$

$$= \int_{\left\{\xi \in \mathcal{O}_E^\times \mid N_{E/F}(\xi) \in a(1+\varpi^n \mathcal{O}_F)\right\}} \psi_E\left(2\varpi^{-n}\xi\right)$$
$$\left(\int_{y \in \mathcal{O}_F} \psi\left[2\varpi^{-n+m}\mathrm{tr}_{E/F}(\xi\eta)\, y\right] dy\right) d\xi.$$

The inner integral vanishes unless $\mathrm{tr}_{E/F}(\xi\eta) \in \varpi^{n-m}\mathcal{O}_F$, i.e. $\xi = s + \eta\varpi^{n-m}t$ for some $s, t \in \mathcal{O}_F$. Then

$$N_{E/F}(\xi) = s^2 - \eta^2 \varpi^{2(n-m)} t^2 \equiv s^2 \pmod{\varpi^{2(n-m)}}.$$

Hence $\mathcal{H}_n(a)$ vanishes unless $a \equiv s^2 \pmod{\varpi^{2(n-m)}}$ for some $s \in \mathcal{O}_F^\times$. Thus $\mathcal{H}_n(a)$ vanishes unless $a = b^2$ for some $b \in \mathcal{O}_F^\times$.

Suppose that $a = b^2$ for $b \in \mathcal{O}_F^\times$. When n is even, the above computation shows that

$$\mathcal{H}_n(b^2) = \int_{\left\{\xi \in \mathcal{O}_E^\times \mid \xi = s + \eta\varpi^{\frac{n}{2}} t\, (s,t \in \mathcal{O}_F),\ s^2 \equiv b^2 \pmod{\varpi^n}\right\}} \psi_E\left(2\varpi^{-n}\xi\right) d\xi$$
$$= q^{-\frac{3n}{2}}\left\{\psi\left(4\varpi^{-n}b\right) + \psi\left(-4\varpi^{-n}b\right)\right\}.$$

Suppose that n is odd. Then the computation above implies that

$$\mathcal{H}_n(b^2)$$
$$= \int_{\left\{\xi \mid \xi = s + \eta\varpi^{\frac{n-1}{2}} t\, (s,t \in \mathcal{O}_F),\ s^2 \equiv b^2(1+\varpi^{n-1}db^{-2}t^2) \pmod{\varpi^n}\right\}} \psi_E\left(2\varpi^{-n}\xi\right) d\xi.$$

Since

$$\left(1 + \frac{1}{2}\varpi^{n-1}db^{-2}t^2\right)^2 \equiv 1 + \varpi^{n-1}db^{-2}t^2 \pmod{\varpi^n},$$

we have

$$\mathcal{H}_n\left(b^2\right) = q^{-\frac{n-1}{2}} \int_{t\in\mathcal{O}_F} \int_{s\in\pm b\left(1+\frac{1}{2}\varpi^{n-1}db^{-2}t^2\right)+\varpi^n\mathcal{O}_F} \psi\left(4\varpi^{-n}s\right) ds\, dt$$

$$= q^{-\frac{3n-1}{2}} \psi\left(4\varpi^{-n}b\right) \int_{\mathcal{O}_F} \psi\left(2\varpi^{-1}db^{-1}t^2\right) dt$$

$$+ q^{-\frac{3n-1}{2}} \psi\left(-4\varpi^{-n}b\right) \int_{\mathcal{O}_F} \psi\left(-2\varpi^{-1}db^{-1}t^2\right) dt.$$

By Proposition 2.5, we have

$$\int_{\mathcal{O}_F} \psi\left(\pm 2\varpi^{-1}db^{-1}t^2\right) dt$$
$$= C\left(\pm 2\varpi^{-1}db^{-1}\right) = C_1 \cdot \operatorname{sgn}\left(\pm 2db^{-1}\right) = -C_1 \cdot \operatorname{sgn}\left(\pm 2b\right)$$
$$= -C\left(\pm 2\varpi^{-1}b\right).$$

Hence

$$\mathcal{H}_n\left(b^2\right) = -q^{-\frac{3n-1}{2}} \left\{ C\left(2\pi^{-1}b\right) \psi\left(4\varpi^{-n}b\right) + C\left(-2\pi^{-1}b\right) \psi\left(-4\varpi^{-n}b\right) \right\}.$$

Finally we note that

$$C\left(2\varpi^{-n}b\right) = q^{-\frac{n+1}{2}} \mathfrak{G}\left(\operatorname{sgn}\right) \operatorname{sgn}\left(2b\right) = q^{-\frac{n-1}{2}} C\left(2\varpi^{-1}b\right).$$

$\square$

Now let us consider the case when $n=1$.

LEMMA 2.18. *We have*

$$\mathcal{H}_1\left(a\right) = -q^{-1} \cdot \mathcal{K}\ell\left(2\varpi^{-1}, 2\varpi^{-1}a\right).$$

PROOF. Let us denote $\mathcal{O}_E/\varpi\mathcal{O}_E$ by K, and, $\mathcal{O}_F/\varpi\mathcal{O}_F$ by L, respectively. Let

$$S_1 = \sum_{\left\{\zeta\in K^\times \mid N_{K/L}(\zeta)=\bar{a}\right\}} \psi^{(1)}\left[2\operatorname{tr}_{K/L}(\zeta)\right]$$

and

$$S_2 = \sum_{u\in L^\times} \psi^{(1)}\left(u + 4\bar{a}u^{-1}\right).$$

Then we have

$$\mathcal{H}_1\left(a\right) = q^{-2}S_1$$

and

$$\mathcal{K}\ell\left(2\varpi^{-1}, 2\varpi^{-1}a\right) = q^{-1}S_2.$$

Hence it is enough for us to show that

$$S_1 + S_2 = 0.$$

2.3. THE DAVENPORT-HASSE RELATION

Since $N_{K/L}(\zeta) = \gamma^2 - \bar{d}\delta^2$ for $\zeta = \gamma + \bar{\eta}\delta$ where $\gamma, \delta \in L$, we have

$$S_1 = \sum_{\{(\gamma,\delta)\in L\times L | \gamma^2 - \bar{a} = \bar{d}\delta^2, (\gamma,\delta)\neq(0,0)\}} \psi^{(1)}(4\gamma)$$

$$= 2 \sum_{\{\gamma\in L | \gamma^2 - \bar{a}\notin (L)^2\}} \psi^{(1)}(4\gamma)$$

$$+ \begin{cases} 0, & \text{when } \bar{a} \notin (L^\times)^2 \\ \{\psi^{(1)}(4\sqrt{\bar{a}}) + \psi^{(1)}(-4\sqrt{\bar{a}})\}, & \text{when } \bar{a} \in (L^\times)^2. \end{cases}$$

As for S_2, we note that for $\gamma \in L$, there exists $u \in L^\times$ such that $u + 4\bar{a}u^{-1} = 4\gamma$ if and only if the quadratic equation $X^2 - 4\gamma X + 4\bar{a}$ has both roots in L, i.e. $\gamma^2 - \bar{a} = \beta^2$ for some $\beta \in L$. Thus

$$S_2 = \sum_{\{\gamma\in L | \gamma^2 - \bar{a}\in (L)^2\}} \psi^{(1)}(4\gamma) \cdot \#\{u \in L \mid u + 4\bar{a}u^{-1} = 4\gamma\}$$

$$= 2 \sum_{\{\gamma\in L | \gamma^2 - \bar{a}\in (L)^2\}} \psi^{(1)}(4\gamma)$$

$$+ \begin{cases} 0, & \text{when } \bar{a} \notin (L^\times)^2 \\ -\{\psi^{(1)}(4\sqrt{\bar{a}}) + \psi^{(1)}(-4\sqrt{\bar{a}})\}, & \text{when } \bar{a} \in (L^\times)^2. \end{cases}$$

Hence

$$S_1 + S_2 = 2\sum_{\{\gamma\in L | \gamma^2 - \bar{a}\notin (L)^2\}} \psi^{(1)}(4\gamma) + 2\sum_{\{\gamma\in L | \gamma^2 - \bar{a}\in (L)^2\}} \psi^{(1)}(4\gamma)$$

$$= 2\sum_{\gamma\in L} \psi^{(1)}(4\gamma) = 0.$$

□

CHAPTER 3

Matrix Argument Kloosterman Sums

In this chapter we define certain Kloosterman sums, which are generalizations of the one variable Kloosterman sums considered in the previous chapter to n by n symmetric matrices. Then we shall evaluate them explicitly in terms of the classical Kloosterman sum when $n = 2$ and prove a generalization of the Davenport-Hasse relation to our case.

The computation here might be of some independent interest since few of the generalized exponential sums admit explicit evaluation.

3.1. Kloosterman sum over n by n symmetric matrices

For a ring R, let us denote by $\mathrm{Sym}^n(R)$ the set of n by n symmetric matrices with entries in R.

DEFINITION 3.1. Let $A \in \mathrm{GL}_n(F)$, $S \in \mathrm{Sym}^n(F)$ and $\varepsilon \in \mathcal{O}_F^\times$. Then we define $\mathcal{K}\ell^{(n)}(A, S, \varepsilon)$, the Kloosterman sum with respect to A, S and ε, by

$$(3.1) \quad \mathcal{K}\ell^{(n)}(A, S, \varepsilon) = \int_{\mathcal{X}_A} \psi\left[\mathrm{tr}\left\{S\left(X + \varepsilon \cdot A^{-1} X^{-1t} A^{-1}\right)\right\}\right] dX$$

$$= \int_{\mathcal{X}_A} \psi\left[\mathrm{tr}\left(SX + \varepsilon \cdot {}^t A^{-1} S A^{-1} X^{-1}\right)\right] dX$$

where the domain of integration $\mathcal{X}_A$ is defined by

$$(3.2) \quad \mathcal{X}_A = \{X \in \mathrm{Sym}^n(F) \mid XA \in \mathrm{GL}_n(\mathcal{O}_F)\}$$

and dX denotes the additive Haar measure on $\mathrm{Sym}^n(F)$ normalized so that

$$\int_{\mathrm{Sym}^n(\mathcal{O}_F)} dX = 1.$$

PROPOSITION 3.2. 1. Suppose that $u \in \mathrm{GL}_2(\mathcal{O}_F)$. Then we have

$$\mathcal{K}\ell^{(n)}(A, {}^t u S u, \varepsilon) = \mathcal{K}\ell^{(n)}({}^t u^{-1} A u^{-1}, S, \varepsilon).$$

2. Suppose that $k \in \mathrm{GL}_2(\mathcal{O}_F)$ satisfies

$${}^t k S k = \mu S \quad \text{where} \quad \mu \in \mathcal{O}_F^\times.$$

Then we have

$$\mathcal{K}\ell^{(n)}({}^t k A, S, \varepsilon) = \mathcal{K}\ell^{(n)}(Ak, S, \varepsilon) = \mathcal{K}\ell^{(n)}(A, S, \varepsilon \mu^{-1}).$$

PROOF. 1. We have

$$\mathcal{K}\ell^{(n)}(A, {}^t u S u, \varepsilon) = \int_{\mathcal{X}_A} \psi\left[\mathrm{tr}\left({}^t u S u X + \varepsilon \cdot {}^t A^{-1t} u S u A^{-1} X^{-1}\right)\right] dX$$

$$= \int_{\mathcal{X}_A} \psi\left[\mathrm{tr}\left(S u X {}^t u + \varepsilon \cdot {}^t A^{-1 t} u S u A^{-1} X^{-1}\right)\right] dX.$$

Since $u\mathcal{X}_A{}^t u = \mathcal{X}_{{}^t u^{-1} A u^{-1}}$, we have

$$\mathcal{K}\ell^{(n)}\left(A, {}^t uSu, \varepsilon\right) = \int_{\mathcal{X}_{{}^t u^{-1} A u^{-1}}} \psi\left[\operatorname{tr}\left(SY + \varepsilon \cdot {}^t A^{-1\,t} uSuA^{-1\,t} uY^{-1} u\right)\right] dY$$

$$= \int_{\mathcal{X}_{{}^t u^{-1} A u^{-1}}} \psi\left[\operatorname{tr}\left(SY + \varepsilon \cdot u^t A^{-1\,t} uSuA^{-1\,t} uY^{-1}\right)\right] dY = \mathcal{K}\ell^{(n)}\left({}^t u^{-1} A u^{-1}, S, \varepsilon\right).$$

2. From (3.1), it is clear that we have $\mathcal{K}\ell^{(n)}(Ak, S, \varepsilon) = \mathcal{K}\ell^{(n)}(A, S, \varepsilon\mu^{-1})$.
Since $\mathcal{X}_{{}^t kA} = k^{-1}\mathcal{X}_A{}^t k^{-1}$, we have

$$\mathcal{K}\ell^{(n)}\left({}^t kA, S, \varepsilon\right) = \int_{\mathcal{X}_A} \psi\left[\operatorname{tr}\left(Sk^{-1}X^t k^{-1} + \varepsilon \cdot k^{-1\,t} A^{-1} S A^{-1} X^{-1} k\right)\right] dX$$

$$= \int_{\mathcal{X}_A} \psi\left[\operatorname{tr}\left(\mu^{-1} SX + \varepsilon \cdot {}^t A^{-1} S A^{-1} X^{-1}\right)\right] dX$$

$$= \int_{\mathcal{X}_A} \psi\left[\operatorname{tr}\left(SY + \varepsilon\mu^{-1} \cdot {}^t A^{-1} S A^{-1} Y^{-1}\right)\right] dY = \mathcal{K}\ell^{(n)}\left(A, S, \varepsilon\mu^{-1}\right).$$

$\square$

Before we move on to the next section for the explicit evaluation for the case when $n = 2$, here we remark the relationship between our generalized Kloosterman sum and the symmetric matrix argument Kloosterman sum which appears in the Fourier expansion of the Poincaré series for the Hilbert-Siegel modular group in Christian [**Ch**]. We note that it appears implicitly in [**Ch**] and is explicitly written down in Kitaoka [**K**]. The relationship was pointed out to the first author by Siegfried Böcherer. We would like to thank him for his remark.

Generalizing (3.1), for $A \in \operatorname{GL}_n(F) \cap \operatorname{Mat}_{n \times n}(\mathcal{O}_F)$, $S, T \in \operatorname{Sym}^n(\mathcal{O}_F)$ and $\varepsilon \in \mathcal{O}_F^\times$, let

$$(3.3) \qquad \mathcal{K}^{(n)}(A, S, T, \varepsilon) = \int_{\mathcal{X}_A} \psi\left[\operatorname{tr}\left(XS + \varepsilon \cdot TA^{-1}X^{-1\,t}A^{-1}\right)\right] dX.$$

Here we note that we have $\mathcal{K}^{(n)}(A, S, S, \varepsilon) = \mathcal{K}\ell^{(n)}(A, S, \varepsilon)$. Now we assume moreover that

$$(3.4) \qquad\qquad A \in \varpi \cdot \operatorname{Mat}_{n \times n}(\mathcal{O}_F)$$

since this is the essential case. Suppose that $X \in \mathcal{X}_A$ and $X' \in \operatorname{Sym}^n(\mathcal{O}_F)$. Since $(X + X')A \equiv XA \pmod{\varpi}$, we have $\mathcal{X}_A + \operatorname{Sym}^n(\mathcal{O}_F) = \mathcal{X}_A$. We also have $\mathcal{X}_A \subset \operatorname{Sym}^n(F) \cap \operatorname{Mat}_{n \times n}(\mathcal{O}_F)A^{-1}$. Hence $\mathcal{X}_A$ consists of finitely many $\operatorname{Sym}^n(\mathcal{O}_F)$ cosets. Next we note that by (3.4) we have

$$\mathcal{X}_A A = \left\{Y \in \operatorname{GL}_n(\mathcal{O}_F) \mid {}^t AY \in \operatorname{Sym}^n(\mathcal{O}_F)\right\}$$
$$= \left\{Y \in \operatorname{Mat}_{n \times n}(\mathcal{O}_F) \mid \exists\, U, V \in \operatorname{Mat}_{n \times n}(\mathcal{O}_F) \text{ such that } \begin{pmatrix} Y & U \\ A & V \end{pmatrix} \in \operatorname{Sp}_{2n}(\mathcal{O}_F)\right\}$$

and then since $Y\,{}^t V - U\,{}^t A = 1_n$, we have

$$A^{-1}\left(YA^{-1}\right)^{-1}{}^t A^{-1} = Y^{-1}\left(Y\,{}^t V - U\,{}^t A\right){}^t A^{-1} \equiv {}^t V\,{}^t A^{-1} \pmod{\mathcal{O}_F}.$$

Hence we may write (3.3) as

$$\mathcal{K}^{(n)}(A, S, T, \varepsilon) = \sum_{Y \in \mathcal{X}_A A/\mathrm{Sym}^n(\mathcal{O}_F)A} \psi\left[\mathrm{tr}\left(YA^{-1}S + \varepsilon \cdot T\,{}^tV\,{}^tA^{-1}\right)\right]$$

$$= \sum_V \psi\left[\mathrm{tr}\left(YA^{-1}S + \varepsilon \cdot A^{-1}VT\right)\right]$$

where V runs over the set of representatives of

$$\left\{V \in \mathrm{Mat}_{n \times n}(\mathcal{O}_F) \mid \exists\, U, Y \in \mathrm{Mat}_{n \times n}(\mathcal{O}_F) \text{ such that } \begin{pmatrix} Y & U \\ A & V \end{pmatrix} \in \mathrm{Sp}_{2n}(\mathcal{O}_F)\right\}$$

modulo $A\,\mathrm{Sym}^n(\mathcal{O}_F)$. Thus we have

(3.5) $$K(S, T; A) = \mathcal{K}^{(n)}(A, S, T, 1)$$

where the left hand side of (3.5) denotes Kitaoka's Kloosterman sum in [**K**].

In the following two sections, we specify $n = 2$ and the symmetric matrix S to be

$$\begin{pmatrix} 0 & 1 \\ 1 & 0 \end{pmatrix}, \quad \begin{pmatrix} 1 & 0 \\ 0 & -d \end{pmatrix}$$

respectively and evaluate the Kloosterman sum $\mathcal{K}\ell^{(2)}(A, S, \varepsilon)$ explicitly.

3.2. Evaluation of the split Kloosterman sum

DEFINITION 3.3. Let us define $\mathcal{K}_{\mathrm{spl}}(A, \varepsilon)$, the *split Kloosterman sum*, by

$$\mathcal{K}_{\mathrm{spl}}(A, \varepsilon) = \mathcal{K}\ell^{(2)}(A, S_1, \varepsilon) \quad \text{where} \quad S_1 = \begin{pmatrix} 0 & 1 \\ 1 & 0 \end{pmatrix}.$$

This is the Kloosterman sum which appears in the evaluation of the Novodvorsky orbital integrals.

First we note that by Proposition 3.2 we have

LEMMA 3.4. *For* $\varepsilon_1, \varepsilon_2 \in \mathcal{O}_F^\times$, *we have*

$$\mathcal{K}_{\mathrm{spl}}\left(\begin{pmatrix} \varepsilon_1 & 0 \\ 0 & \varepsilon_2 \end{pmatrix} A, \varepsilon\right) = \mathcal{K}_{\mathrm{spl}}\left(A \begin{pmatrix} \varepsilon_1 & 0 \\ 0 & \varepsilon_2 \end{pmatrix}, \varepsilon\right) = \mathcal{K}_{\mathrm{spl}}(A, \varepsilon \varepsilon_1^{-1} \varepsilon_2^{-1})$$

and we also have

$$\mathcal{K}_{\mathrm{spl}}\left(\begin{pmatrix} 0 & 1 \\ 1 & 0 \end{pmatrix} A, \varepsilon\right) = \mathcal{K}_{\mathrm{spl}}\left(A \begin{pmatrix} 0 & 1 \\ 1 & 0 \end{pmatrix}, \varepsilon\right) = \mathcal{K}_{\mathrm{spl}}(A, \varepsilon).$$

The rest of this section is devoted to the proof of the following theorem.

THEOREM 3.5. *For*

$$A = \begin{pmatrix} \alpha & \beta \\ \gamma & \delta \end{pmatrix} \in \varpi \cdot \mathrm{Mat}_{2 \times 2}(\mathcal{O}_F) \cap \mathrm{GL}_2(F) \quad \text{and} \quad \varepsilon \in \mathcal{O}_F^\times,$$

we have

(3.6) $$\mathcal{K}_{\mathrm{spl}}(A, \varepsilon) = |\Delta|^{-1}\left\{\mathcal{K}\ell\left(2\alpha\Delta^{-1}, 2\varepsilon\delta\Delta^{-1}\right) + \mathcal{K}\ell\left(2\beta\Delta^{-1}, 2\varepsilon\gamma\Delta^{-1}\right)\right\}$$

where $\Delta = \det A$.

By dividing into cases, we prove the equality (3.6) by brute force computation. Since the both sides of (3.6) is invariant under the multiplication by a matrix $\begin{pmatrix} 0 & 1 \\ 1 & 0 \end{pmatrix}$ from both left and right by (2.8), (2.9) and Lemma 3.4, we may assume that $A = \begin{pmatrix} \alpha & \beta \\ \gamma & \delta \end{pmatrix} \in \varpi \cdot \mathrm{Mat}_{2\times 2}(\mathcal{O}_F) \cap \mathrm{GL}_2(F)$ satisfies

$$(3.7) \qquad \max\{|\alpha|, |\beta|, |\gamma|, |\delta|\} = |\delta| < 1,$$

without loss of generality. Then we have the Cartan decomposition

$$A = \begin{pmatrix} 1 & \beta\delta^{-1} \\ 0 & 1 \end{pmatrix} \begin{pmatrix} \Delta\delta^{-1} & 0 \\ 0 & \delta \end{pmatrix} \begin{pmatrix} 1 & 0 \\ \gamma\delta^{-1} & 1 \end{pmatrix}.$$

Let

$$\kappa_1 = \begin{pmatrix} 1 & \beta\delta^{-1} \\ 0 & 1 \end{pmatrix}, \quad \kappa_2 = \begin{pmatrix} 1 & 0 \\ \gamma\delta^{-1} & 1 \end{pmatrix} \quad \text{and} \quad T_A = \begin{pmatrix} \Delta\delta^{-1} & 0 \\ 0 & \delta \end{pmatrix}.$$

Then since $\mathcal{X}_A = {}^t\kappa_1^{-1}\mathcal{X}_{T_A}\kappa_1^{-1}$, we have

$$(3.8) \quad \mathcal{K}_{\mathrm{spl}}(A, \varepsilon) = \int_{Y \in \mathcal{X}_{T_A}}$$
$$\cdot \psi\left[\mathrm{tr}\left(\kappa_1^{-1}\begin{pmatrix} 0 & 1 \\ 1 & 0 \end{pmatrix}{}^t\kappa_1^{-1}Y + \varepsilon \cdot T_A^{-1}{}^t\kappa_2^{-1}\begin{pmatrix} 0 & 1 \\ 1 & 0 \end{pmatrix}\kappa_2^{-1}T_A^{-1}Y^{-1}\right)\right]dY$$
$$= \int_{\binom{r\ s}{s\ t} \in \mathrm{GL}_2(\mathcal{O}_F)\begin{pmatrix}\Delta\delta^{-1} & 0 \\ 0 & \delta\end{pmatrix}^{-1}} \psi\left[-2\beta\delta^{-1}r + 2s - \frac{2\varepsilon\Delta^{-2}(\gamma\delta t + \Delta s)}{rt - s^2}\right]dr\,ds\,dt.$$

3.2.1. The case when $|\Delta| < |\delta|^2$. Then it is easily seen that we have

$$\begin{pmatrix} r & s \\ s & t \end{pmatrix}\begin{pmatrix} \Delta\delta^{-1} & 0 \\ 0 & \delta \end{pmatrix} = \begin{pmatrix} r\Delta\delta^{-1} & s\delta \\ s\Delta\delta^{-1} & t\delta \end{pmatrix} \in \mathrm{GL}_2(\mathcal{O}_F)$$

if and only if

$$(3.9) \qquad r \in \Delta^{-1}\delta\mathcal{O}_F^\times, \quad s \in \delta^{-1}\mathcal{O}_F, \quad t \in \delta^{-1}\mathcal{O}_F^\times.$$

Thus let us write

$$r = \Delta^{-1}\delta\varepsilon_1, \quad s = \delta^{-1}\xi, \quad t = \delta^{-1}\varepsilon_2.$$

Then we have

$$\mathcal{K}_{\mathrm{spl}}(A, \varepsilon) = |\Delta\delta|^{-1}\int_{\mathcal{O}_F^\times}\int_{\mathcal{O}_F^\times}\int_{\mathcal{O}_F}$$
$$\cdot \psi\left[-2\beta\Delta^{-1}\varepsilon_1 + 2\delta^{-1}\xi - \frac{2\varepsilon\Delta^{-2}(\gamma\varepsilon_2 + \Delta\delta^{-1}\xi)}{\Delta^{-1}\varepsilon_1\varepsilon_2 - \delta^{-2}\xi^2}\right]d\varepsilon_1\,d\varepsilon_2\,d\xi.$$

By a change of variable $\varepsilon_2 \mapsto \varepsilon_1^{-1}\varepsilon_2$, we have

$$\mathcal{K}_{\mathrm{spl}}(A, \varepsilon) = |\Delta\delta|^{-1}\int_{\mathcal{O}_F^\times}\int_{\mathcal{O}_F^\times}\int_{\mathcal{O}_F}$$
$$\cdot \psi\left[-2\beta\Delta^{-1}\varepsilon_1 + 2\delta^{-1}\xi - \frac{2\varepsilon\Delta^{-1}(\gamma\varepsilon_1^{-1}\varepsilon_2 + \Delta\delta^{-1}\xi)}{\varepsilon_2 - \Delta\delta^{-2}\xi^2}\right]d\varepsilon_1\,d\varepsilon_2\,d\xi.$$

Since $|\Delta\delta^{-2}| < 1$, by a further change of variable $\varepsilon_2 \mapsto \varepsilon_2 + \Delta\delta^{-2}\xi^2$, followed by another change of variable $\varepsilon_2 \mapsto \varepsilon_2^{-1}$, we have

$$\mathcal{K}_{\mathrm{spl}}(A,\varepsilon) = |\Delta\delta|^{-1} \int_{\mathcal{O}_F^\times} \int_{\mathcal{O}_F^\times} \int_{\mathcal{O}_F}$$
$$\psi\left[-2\beta\Delta^{-1}\varepsilon_1 + 2\delta^{-1}\xi - 2\varepsilon\Delta^{-1}\varepsilon_2^{-1}\left(\gamma\varepsilon_1^{-1}\varepsilon_2 + \Delta\gamma\delta^{-2}\varepsilon_1^{-1}\xi^2 + \Delta\delta^{-1}\xi\right)\right] d\varepsilon_1 d\varepsilon_2 d\xi$$
$$= |\Delta\delta|^{-1} \int_{\mathcal{O}_F^\times} \psi\left(-2\beta\Delta^{-1}\varepsilon_1 - 2\varepsilon\gamma\Delta^{-1}\varepsilon_1^{-1}\right) \int_{\mathcal{O}_F} \psi\left(2\delta^{-1}\xi\right)$$
$$\left(\int_{\mathcal{O}_F^\times} \psi\left[-2\varepsilon\delta^{-1}\xi\left(1 + \varepsilon_1^{-1}\gamma\delta^{-1}\xi\right)\varepsilon_2\right] d\varepsilon_2\right) d\xi\, d\varepsilon_1.$$

3.2.1.1. *When $|\gamma| < |\delta|$.* Then since $1 + \varepsilon_1^{-1}\gamma\delta\xi \in \mathcal{O}_F^\times$, by a change of variable $\varepsilon_2 \mapsto \left(1 + \varepsilon_1^{-1}\gamma\delta\xi\right)^{-1}\varepsilon_2$, we have

$$\mathcal{K}_{\mathrm{spl}}(A,\varepsilon) = |\Delta\delta|^{-1} \cdot \mathcal{K}\ell\left(2\beta\Delta^{-1}, 2\varepsilon\gamma\Delta^{-1}\right) \int_{\mathcal{O}_F^\times} \left(\int_{\mathcal{O}_F} \psi\left[2\delta^{-1}\left(1 - \varepsilon\varepsilon_2\right)\xi\right] d\xi\right) d\varepsilon_2.$$

Here

$$\int_{\mathcal{O}_F^\times} \left(\int_{\mathcal{O}_F} \psi\left[2\delta^{-1}\left(1 - \varepsilon\varepsilon_2\right)\xi\right] d\xi\right) d\varepsilon_2 = \int_{\varepsilon^{-1}+\delta\mathcal{O}_F} d\varepsilon_2 = |\delta|.$$

Hence

$$\mathcal{K}_{\mathrm{spl}}(A,\varepsilon) = |\Delta|^{-1} \cdot \mathcal{K}\ell\left(2\beta\Delta^{-1}, 2\varepsilon\gamma\Delta^{-1}\right).$$

If $|\alpha| = |\delta|$, then $|\Delta| = |\alpha\delta - \beta\gamma| = |\delta|^2$ and this is a contradiction. Thus we have $|\alpha| < |\delta|$. Hence

$$|\alpha\Delta^{-1}| \neq |\delta\Delta^{-1}| > |\delta|^{-1} > 1.$$

Thus by Proposition 2.7 we have

$$\mathcal{K}\ell\left(2\alpha\Delta^{-1}, 2\varepsilon\delta\Delta^{-1}\right) = 0$$

in this case. Therefore the equality (3.6) holds in this case.

3.2.1.2. *When $|\gamma| = |\delta|$.* Let us write

$$\vartheta = -2\varpi\varepsilon\delta^{-1}\xi\left(1 + \varepsilon_1^{-1}\gamma\delta^{-1}\xi\right) = -2\varepsilon\varepsilon_1^{-1}\gamma\delta^{-1} \cdot \varpi\delta^{-1}\xi\left(\xi + \gamma^{-1}\delta\varepsilon_1\right).$$

We recall that

$$\int_{\mathcal{O}_F^\times} \psi\left(\varpi^{-1}\vartheta\varepsilon_2\right) d\varepsilon_2 = \begin{cases} 1 - q^{-1}, & \text{when } |\vartheta| < 1 \\ -q^{-1}, & \text{when } |\vartheta| = 1 \\ 0, & \text{otherwise.} \end{cases}$$

Here we have

$$|\vartheta| < 1 \iff \xi \in \delta\mathcal{O}_F, \text{ or, } \xi \in -\gamma^{-1}\delta\varepsilon_1 + \delta\mathcal{O}_F$$

and

$$|\vartheta| = 1 \iff \begin{cases} \xi \in \varpi^{-1}\delta\mathcal{O}_F^\times, \text{ or, } \xi \in -\gamma^{-1}\delta\varepsilon_1 + \varpi^{-1}\delta\mathcal{O}_F^\times, & \text{when } \mathrm{ord}\,(\delta) \geq 2 \\ \xi \in \mathcal{O}_F \setminus \left(\varpi\mathcal{O}_F \cup \left(-\gamma^{-1}\delta\varepsilon_1 + \varpi\mathcal{O}_F\right)\right), & \text{when } \mathrm{ord}\,(\delta) = 1. \end{cases}$$

Therefore

$$\int_{\mathcal{O}_F} \psi\left(2\delta^{-1}\xi\right) \left(\int_{\mathcal{O}_F^\times} \psi\left[-2\varepsilon\delta^{-1}\xi\left(1+\varepsilon_1^{-1}\gamma\delta^{-1}\xi\right)\varepsilon_2\right] d\varepsilon_2\right) d\xi$$

$$= (1-q^{-1}) \int_{\delta\mathcal{O}_F \cup (-\gamma^{-1}\delta\varepsilon_1+\delta\mathcal{O}_F)} \psi\left(2\delta^{-1}\xi\right) d\xi$$

$$- q^{-1} \cdot \begin{cases} \int_{\varpi^{-1}\delta\mathcal{O}_F^\times \cup (-\gamma^{-1}\delta\varepsilon_1+\varpi^{-1}\delta\mathcal{O}_F^\times)} \psi\left(2\delta^{-1}\xi\right) d\xi, & \text{when ord}(\delta) \geq 2 \\ \int_{\mathcal{O}_F \setminus (\varpi\mathcal{O}_F \cup -\gamma^{-1}\delta\varepsilon_1+\varpi\mathcal{O}_F)} \psi\left(2\delta^{-1}\xi\right) d\xi, & \text{when ord}(\delta) = 1 \end{cases}$$

$$= |\delta| \left\{1 + \psi\left(-2\gamma^{-1}\varepsilon_1\right)\right\}.$$

Hence

$$\mathcal{K}_{\mathrm{spl}}(A,\varepsilon) = |\Delta|^{-1} \int_{\mathcal{O}_F^\times}$$
$$\cdot \left\{\psi\left(-2\beta\Delta^{-1}\varepsilon_1 - 2\varepsilon\gamma\Delta^{-1}\varepsilon_1^{-1}\right) + \psi\left[-2\left(\beta\Delta^{-1}+\gamma^{-1}\right)\varepsilon_1 - 2\varepsilon\gamma\Delta^{-1}\varepsilon_1^{-1}\right]\right\} d\varepsilon_1$$
$$= |\Delta|^{-1} \left\{\mathcal{K}\ell\left(2\beta\Delta^{-1}, 2\varepsilon\gamma\Delta^{-1}\right) + \mathcal{K}\ell\left(2\gamma^{-1}\delta\alpha\Delta^{-1}, 2\varepsilon\gamma\Delta^{-1}\right)\right\}$$
$$= |\Delta|^{-1} \left\{\mathcal{K}\ell\left(2\beta\Delta^{-1}, 2\varepsilon\gamma\Delta^{-1}\right) + \mathcal{K}\ell\left(2\alpha\Delta^{-1}, 2\varepsilon\delta\Delta^{-1}\right)\right\},$$

since $\gamma^{-1}\delta \in \mathcal{O}_F^\times$. Thus the equality (3.6) holds in this case also.

3.2.2. The case when $|\Delta| = |\delta|^2$. Then since

$$\begin{pmatrix} \Delta\delta^{-1} & 0 \\ 0 & \delta \end{pmatrix} \in \delta \cdot \mathrm{GL}_2(\mathcal{O}_F),$$

we have

$$\mathrm{Sym}^2(F) \cap \mathrm{GL}_2(\mathcal{O}_F) \begin{pmatrix} \Delta\delta^{-1} & 0 \\ 0 & \delta \end{pmatrix}^{-1} = \mathrm{Sym}^2(F) \cap \delta^{-1}\mathrm{GL}_2(\mathcal{O}_F).$$

Hence (3.8) becomes

$$\mathcal{K}_{\mathrm{spl}}(A,\varepsilon) = |\delta|^{-3} \int_{\{(r,s,t)\in\mathcal{O}_F^3 \mid |rt-s^2|=1\}}$$
$$\cdot \psi\left[-2\beta\delta^{-2}r + 2\delta^{-1}s - \frac{2\varepsilon\Delta^{-2}\left(\gamma\delta^2 t + \Delta\delta s\right)}{rt-s^2}\right] dr\,ds\,dt.$$

Thus we have

$$\mathcal{K}_{\mathrm{spl}}(A,\varepsilon) = \mathcal{K}_{\mathrm{spl}}^{(1)}(A,\varepsilon) + \mathcal{K}_{\mathrm{spl}}^{(2)}(A,\varepsilon)$$

where

$$(3.10) \quad \mathcal{K}_{\mathrm{spl}}^{(1)}(A,\varepsilon) = |\delta|^{-3} \int_{(\varepsilon_1,s,\varepsilon_2)\in\mathcal{O}_F^\times \times \varpi\mathcal{O}_F \times \mathcal{O}_F^\times}$$
$$\cdot \psi\left[-2\beta\delta^{-2}\varepsilon_1 + 2\delta^{-1}s - \frac{2\varepsilon\Delta^{-2}\left(\gamma\delta^2\varepsilon_2 + \Delta\delta s\right)}{\varepsilon_1\varepsilon_2 - s^2}\right] d\varepsilon_1\,d\varepsilon_2\,ds$$

and

$$(3.11) \quad \mathcal{K}_{\text{spl}}^{(2)}(A,\varepsilon) = |\delta|^{-3} \int_{\{(r,s,t)\in\mathcal{O}_F\times\mathcal{O}_F^\times\times\mathcal{O}_F|\ |rt-s^2|=1\}}$$
$$\cdot \psi\left[-2\beta\delta^{-2}r + 2\delta^{-1}s - \frac{2\varepsilon\Delta^{-2}\left(\gamma\delta^2 t + \Delta\delta s\right)}{rt-s^2}\right] dr\, ds\, dt.$$

We shall show that

$$(3.12) \quad \mathcal{K}_{\text{spl}}^{(1)}(A,\varepsilon) = \begin{cases} |\Delta|^{-1}\cdot\mathcal{K}\ell\left(2\beta\Delta^{-1}, 2\varepsilon\gamma\Delta^{-1}\right), & \text{when ord}(\delta) > 1 \\ \left(1-q^{-1}\right)|\Delta|^{-1}\cdot\mathcal{K}\ell\left(2\beta\Delta^{-1}, 2\varepsilon\gamma\Delta^{-1}\right), & \text{when ord}(\delta) = 1 \end{cases}$$

and

$$(3.13) \quad \mathcal{K}_{\text{spl}}^{(2)}(A,\varepsilon) = |\Delta|^{-1}\cdot\mathcal{K}\ell\left(2\alpha\Delta^{-1}, 2\varepsilon\delta\Delta^{-1}\right)$$
$$+ \begin{cases} 0, & \text{when ord}(\delta) > 1 \\ q^{-1}|\Delta|^{-1}\cdot\mathcal{K}\ell\left(2\beta\Delta^{-1}, 2\varepsilon\gamma\Delta^{-1}\right), & \text{when ord}(\delta) = 1. \end{cases}$$

Now let us write

$$(3.14) \quad A = \begin{pmatrix} \alpha & \beta \\ \gamma & \delta \end{pmatrix} = \delta\begin{pmatrix} a & b \\ c & 1 \end{pmatrix}.$$

Then by assumption, we have

$$(3.15) \quad a,b,c \in \mathcal{O}_F,\ |a-bc|=1.$$

3.2.2.1. *Proof of* (3.12). By a change of variable $\varepsilon_2 \mapsto \varepsilon_1^{-1}\varepsilon_2$, we have

$$\mathcal{K}_{\text{spl}}^{(1)}(A,\varepsilon) = |\delta|^{-3}\int_{(\varepsilon_1,s,\varepsilon_2)\in\mathcal{O}_F^\times\times\varpi\mathcal{O}_F\times\mathcal{O}_F^\times}$$
$$\cdot \psi\left[-2b\delta^{-1}\varepsilon_1 + 2\delta^{-1}s - \frac{2\varepsilon\Delta^{-2}\delta\left(c\delta^2\varepsilon_1^{-1}\varepsilon_2 + \Delta s\right)}{\varepsilon_2 - s^2}\right] d\varepsilon_1\, d\varepsilon_2\, ds.$$

Further, by a change of variable $\varepsilon_2 \mapsto \varepsilon_2 + s^2$, followed by another change of variable $\varepsilon_2 \mapsto \varepsilon_2^{-1}$, we have

$$\mathcal{K}_{\text{spl}}^{(1)}(A,\varepsilon) = |\delta|^{-3}\int_{\mathcal{O}_F^\times} \psi\left(-2b\delta^{-1}\varepsilon_1 - 2\varepsilon c\Delta^{-2}\delta^3\varepsilon_1^{-1}\right) \int_{\varpi\mathcal{O}_F} \psi\left(2\delta^{-1}s\right)$$
$$\int_{\mathcal{O}_F^\times} \psi\left[-2\varepsilon\Delta^{-2}\delta^4\varepsilon_1^{-1}\left(cs+\Delta\delta^{-2}\varepsilon_1\right)\cdot\delta^{-1}s\varepsilon_2\right] d\varepsilon_2\, ds\, d\varepsilon_1.$$

Since $s\in\varpi\mathcal{O}_F$ and $\Delta\delta^{-2}\in\mathcal{O}_F^\times$, we have
$$-2\varepsilon\Delta^{-2}\delta^4\varepsilon_1^{-1}\left(cs+\Delta\delta^{-2}\varepsilon_1\right)\in\mathcal{O}_F^\times.$$

Hence we have

$$\int_{\mathcal{O}_F^\times} \psi\left[-2\varepsilon\Delta^{-2}\delta^4\varepsilon_1^{-1}\left(cs+\Delta\delta^{-2}\varepsilon_1\right)\cdot\delta^{-1}s\varepsilon_2\right] d\varepsilon_2$$
$$= \begin{cases} 1-q^{-1}, & \text{when } s\in\delta\mathcal{O}_F \\ -q^{-1}, & \text{when } s\in\varpi^{-1}\delta\mathcal{O}_F^\times \\ 0, & \text{otherwise}. \end{cases}$$

Thus

$$(1-q^{-1})\int_{\varpi\mathcal{O}_F\cap\delta\mathcal{O}_F}\psi\left(2\delta^{-1}s\right)ds - q^{-1}\int_{\varpi\mathcal{O}_F\cap\delta\varpi^{-1}\mathcal{O}_F^\times}\psi\left(2\delta^{-1}s\right)ds$$

$$= \begin{cases} |\delta|, & \text{when ord}(\delta) > 1 \\ (1-q^{-1})|\delta|, & \text{when ord}(\delta) = 1. \end{cases}$$

On the other hand, since $\Delta^{-1}\delta^2 \in \mathcal{O}_F^\times$, we have

$$\int_{\mathcal{O}_F^\times}\psi\left(-2b\delta^{-1}\varepsilon_1 - 2\varepsilon c\Delta^{-2}\delta^3\varepsilon_1^{-1}\right)d\varepsilon_1 = \mathcal{K}\ell\left(2\beta\delta^{-2}, 2\varepsilon\gamma\Delta^{-2}\delta^2\right)$$
$$=\mathcal{K}\ell\left(2\beta\delta^{-2}\cdot\Delta^{-1}\delta^2, 2\varepsilon\gamma\Delta^{-2}\delta^2\cdot\Delta\delta^{-2}\right) = \mathcal{K}\ell\left(2\beta\Delta^{-1}, 2\varepsilon\gamma\Delta^{-1}\right)$$

by (2.9). Thus we have proved the equality (3.12).

3.2.2.2. *Proof of* (3.13). By a change of variable $r \mapsto rs$, $t \mapsto st$, we have

(3.16) $$\mathcal{K}_{\text{spl}}^{(2)}(A,\varepsilon) = |\delta|^{-3}\int_{\mathcal{O}_F^\times}\psi\left(2\delta^{-1}s\right)$$
$$\left(\int_{\{(r,t)\mid r,t\in\mathcal{O}_F,|rt-1|=1\}}\psi\left[-2b\delta^{-1}rs - \frac{2\varepsilon\Delta^{-2}\delta\left(c\delta^2 t + \Delta\right)}{rt-1}s^{-1}\right]dr\,dt\right)ds.$$

Let us denote the inner integral of (3.16) by $\mathcal{I}(s)$, i.e.

(3.17)
$$\mathcal{I}(s) = \int_{\{(r,t)\mid r,t\in\mathcal{O}_F,|rt-1|=1\}}\psi\left[-2b\delta^{-1}rs - \frac{2\varepsilon\Delta^{-2}\delta\left(c\delta^2 t + \Delta\right)}{rt-1}s^{-1}\right]dr\,dt.$$

Here we note that

$$\{(r,t) \mid r,t \in \mathcal{O}_F, |rt-1| = 1\}$$
$$= \{(r,t) \mid r,t \in \mathcal{O}_F, |rt-1| = 1, r \neq 0\} \cup \{(0,t) \mid t \in \mathcal{O}_F\}$$

and the latter set is of measure zero. Hence

$$\mathcal{I}(s) = \int_{\{(r,t)\mid r,t\in\mathcal{O}_F,|rt-1|=1,r\neq 0\}}\psi\left[-2b\delta^{-1}rs - \frac{2\varepsilon\Delta^{-2}\delta\left(c\delta^2 t + \Delta\right)}{rt-1}s^{-1}\right]dr\,dt.$$

Let $v = r$ and $w = rt - 1$. Then we have

$$\mathcal{I}(s) = \int_{\{(v,w)\mid w\in\mathcal{O}_F^\times,|w+1|\leq|v|\leq 1,v\neq 0\}}$$
$$|v|^{-1}\cdot\psi\left[-2b\delta^{-1}vs - \frac{2\varepsilon\Delta^{-2}\delta\left(c\delta^2 v^{-1}(w+1) + \Delta\right)}{w}s^{-1}\right]dv\,dw.$$

Here we note that

$$\{(v,w) \mid w \in \mathcal{O}_F^\times, |w+1| \leq |v| \leq 1, v \neq 0\}$$
$$= \{(v,w) \mid |v| = |w| = 1\} \cup \left(\bigcup_{m\geq 1}\{(v,w) \mid v \in \varpi^m\mathcal{O}_F^\times, w \in -1+\varpi^m\mathcal{O}_F\}\right).$$

Hence let

$$
(3.18) \quad \mathcal{I}^{(0)}(s) = \int_{\{(v,w)\,|\,|v|=|w|=1\}} \psi\left[-2b\delta^{-1}vs - \frac{2\varepsilon\Delta^{-2}\delta\left(c\delta^2 v^{-1}(w+1)+\Delta\right)}{w}s^{-1}\right] dv\, dw
$$

and let

$$
(3.19) \quad \mathcal{I}^{(m)}(s) = q^m \int_{\{(v,w)\,|\,v\in\varpi^m\mathcal{O}_F^\times,\, w\in -1+\varpi^m\mathcal{O}_F\}} \psi\left[-2b\delta^{-1}vs - \frac{2\varepsilon\Delta^{-2}\delta\left(c\delta^2 v^{-1}(w+1)+\Delta\right)}{w}s^{-1}\right] dv\, dw
$$

for $m \geq 1$.

Let us compute $\mathcal{I}^{(0)}(s)$ first. By a change of variable $w \mapsto w^{-1}$, we have

$$
\mathcal{I}^{(0)}(s) = \int_{\mathcal{O}_F^\times} \psi\left(-2b\delta^{-1}sv - 2\varepsilon\Delta^{-2}\delta^3 cs^{-1}v^{-1}\right)
$$
$$
\left(\int_{\mathcal{O}_F^\times} \psi\left[-2\varepsilon\Delta^{-1}\delta^2 v^{-1}s^{-1}\cdot \delta^{-1}\left(v + c\Delta^{-1}\delta^2\right)w\right] dw\right) dv.
$$

Since
$$
-2\varepsilon\Delta^{-1}\delta^2 v^{-1}s^{-1} \in \mathcal{O}_F^\times,
$$
we have

$$
(3.20) \quad \mathcal{I}^{(0)}(s) = \int_{\mathcal{O}_F^\times} \psi\left(-2b\delta^{-1}sv - 2\varepsilon\Delta^{-2}\delta^3 cs^{-1}v^{-1}\right)
$$
$$
\left(\int_{\mathcal{O}_F^\times} \psi\left[\delta^{-1}\left(v + c\Delta^{-1}\delta^2\right)w\right] dw\right) dv.
$$

When $|c| < 1$, we have
$$
v + c\Delta^{-1}\delta^2 \in \mathcal{O}_F^\times.
$$

Hence
$$
\int_{\mathcal{O}_F^\times} \psi\left[\delta^{-1}\left(v + c\Delta^{-1}\delta^2\right)w\right] dw = \begin{cases} 0, & \text{when } \mathrm{ord}(\delta) > 1 \\ -q^{-1}, & \text{when } \mathrm{ord}(\delta) = 1. \end{cases}
$$

Thus

$$
(3.21) \quad \mathcal{I}^{(0)}(s) = \begin{cases} 0, & \text{when } \mathrm{ord}(\delta) > 1 \\ -q^{-1}\cdot \mathcal{K}\ell\left(2\beta\Delta^{-1}, 2\varepsilon\gamma\Delta^{-1}\right), & \text{when } \mathrm{ord}(\delta) = 1. \end{cases}
$$

Suppose that $|c| = 1$. Then we have

$$
(3.22) \quad \mathcal{I}^{(0)}(s) = \left(1 - q^{-1}\right) \int_{-c\Delta^{-1}\delta^2 + \delta\mathcal{O}_F} \psi\left(-2b\delta^{-1}sv - 2\varepsilon\Delta^{-2}\delta^3 cs^{-1}v^{-1}\right) dv
$$
$$
- q^{-1} \int_{\mathcal{O}_F^\times \cap \left(-c\Delta^{-1}\delta^2 + \varpi^{-1}\delta\mathcal{O}_F^\times\right)} \psi\left(-2b\delta^{-1}sv - 2\varepsilon\Delta^{-2}\delta^3 cs^{-1}v^{-1}\right) dv.
$$

3.2. EVALUATION OF THE SPLIT KLOOSTERMAN SUM

When $\text{ord}(\delta) > 1$, we have

$$\mathcal{I}^{(0)}(s) = \int_{-c\Delta^{-1}\delta^2 + \delta\mathcal{O}_F} \psi\left(-2b\delta^{-1}sv - 2\varepsilon\Delta^{-2}\delta^3 cs^{-1}v^{-1}\right) dv$$
$$- q^{-1} \int_{-c\Delta^{-1}\delta^2 + \varpi^{-1}\delta\mathcal{O}_F} \psi\left(-2b\delta^{-1}sv - 2\varepsilon\Delta^{-2}\delta^3 cs^{-1}v^{-1}\right) dv.$$

For $v = -c\Delta^{-1}\delta^2 + \varpi^{-1}\delta x$ where $x \in \mathcal{O}_F$, we have

$$v^{-1} \equiv -c^{-1}\Delta\delta^{-2} - \varpi^{-1}c^{-2}\Delta^2\delta^{-3}x \pmod{\delta}$$

since $\text{ord}(\delta) > 1$. Hence we have

$$- 2b\delta^{-1}sv - 2\varepsilon\Delta^{-2}\delta^3 cs^{-1}v^{-1}$$
$$\equiv 2bc\delta\Delta^{-1}s + 2\varepsilon\delta\Delta^{-1}s^{-1} - 2\varpi^{-1}s^{-1}\left(bs^2 - \varepsilon c^{-1}\right)x \pmod{\mathcal{O}_F}.$$

Therefore

(3.23) $$\mathcal{I}^{(0)}(s) = \int_{-c\Delta^{-1}\delta^2 + \delta\mathcal{O}_F} \psi\left(-2b\delta^{-1}sv - 2\varepsilon\Delta^{-2}\delta^3 cs^{-1}v^{-1}\right) dv$$
$$- q^{-1} \int_{-c\Delta^{-1}\delta^2 + \varpi^{-1}\delta\mathcal{O}_F} \psi\left(-2b\delta^{-1}sv - 2\varepsilon\Delta^{-2}\delta^3 cs^{-1}v^{-1}\right) dv$$
$$= |\delta| \cdot \psi\left(2bc\delta\Delta^{-1}s + 2\varepsilon\delta\Delta^{-1}s^{-1}\right)$$
$$\cdot \left\{1 - \int_{\mathcal{O}_F} \psi\left[-2\varpi^{-1}s^{-1}\left(bs^2 - \varepsilon c^{-1}\right)x\right] dx\right\}$$
$$= |\delta| \cdot \psi\left(2bc\delta\Delta^{-1}s + 2\varepsilon\delta\Delta^{-1}s^{-1}\right) \cdot (1 - \varrho(s)),$$

where $\varrho(s)$ denotes the characteristic function of the set

$$\Upsilon = \left\{s \in \mathcal{O}_F^\times \mid bs^2 - \varepsilon c^{-1} \in \varpi\mathcal{O}_F\right\}.$$

We note that Υ is empty unless $b \in \mathcal{O}_F^\times$ and $\varepsilon(bc)^{-1} \in \left(\mathcal{O}_F^\times\right)^2$, and then

(3.24)
$$\Upsilon = \zeta(1 + \varpi\mathcal{O}_F) \cup \{-\zeta(1 + \varpi\mathcal{O}_F)\}, \text{ where } \zeta \in \mathcal{O}_F^\times \text{ such that } \zeta^2 = \varepsilon(bc)^{-1}.$$

When $\text{ord}(\delta) = 1$, we have

$$\mathcal{I}^{(0)}(s) = (1 - q^{-1}) \int_{-c\Delta^{-1}\delta^2 + \delta\mathcal{O}_F} \psi\left(-2b\delta^{-1}sv - 2\varepsilon\Delta^{-2}\delta^3 cs^{-1}v^{-1}\right) dv$$
$$q^{-1} \int_{\mathcal{O}_F^\times \setminus (-c\Delta^{-1}\delta^2 + \varpi\mathcal{O}_F)} \psi\left(-2b\delta^{-1}sv - 2\varepsilon\Delta^{-2}\delta^3 cs^{-1}v^{-1}\right) dv.$$

Hence

(3.25) $$\mathcal{I}^{(0)}(s) = \int_{-c\Delta^{-1}\delta^2 + \delta\mathcal{O}_F} \psi\left(-2b\delta^{-1}sv - 2\varepsilon\Delta^{-2}\delta^3 cs^{-1}v^{-1}\right) dv$$
$$- q^{-1} \int_{\mathcal{O}_F^\times} \psi\left(-2b\delta^{-1}sv - 2\varepsilon\Delta^{-2}\delta^3 cs^{-1}v^{-1}\right) dv$$
$$= |\delta| \cdot \psi\left(2bc\delta\Delta^{-1}s + 2\varepsilon\delta\Delta^{-1}s^{-1}\right) - |\delta| \cdot \mathcal{K}\ell\left(2\beta\Delta^{-1}, 2\varepsilon\gamma\Delta^{-1}\right).$$

Now let us compute $\mathcal{I}^{(m)}(s)$ for $m \geq 1$. By a change of variable $w \mapsto w^{-1}$, (3.19) becomes

$$(3.26) \quad \mathcal{I}^{(m)}(s) = q^m \int_{\varpi^m \mathcal{O}_F^\times} \psi\left(-2b\delta^{-1}sv - 2\varepsilon\Delta^{-2}\delta^3 cs^{-1}v^{-1}\right)$$

$$\left(\int_{-1+\varpi^m \mathcal{O}_F} \psi\left[-2\varepsilon\Delta^{-1}\delta^2 s^{-1} \cdot \delta^{-1}v^{-1}\left(v + c\Delta^{-1}\delta^2\right)w\right]dw\right)dv.$$

Suppose that $|c| = 1$. Then the inner integral vanishes, since

$$v + c\Delta^{-1}\delta^2 \in \mathcal{O}_F^\times$$

and it in turn implies that

$$|-2\varepsilon\Delta^{-1}\delta^2 s^{-1} \cdot \delta^{-1}v^{-1}\left(v + c\Delta^{-1}\delta^2\right)| = |\delta^{-1}v^{-1}| > q^m.$$

Suppose that $|c| < 1$. Let us write $w = -1 + vy$ in (3.26). Then we have

$$(3.27) \quad \mathcal{I}^{(m)}(s) = \psi\left(2\varepsilon\Delta^{-1}\delta s^{-1}\right) \int_{\mathcal{O}_F} \psi\left(-2\varepsilon c\Delta^{-2}\delta^3 s^{-1}y\right)$$

$$\left(\int_{\varpi^m \mathcal{O}_F^\times} \psi\left[-2\varepsilon\Delta^{-1}\delta^2 s^{-1} \cdot \delta^{-1}\left(y + \varepsilon^{-1}b\Delta\delta^{-2}s^2\right)v\right]dv\right)dy.$$

Thus by a change of variable $y \mapsto y - \varepsilon^{-1}b\Delta\delta^{-2}s^2$ and $v \mapsto \left(-2\varepsilon\Delta^{-1}\delta^2 s^{-1}\right)^{-1}v$, we have

$$(3.28) \quad \mathcal{I}^{(m)}(s) = \psi\left(2\varepsilon\Delta^{-1}\delta s^{-1} + 2bc\Delta^{-1}\delta s\right)$$

$$\cdot \int_{\varpi^m \mathcal{O}_F^\times} \left(\int_{\mathcal{O}_F} \psi\left[\delta^{-1}\left(v - 2\varepsilon c\Delta^{-2}\delta^4 s^{-1}\right)y\right]dy\right)dv$$

$$= \psi\left(2\varepsilon\Delta^{-1}\delta s^{-1} + 2bc\Delta^{-1}\delta s\right) \int_{\varpi^m \mathcal{O}_F^\times \cap (2\varepsilon c\Delta^{-2}\delta^4 s^{-1} + \delta\mathcal{O}_F)} dv.$$

Hence we have

$$\sum_{m=1}^\infty \mathcal{I}^{(m)}(s) = \psi\left(2\varepsilon\Delta^{-1}\delta s^{-1} + 2bc\Delta^{-1}\delta s\right) \sum_{m=1}^\infty \int_{\varpi^m \mathcal{O}_F^\times \cap (2\varepsilon c\Delta^{-2}\delta^4 s^{-1} + \delta\mathcal{O}_F)} dv$$

$$= \psi\left(2\varepsilon\Delta^{-1}\delta s^{-1} + 2bc\Delta^{-1}\delta s\right) \int_{\varpi\mathcal{O}_F \cap (2\varepsilon c\Delta^{-2}\delta^4 s^{-1} + \delta\mathcal{O}_F)} dv$$

$$= |\delta| \cdot \psi\left(2\varepsilon\Delta^{-1}\delta s^{-1} + 2bc\Delta^{-1}\delta s\right),$$

since

$$2\varepsilon c\Delta^{-2}\delta^4 s^{-1} \in \varpi\mathcal{O}_F.$$

Thus we have shown that for

$$\mathcal{I}(s) = \mathcal{I}^{(0)}(s) + \sum_{m=1}^\infty \mathcal{I}^{(m)}(s),$$

we have

1. When $\mathrm{ord}(\delta) > 1$ and $|c| < 1$,

$$\mathcal{I}(s) = |\delta| \cdot \psi\left(2\varepsilon\Delta^{-1}\delta s^{-1} + 2\beta\gamma\Delta^{-1}\delta^{-1}s\right).$$

2. When $\mathrm{ord}(\delta) > 1$ and $|c| = 1$,

$$\mathcal{I}(s) = |\delta| \cdot \psi\left(2\varepsilon\Delta^{-1}\delta s^{-1} + 2\beta\gamma\Delta^{-1}\delta^{-1}s\right) \cdot (1 - \varrho(s)).$$

3.2. EVALUATION OF THE SPLIT KLOOSTERMAN SUM

3. When $\text{ord}(\delta) = 1$,
$$\mathcal{I}(s) = |\delta| \cdot \psi\left(2\varepsilon\Delta^{-1}\delta s^{-1} + 2\beta\gamma\Delta^{-1}\delta^{-1}s\right) - |\delta| \cdot \mathcal{K}\ell\left(2\beta\Delta^{-1}, 2\varepsilon\gamma\Delta^{-1}\right).$$

Here we have
$$|\delta|^{-3} \int_{\mathcal{O}_F^\times} \psi\left(2\delta^{-1}s\right) |\delta| \cdot \psi\left(2\varepsilon\Delta^{-1}\delta s^{-1} + 2\beta\gamma\Delta^{-1}\delta^{-1}s\right) ds$$
$$= |\Delta|^{-1} \int_{\mathcal{O}_F^\times} \psi\left(2\alpha\Delta^{-1}s + 2\varepsilon\delta\Delta^{-1}s^{-1}\right) ds$$
$$= |\Delta|^{-1} \cdot \mathcal{K}\ell\left(2\alpha\Delta^{-1}, 2\varepsilon\delta\Delta^{-1}\right)$$

and when $\text{ord}(\delta) = 1$
$$- |\delta|^{-2} \cdot \mathcal{K}\ell\left(2\beta\Delta^{-1}, 2\varepsilon\gamma\Delta^{-1}\right) \int_{\mathcal{O}_F^\times} \psi\left(2\delta^{-1}s\right) ds$$
$$= q^{-1}|\Delta|^{-1} \cdot \mathcal{K}\ell\left(2\beta\Delta^{-1}, 2\varepsilon\gamma\Delta^{-1}\right).$$

Finally suppose that $\text{ord}(\delta) > 1$, $b \in \mathcal{O}_F^\times$ and $\varepsilon(bc)^{-1} \in \left(\mathcal{O}_F^\times\right)^2$. Let $\zeta \in \mathcal{O}_F^\times$ such that $\zeta^2 = \varepsilon(bc)^{-1}$ and let $\ell = \text{ord}(\delta)$. Then we have

$$\int_{\mathcal{O}_F^\times} \psi\left(2\delta^{-1}s\right) \cdot \psi\left(2\varepsilon\Delta^{-1}\delta s^{-1} + 2\beta\gamma\Delta^{-1}\delta^{-1}s\right) \varrho(s) ds$$
$$= \int_{\pm\zeta(1+\varpi\mathcal{O}_F)} \psi\left(2\alpha\Delta^{-1}s + 2\varepsilon\delta\Delta^{-1}s^{-1}\right) ds$$
$$= \int_{\mathcal{O}_F} \int_{\pm\zeta(1+\varpi\mathcal{O}_F)} \psi\left[2\alpha\Delta^{-1}s\left(1+\varpi^{\ell-1}x\right)^{-1} + 2\varepsilon\delta\Delta^{-1}s^{-1}\left(1+\varpi^{\ell-1}x\right)\right] ds\, dx.$$

Here we have
$$\left(1+\varpi^{\ell-1}x\right)^{-1} \in \left(1 - \varpi^{\ell-1}x\right) + \delta\mathcal{O}_F$$

and hence
$$2\alpha\Delta^{-1}s\left(1+\varpi^{\ell-1}x\right)^{-1} \in \left(2\alpha\Delta^{-1}s - 2\varpi^{\ell-1}\alpha\Delta^{-1}sx\right) + \mathcal{O}_F.$$

Thus
(3.29)
$$\int_{\mathcal{O}_F^\times} \psi\left(2\delta^{-1}s\right) \cdot \psi\left(2\varepsilon\Delta^{-1}\delta s^{-1} + 2\beta\gamma\Delta^{-1}\delta^{-1}s\right) \varrho(s) ds$$
$$= \int_{\pm\zeta(1+\varpi\mathcal{O}_F)} \psi\left(2\alpha\Delta^{-1}s + 2\varepsilon\delta\Delta^{-1}s^{-1}\right) \left(\int_{\mathcal{O}_F} \psi\left[-2\varpi^{\ell-1}\Delta^{-1}s^{-1}\left(\alpha s^2 - \varepsilon\delta\right)x\right] dx\right) ds.$$

For $s \in \pm\zeta(1+\varpi\mathcal{O}_F)$, we have
$$\alpha s^2 - \varepsilon\delta \equiv \alpha\varepsilon(bc)^{-1} - \varepsilon\delta \pmod{\varpi\delta\mathcal{O}_F}$$

and here
$$\alpha\varepsilon(bc)^{-1} - \varepsilon\delta = \varepsilon(bc)^{-1}\delta^{-2}\Delta \cdot \delta.$$

Hence
$$\left|-2\varpi^{\ell-1}\Delta^{-1}s^{-1}\left(\alpha s^2 - \varepsilon\delta\right)\right| = |\delta\varpi^{-1}\Delta^{-1}\delta| = q > 1$$

and the inner integral of (3.29) vanishes. Thus we have proved the equality (3.13).

3.3. Evaluation of the anisotropic Kloosterman sum

We recall that $E = F(\eta)$ where $\eta^2 = d \in \mathcal{O}_F^\times \setminus (\mathcal{O}_F^\times)^2$. Let

$$S_2 = \begin{pmatrix} 1 & 0 \\ 0 & -d \end{pmatrix}.$$

In this section we shall consider $\mathcal{K}\ell^{(2)}(A, S_2, \varepsilon)$. For our later use, we shall formulate our evaluation utilizing the realization of $\mathrm{Mat}_{2\times 2}(F)$ as D_1.

We recall that

$$D_1 = \left\{ \begin{pmatrix} a & b \\ b^\sigma & a^\sigma \end{pmatrix} \mid a, b \in E \right\}$$

and we have

$$\mathrm{Mat}_{2\times 2}(F) = C D_1 C^{-1} \quad \text{where} \quad C = \begin{pmatrix} 1 & 1 \\ \eta & -\eta \end{pmatrix}.$$

Since

$$\mathrm{Sym}^2(F) = w_0 \{T \in \mathrm{Mat}_{2\times 2}(F) \mid \mathrm{tr}(T) = 0\}$$

where $w_0 = \begin{pmatrix} 0 & 1 \\ -1 & 0 \end{pmatrix}$, we have

$$\mathrm{Sym}^2(F) = w_0 C D_1^- C^{-1}$$

where $D_1^- = \{x \in D_1 \mid \mathrm{tr}(x) = 0\}$. Thus

$$\mathcal{K}\ell^{(2)}(A, S_2, \varepsilon) = \int_{\{Y \in D_1^- \mid YC^{-1}A \in \mathrm{GL}_2(\mathcal{O}_E)\}} \psi\left[\mathrm{tr}\left(S_2 w_0 C Y C^{-1} + \varepsilon \cdot {}^t A^{-1} S_2 A^{-1} C Y^{-1} C^{-1} w_0^{-1}\right)\right] dY.$$

We note that

$$C^{-1} S_2 w_0 C = C^{-1} \begin{pmatrix} 0 & 1 \\ d & 0 \end{pmatrix} C = \begin{pmatrix} \eta & 0 \\ 0 & -\eta \end{pmatrix}.$$

Then we write

$$C^{-1} w_0^{-1}\, {}^t A^{-1} S_2 A^{-1} C = \left(C^{-1} w_0^{-1}\, {}^t A^{-1} C\right) \left(C^{-1} S_2 w_0 C\right) \left(C^{-1} w_0^{-1} A^{-1} C\right).$$

Let $B = C^{-1} A w_0 C$. Since

$$w_0^{-1}\, {}^t A^{-1} = \left(w_0^{-1}\, {}^t A^{-1} w_0\right) w_0^{-1} = -(\det A)^{-1} A w_0,$$

we have

$$C^{-1} w_0^{-1}\, {}^t A^{-1} C = -(\det B)^{-1} B.$$

Hence we have
(3.30)
$$\mathcal{K}\ell^{(2)}(A, S_2, \varepsilon) = \int_{\mathcal{Y}_B} \psi\left[\mathrm{tr}\left\{\begin{pmatrix} \eta & 0 \\ 0 & -\eta \end{pmatrix} Y - \frac{\varepsilon}{\det B} \cdot B \begin{pmatrix} \eta & 0 \\ 0 & -\eta \end{pmatrix} B^{-1} Y^{-1}\right\}\right] dY$$

where
(3.31)
$$\mathcal{Y}_B = \left\{Y \in D_1^- \mid YB \in \mathrm{GL}_2(\mathcal{O}_E)\right\}$$

for $B = C^{-1} A w_0 C$.

3.3. EVALUATION OF THE ANISOTROPIC KLOOSTERMAN SUM

DEFINITION 3.6. For $B \in D_1^\times$ and $\varepsilon \in \mathcal{O}_F^\times$, we define the *anisotropic Kloosterman sum* $\mathcal{K}_{\mathrm{an}}(B, \varepsilon)$ by the right hand side of (3.30), i.e.

$$\mathcal{K}_{\mathrm{an}}(B, \varepsilon) = \int_{\mathcal{Y}_B} \psi \left[\mathrm{tr} \left\{ \begin{pmatrix} \eta & 0 \\ 0 & -\eta \end{pmatrix} \left(Y - \varepsilon (\det B)^{-1} \cdot B^{-1} Y^{-1} B \right) \right\} \right] dY.$$

By Proposition 3.2 we have

LEMMA 3.7. *For* $\mu \in \mathcal{O}_E^\times$ *we have*

$$\mathcal{K}_{\mathrm{an}} \left(\begin{pmatrix} \mu & 0 \\ 0 & \mu^\sigma \end{pmatrix} B, \varepsilon \right) = \mathcal{K}_{\mathrm{an}} \left(B \begin{pmatrix} \mu & 0 \\ 0 & \mu^\sigma \end{pmatrix}, \varepsilon \right) = \mathcal{K}_{\mathrm{an}} \left(B, \varepsilon (\mu \mu^\sigma)^{-1} \right)$$

and we also have

$$\mathcal{K}_{\mathrm{an}} \left(\begin{pmatrix} 0 & 1 \\ 1 & 0 \end{pmatrix} B, \varepsilon \right) = \mathcal{K}_{\mathrm{an}} \left(B \begin{pmatrix} 0 & 1 \\ 1 & 0 \end{pmatrix}, \varepsilon \right) = \mathcal{K}_{\mathrm{an}}(B, \varepsilon).$$

Thus without loss of generality we may assume that the matrix B is of the form $B = \varpi^n \begin{pmatrix} 1 & u \\ u^\sigma & 1 \end{pmatrix}$ where n is an integer and $u \in E$ such that $uu^\sigma \neq 1$.

For $u \in E$ such that $uu^\sigma \neq 1$, let us define a matrix A_u by

(3.32) $$A_u = \begin{pmatrix} 1 & u \\ u^\sigma & 1 \end{pmatrix}.$$

LEMMA 3.8. *Suppose that* $u, v \in E^\times$ *satisfy* $uu^\sigma = vv^\sigma (\neq 1)$. *Then we have*

$$\mathcal{K}_{\mathrm{an}}(\varpi^n A_u, \varepsilon) = \mathcal{K}_{\mathrm{an}}(\varpi^n A_v, \varepsilon).$$

In particular we have

$$\mathcal{K}_{\mathrm{an}}(\varpi^n A_u, \varepsilon) = \mathcal{K}_{\mathrm{an}}(\varpi^n A_{u^\sigma}, \varepsilon).$$

PROOF. By the assumption there exists $\zeta \in \mathcal{O}_E^\times$ such that $v = \zeta \zeta^{-\sigma} u$. Thus

$$A_v = \begin{pmatrix} \zeta & 0 \\ 0 & \zeta^\sigma \end{pmatrix} A_u \begin{pmatrix} \zeta & 0 \\ 0 & \zeta^\sigma \end{pmatrix}^{-1}$$

and the assertion follows from Lemma 3.7. $\square$

We have the following *functional equation* for $\mathcal{K}_{\mathrm{an}}(\varpi^n A_u, \varepsilon)$.

PROPOSITION 3.9. *Let* $u \in E^\times$ *such that* $uu^\sigma \neq 1$. *Let us write* $u = \varpi^m \varepsilon_u$ *where* m *is an integer and* $\varepsilon_u \in \mathcal{O}_E^\times$.

Then we have

(3.33) $$\mathcal{K}_{\mathrm{an}}(\varpi^n A_u, \varepsilon) = \mathcal{K}_{\mathrm{an}} \left(\varpi^{n+m} A_{u^{-1}}, \varepsilon (\varepsilon_u \varepsilon_u^\sigma)^{-1} \right).$$

PROOF. We have

(3.34) $$A_u \begin{pmatrix} 1 & 0 \\ 0 & -1 \end{pmatrix} A_u^{-1} = \frac{1}{1 - uu^\sigma} \begin{pmatrix} 1 & u \\ u^\sigma & 1 \end{pmatrix} \begin{pmatrix} 1 & 0 \\ 0 & -1 \end{pmatrix} \begin{pmatrix} 1 & -u \\ -u^\sigma & 1 \end{pmatrix}$$

$$= \frac{1}{1 - uu^\sigma} \begin{pmatrix} 1 + uu^\sigma & -2u \\ 2u^\sigma & -1 - uu^\sigma \end{pmatrix}$$

$$= \frac{-1}{1 - u^{-1}u^{-\sigma}} \begin{pmatrix} 1 + u^{-1}u^{-\sigma} & -2u^{-\sigma} \\ 2u^{-1} & -1 - u^{-1}u^{-\sigma} \end{pmatrix}$$

$$= -A_{u^{-\sigma}} \begin{pmatrix} 1 & 0 \\ 0 & -1 \end{pmatrix} A_{u^{-\sigma}}^{-1}.$$

Hence

$$\mathcal{K}_{\mathrm{an}}\left(\varpi^{n}A_{u},\varepsilon\right)=\int_{\mathcal{Y}_{\varpi^{n}A_{u}}}\psi\left[\operatorname{tr}\left\{\begin{pmatrix}\eta&0\\0&-\eta\end{pmatrix}Y-\frac{\varepsilon}{\varpi^{2n}\left(1-uu^{\sigma}\right)}A_{u}\begin{pmatrix}\eta&0\\0&-\eta\end{pmatrix}A_{u}^{-1}Y^{-1}\right\}\right]dY$$

$$=\int_{\mathcal{Y}_{\varpi^{n}A_{u}}}\psi\left[\operatorname{tr}\left\{\begin{pmatrix}\eta&0\\0&-\eta\end{pmatrix}Y+\frac{\varepsilon}{\varpi^{2n}\left(1-uu^{\sigma}\right)}A_{u^{-\sigma}}\begin{pmatrix}\eta&0\\0&-\eta\end{pmatrix}A_{u^{-\sigma}}^{-1}Y^{-1}\right\}\right]dY$$

$$=\int_{\mathcal{Y}_{\varpi^{n}A_{u}}}\psi\left[\operatorname{tr}\left\{\begin{pmatrix}\eta&0\\0&-\eta\end{pmatrix}Y-\frac{\varepsilon\left(\varepsilon_{u}\varepsilon_{u}^{\sigma}\right)^{-1}}{\varpi^{2(n+m)}\det\left(A_{u^{-\sigma}}\right)}A_{u^{-\sigma}}\begin{pmatrix}\eta&0\\0&-\eta\end{pmatrix}A_{u^{-\sigma}}^{-1}Y^{-1}\right\}\right]dY.$$

Moreover since

$$\varpi^{n+m}A_{u^{-\sigma}}=\varepsilon_{u}^{-1}\varpi^{n}uA_{u^{-\sigma}}=\varepsilon_{u}^{-1}\varpi^{n}\begin{pmatrix}u&uu^{-\sigma}\\1&u\end{pmatrix}=\varepsilon_{u}^{-1}\varpi^{n}A_{u}\begin{pmatrix}0&uu^{-\sigma}\\1&0\end{pmatrix}$$

where $\begin{pmatrix}0&uu^{-\sigma}\\1&0\end{pmatrix}\in\mathrm{GL}_{2}\left(\mathcal{O}_{E}\right)$, we have

$$\mathcal{Y}_{\varpi^{n}A_{u}}=\mathcal{Y}_{\varpi^{n+m}A_{u^{-\sigma}}}.$$

Hence we have

$$\mathcal{K}_{\mathrm{an}}\left(\varpi^{n}A_{u},\varepsilon\right)=\mathcal{K}_{\mathrm{an}}\left(\varpi^{n+m}A_{u^{-\sigma}},\varepsilon\left(\varepsilon_{u}\varepsilon_{u}^{\sigma}\right)^{-1}\right)$$
$$=\mathcal{K}_{\mathrm{an}}\left(\varpi^{n+m}A_{u^{-1}},\varepsilon\left(\varepsilon_{u}\varepsilon_{u}^{\sigma}\right)^{-1}\right)$$

by Lemma 3.8. $\square$

The rest of this section is devoted to the proof of the following theorem.

THEOREM 3.10. *Suppose that $n>0$ and $u\in\mathcal{O}_{E}$ such that $u\neq 0$ and $uu^{\sigma}\neq 1$. Let $m=\mathrm{ord}\,(u)$ and let us write $u=\varpi^{m}\varepsilon_{u}$.*
1. *When $m\geq n$, we have*

(3.35) $$\mathcal{K}_{\mathrm{an}}\left(\varpi^{n}A_{u},\varepsilon\right)=q^{2n}\left\{(-1)^{n}\mathcal{K}\ell\left(2\varpi^{-n},-2\varpi^{-n}d\varepsilon\right)+1+q^{-1}\right\}.$$

2. *When $0\leq m<n$, we have*

(3.36) $$\mathcal{K}_{\mathrm{an}}\left(\varpi^{n}A_{u},\varepsilon\right)=(-1)^{n}q^{2n}\left|1-uu^{\sigma}\right|^{-1}$$
$$\cdot\left\{\mathcal{K}\ell\left(\frac{2\varpi^{-n}}{1-uu^{\sigma}},\frac{-2\varpi^{-n}d\varepsilon}{1-uu^{\sigma}}\right)+(-1)^{m}\mathcal{K}\ell\left(\frac{2\varpi^{m-n}}{1-uu^{\sigma}},\frac{-2\varpi^{m-n}d\varepsilon\varepsilon_{u}\varepsilon_{u}^{\sigma}}{1-uu^{\sigma}}\right)\right\}.$$

By Lemma 3.7 we may state the assertion of the theorem in the general setting.

COROLLARY 3.11. *Let $a,b\in\varpi\mathcal{O}_{E}\setminus\{0\}$ and suppose that $\Delta=aa^{\sigma}-bb^{\sigma}\neq 0$.*
1. *We have*

$$\mathcal{K}_{\mathrm{an}}\left(\begin{pmatrix}a&b\\b^{\sigma}&a^{\sigma}\end{pmatrix},\varepsilon\right)=\mathcal{K}_{\mathrm{an}}\left(\begin{pmatrix}b&a\\a^{\sigma}&b^{\sigma}\end{pmatrix},\varepsilon\right).$$

2. When $\operatorname{ord}(b) \geq 2\operatorname{ord}(a)$, we have

$$\mathcal{K}_{\mathrm{an}}\left(\begin{pmatrix} a & b \\ b^\sigma & a^\sigma \end{pmatrix}, \varepsilon\right)$$
$$= |\Delta|^{-1}\left\{(-1)^{\operatorname{ord}(a)} \mathcal{K}\ell\left(\frac{2\varpi^{-\operatorname{ord}(a)}aa^\sigma}{\Delta}, -\frac{2\varpi^{\operatorname{ord}(a)}d\varepsilon}{\Delta}\right) + 1 + q^{-1}\right\}.$$

3. When $\operatorname{ord}(a) \leq \operatorname{ord}(b) < 2\operatorname{ord}(a)$, we have

$$\mathcal{K}_{\mathrm{an}}\left(\begin{pmatrix} a & b \\ b^\sigma & a^\sigma \end{pmatrix}, \varepsilon\right) = |\Delta|^{-1}(-1)^{\operatorname{ord}(a)}\mathcal{K}\ell\left(\frac{2\varpi^{-\operatorname{ord}(a)}aa^\sigma}{\Delta}, -\frac{2\varpi^{\operatorname{ord}(a)}d\varepsilon}{\Delta}\right)$$
$$+ |\Delta|^{-1}(-1)^{\operatorname{ord}(b)}\mathcal{K}\ell\left(\frac{2\varpi^{-\operatorname{ord}(b)}bb^\sigma}{\Delta}, -\frac{2\varpi^{\operatorname{ord}(b)}d\varepsilon}{\Delta}\right).$$

Before delving into the proof of Theorem 3.10, we remark the following lemma.

LEMMA 3.12. *Suppose that $u \in \varpi^m \mathcal{O}_E^\times$. Then there exists $v \in \varpi^m \mathcal{O}_E^\times$ such that $uu^\sigma = vv^\sigma$ and $v + v^\sigma \in \varpi^m \mathcal{O}_F^\times$.*

PROOF. Let us write $u = \varpi^m(a + \eta b)$ where $a, b \in \mathcal{O}_F$. Then $u + u^\sigma = 2\varpi^m a$. Hence when $a \in \mathcal{O}_F^\times$, there is nothing to prove. Thus suppose that $a \in \varpi\mathcal{O}_F$. Then we have $b \in \mathcal{O}_F^\times$ since $a + \eta b \in \mathcal{O}_E^\times$. Let $\xi = 1 + \eta$. Since $\xi \in \mathcal{O}_E$ and $\xi + \xi^\sigma = 2 \in \mathcal{O}_F^\times$, we have $\xi \in \mathcal{O}_E^\times$. Then we have

$$u\frac{\xi}{\xi^\sigma} = \varpi^m \frac{(1+\eta)^2(a+\eta b)}{(1-\eta)(1+\eta)} = \varpi^m \frac{\{a(1+d) + 2bd\} + \eta\{2a + b(1+d)\}}{(1-\eta)(1+\eta)}$$

where

$$a(1+d) + 2bd \in 2bd + \varpi\mathcal{O}_F \subset \mathcal{O}_F^\times.$$

□

Hence by Lemma 3.8, in the proof of Theorem 3.10, we may assume that $u \in \varpi^m \mathcal{O}_E^\times$ satisfies

(3.37) $$u + u^\sigma \in \varpi^m \mathcal{O}_F^\times.$$

Moreover when $uu^\sigma \in (F^\times)^2$, we may take $u \in F^\times$.

Suppose that $uu^\sigma \notin (F^\times)^2$. If $|\varepsilon_u - \varepsilon_u^\sigma| < 1$, then $\varepsilon_u \in \mu(1 + \varpi\mathcal{O}_E)$ for some $\mu \in \mathcal{O}_F^\times$ and we have $\varepsilon_u\varepsilon_u^\sigma \in \mu^2(1 + \varpi\mathcal{O}_F) \subset (\mathcal{O}_F^\times)^2$. This is a contradiction. Thus in this case we have

(3.38) $$\varepsilon_u - \varepsilon_u^\sigma \in \mathcal{O}_E^\times.$$

Let us start the computation. The computation is somewhat lengthy and we divide into three subsections depending on the cases.

We recall that

(3.39) $\mathcal{K}_{\mathrm{an}}(\varpi^n A_u, \varepsilon)$
$$= \int_{\mathcal{Y}_{\varpi^n A_u}} \psi\left[\operatorname{tr}\left(\begin{pmatrix} \eta & 0 \\ 0 & -\eta \end{pmatrix}Y - \frac{\varpi^{-2n}\varepsilon}{1 - uu^\sigma}A_u\begin{pmatrix} \eta & 0 \\ 0 & -\eta \end{pmatrix}A_u^{-1}Y^{-1}\right)\right] dY$$

where

$$\mathcal{Y}_{\varpi^n A_u} = \{Y \in D_1^- \mid \varpi^n Y A_u \in \operatorname{GL}_2(\mathcal{O}_E)\}.$$

3.3.1. The case when $|1 - uu^\sigma| < 1$. Then we have
$$uu^\sigma = 1 - (1 - uu^\sigma) \in 1 + \varpi \mathcal{O}_F \subset \left(\mathcal{O}_F^\times\right)^2.$$
Hence by Lemma 3.8, we may assume that $u = y \in \mathcal{O}_F^\times$ such that
$$(3.40) \qquad 1 + y \in \mathcal{O}_F^\times, \quad 1 - y \in \varpi \mathcal{O}_F.$$

We employ the $\mathrm{Mat}_{2\times 2}(F)$ realization for our computation in this case. For $C = \begin{pmatrix} 1 & 1 \\ \eta & -\eta \end{pmatrix}$, we note that
$$CA_y C^{-1} = \begin{pmatrix} 1+y & 0 \\ 0 & 1-y \end{pmatrix}, \quad C\begin{pmatrix} \eta & 0 \\ 0 & -\eta \end{pmatrix} C^{-1} = \begin{pmatrix} 0 & 1 \\ d & 0 \end{pmatrix},$$
and, for $\begin{pmatrix} r & s \\ t & -r \end{pmatrix} \in \mathrm{Mat}_{2\times 2}(F)$, we have
$$\varpi^n \begin{pmatrix} r & s \\ t & -r \end{pmatrix} \begin{pmatrix} 1+y & 0 \\ 0 & 1-y \end{pmatrix} \in \mathrm{GL}_2(\mathcal{O}_F) \iff \begin{cases} |r| \leq q^n \\ |s(1-y)| = q^n \\ |t| = q^n. \end{cases}$$

Thus (3.39) becomes
$$\mathcal{K}_{\mathrm{an}}(\varpi^n A_y, \varepsilon) = \int_{S = \binom{r\ s}{t\ -r} \in \varpi^{-n} \mathrm{GL}_2(\mathcal{O}_F) \binom{1+y\ 0}{0\ 1-y}^{-1}}$$
$$\psi\left[\mathrm{tr}\left\{C\begin{pmatrix} \eta & 0 \\ 0 & -\eta \end{pmatrix}C^{-1}S - \frac{\varpi^{-2n}\varepsilon}{1-y^2} CA_y \begin{pmatrix} \eta & 0 \\ 0 & -\eta \end{pmatrix} A_y^{-1} C^{-1} S^{-1}\right\}\right] dS$$
$$= \int_{|r| \leq q^n, |(1-y)s| = q^n, |t| = q^n} \psi\left[ds + t - \frac{\varpi^{-2n}\varepsilon}{r^2 + st}\left\{\frac{t}{(1-y)^2} + \frac{ds}{(1+y)^2}\right\}\right] dr\, ds\, dt.$$

Let $r = \varpi^{-n}\xi$, $s = \varpi^{-n}(1-y)^{-1}\varepsilon_1$ and $t = \varpi^{-n}\varepsilon_2$. Then
$$\mathcal{K}_{\mathrm{an}}(\varpi^n A_y, \varepsilon) = |1 - y^2|^{-1} q^{3n} \int_{\mathcal{O}_F^\times} \int_{\mathcal{O}_F^\times} \int_{\mathcal{O}_F}$$
$$\psi\left[\varpi^{-n}\left\{\frac{d\varepsilon_1}{1-y} + \varepsilon_2 - \frac{\varepsilon}{\varepsilon_1\varepsilon_2 + (1-y)\xi^2}\left(\frac{d\varepsilon_1}{(1+y)^2} + \frac{\varepsilon_2}{1-y}\right)\right\}\right] d\varepsilon_1\, d\varepsilon_2\, d\xi.$$

Then by a change of variable $\varepsilon_1 \mapsto \varepsilon_1 \varepsilon_2^{-1}$, followed by another change of variable $\varepsilon_1 \mapsto \varepsilon_1 - (1-y)\xi^2$, we have
$$(3.41)$$
$$\mathcal{K}_{\mathrm{an}}(\varpi^n A_y, \varepsilon) = |1 - y^2|^{-1} q^{3n} \int_{\mathcal{O}_F^\times} \int_{\mathcal{O}_F^\times} \int_{\mathcal{O}_F}$$
$$\psi\left[\varpi^{-n}\left\{\frac{d\varepsilon_1 \varepsilon_2^{-1}}{1-y} + \varepsilon_2 - \frac{\varepsilon}{\varepsilon_1 + (1-y)\xi^2}\left(\frac{d\varepsilon_1 \varepsilon_2^{-1}}{(1+y)^2} + \frac{\varepsilon_2}{1-y}\right)\right\}\right] d\varepsilon_1\, d\varepsilon_2\, d\xi$$
$$= |1 - y^2|^{-1} q^{3n} \int_{\mathcal{O}_F^\times} \int_{\mathcal{O}_F^\times} \psi\left[\varpi^{-n}\left\{\frac{d\varepsilon_1\varepsilon_2^{-1}}{1-y} + \varepsilon_2 - \frac{d\varepsilon\varepsilon_2^{-1}}{(1+y)^2} - \frac{\varepsilon\varepsilon_1^{-1}\varepsilon_2}{1-y}\right\}\right]$$
$$\left(\int_{\mathcal{O}_F} \psi\left[-\varpi^{-n} d\varepsilon_2 \cdot \varepsilon_2^{-2}\left\{1 - (1+y)^{-2}(1-y)d^{-1}\varepsilon\varepsilon_1^{-1}\right\}\xi^2\right] d\xi\right) d\varepsilon_1\, d\varepsilon_2.$$

3.3. EVALUATION OF THE ANISOTROPIC KLOOSTERMAN SUM

Since $1 - (1+y)^{-2}(1-y)d^{-1}\varepsilon\varepsilon_1^{-1} \in 1 + \varpi\mathcal{O}_F \subset (\mathcal{O}_F^\times)^2$, the inner integral of (3.41) simplifies to

$$\int_{\mathcal{O}_F} \psi\left(-\varpi^{-n}d\varepsilon_2\xi^2\right)d\xi.$$

Thus after a change of variable $\varepsilon_1 \mapsto \varepsilon_1\varepsilon_2$, we have

(3.42)
$$\mathcal{K}_{\mathrm{an}}\left(\varpi^n A_y, \varepsilon\right) = |1-y^2|^{-1}q^{3n}\int_{\mathcal{O}_F^\times}\psi\left(\frac{\varpi^{-n}d\varepsilon_1}{1-y} - \frac{\varpi^{-n}\varepsilon\varepsilon_1^{-1}}{1-y}\right)d\varepsilon_1$$
$$\int_{\mathcal{O}_F^\times}\int_{\mathcal{O}_F}\psi\left[\varpi^{-n}\varepsilon_2 - \varpi^{-n}(1+y)^{-2}d\varepsilon\varepsilon_2^{-1} - \varpi^{-n}d\varepsilon_2\xi^2\right]d\varepsilon_2\,d\xi$$
$$=|1-y^2|^{-1}q^{3n}\cdot \mathcal{K}\ell\left(\varpi^{-n}d(1-y)^{-1}, -\varpi^{-n}\varepsilon(1-y)^{-1}\right)$$
$$\int_{\mathcal{O}_F^\times}\psi\left[\varpi^{-n}\varepsilon_2 - \varpi^{-n}(1+y)^{-2}d\varepsilon\varepsilon_2^{-1}\right]\left(\int_{\mathcal{O}_F}\psi\left(-\varpi^{-n}d\varepsilon_2\xi^2\right)d\xi\right)d\varepsilon_2.$$

By Proposition 2.11, unless $-d\varepsilon \in (\mathcal{O}_F^\times)^2$, we have

$$\mathcal{K}\ell\left(\varpi^{-n}d(1-y)^{-1}, -\varpi^{-n}\varepsilon(1-y)^{-1}\right) = 0$$

and also the right hand side of (3.36) vanishes. Now suppose that $-d\varepsilon \in (\mathcal{O}_F^\times)^2$ and take $\zeta \in \mathcal{O}_F^\times$ such that $\zeta^2 = -d\varepsilon$.

First suppose that n is even. Let us write $n = 2r$. Then by Proposition 2.5 and Proposition 2.11, we have

$$\int_{\mathcal{O}_F^\times}\psi\left[\varpi^{-2r}\varepsilon_2 - \varpi^{-2r}(1+y)^{-2}d\varepsilon\varepsilon_2^{-1}\right]\left(\int_{\mathcal{O}_F}\psi\left(-\varpi^{-2r}d\varepsilon_2\xi^2\right)d\xi\right)d\varepsilon_2$$
$$=q^{-r}\int_{\mathcal{O}_F^\times}\psi\left(\varpi^{-2r}\varepsilon_2 - \varpi^{-2r}(1+y)^{-2}d\varepsilon\varepsilon_2^{-1}\right)d\varepsilon_2$$
$$=q^{-r}\cdot \mathcal{K}\ell\left(\varpi^{-2r}, -\varpi^{-2r}d\varepsilon(1+y)^{-2}\right)$$
$$=q^{-r}\left\{\psi\left(\frac{2\varpi^{-2r}\zeta}{1+y}\right) + \psi\left(-\frac{2\varpi^{-2r}\zeta}{1+y}\right)\right\}.$$

Suppose that n is odd. Let us write $n = 2r+1$ where $r \geq 0$. Then by Proposition 2.5 and Proposition 2.14, we have

$$\int_{\mathcal{O}_F^\times}\psi\left[\varpi^{-2r-1}\varepsilon_2 - \varpi^{-2r-1}(1+y)^{-2}d\varepsilon\varepsilon_2^{-1}\right]\left(\int_{\mathcal{O}_F}\psi\left(-\varpi^{-2r-1}d\varepsilon_2\xi^2\right)d\xi\right)d\varepsilon_2$$
$$=q^{-r-1}\mathrm{sgn}(-d)\,\mathfrak{G}(\mathrm{sgn})\int_{\mathcal{O}_F^\times}\mathrm{sgn}(\varepsilon_2)\cdot\psi\left[\varpi^{-2r-1}\varepsilon_2 - \varpi^{-2r-1}(1+y)^{-2}d\varepsilon\varepsilon_2^{-1}\right]d\varepsilon_2$$
$$=q^{-r-1}\mathrm{sgn}(-d)\,\mathfrak{G}(\mathrm{sgn})\cdot \mathcal{S}\left(\varpi^{-2r-1}, -\varpi^{-2r-1}d\varepsilon(1+y)^{-2}\right)$$
$$=-q^{-2r-1}\left\{\psi\left(2\varpi^{-2r-1}\zeta(1+y)^{-1}\right) + \psi\left(-2\varpi^{-2r-1}\zeta(1+y)^{-1}\right)\right\}.$$

Thus we have shown that

$$(3.43) \quad \mathcal{K}_{\mathrm{an}}(\varpi^n A_u, \varepsilon) = (-1)^n q^{2n} |1-y^2|^{-1}$$
$$\left\{ \psi\left(\frac{2\varpi^{-n}\zeta}{1+y}\right) + \psi\left(-\frac{2\varpi^{-n}\zeta}{1+y}\right) \right\} \cdot \mathcal{K}\ell\left(\varpi^{-n}(1-y)^{-1}, \varpi^{-n}\zeta^2(1-y)^{-1}\right).$$

Finally let us compare (3.43) with (3.36). By Proposition 2.11, we have

$$\left\{ \psi\left(\frac{2\varpi^{-n}\zeta}{1+y}\right) + \psi\left(-\frac{2\varpi^{-n}\zeta}{1+y}\right) \right\} \cdot \mathcal{K}\ell\left(\varpi^{-n}(1-y)^{-1}, \varpi^{-n}\zeta^2(1-y)^{-1}\right)$$
$$= \left\{ \psi\left(\frac{2\varpi^{-n}\zeta}{1+y}\right) + \psi\left(-\frac{2\varpi^{-n}\zeta}{1+y}\right) \right\}$$
$$\cdot \left\{ C\left(\frac{\varpi^{-n}\zeta}{1-y}\right)\psi\left(\frac{2\varpi^{-n}\zeta}{1-y}\right) + C\left(-\frac{\varpi^{-n}\zeta}{1-y}\right)\psi\left(-\frac{2\varpi^{-n}\zeta}{1-y}\right) \right\}$$
$$= \left\{ C\left(\frac{2\varpi^{-n}\zeta}{1-y^2}\right)\psi\left(\frac{4\varpi^{-n}\zeta}{1-y^2}\right) + C\left(-\frac{2\varpi^{-n}\zeta}{1-y^2}\right)\psi\left(-\frac{4\varpi^{-n}\zeta}{1-y^2}\right) \right\}$$
$$+ \left\{ C\left(\frac{2\varpi^{-n}y\zeta}{1-y^2}\right)\psi\left(\frac{4\varpi^{-n}y\zeta}{1-y^2}\right) + C\left(-\frac{2\varpi^{-n}y\zeta}{1-y^2}\right)\psi\left(-\frac{4\varpi^{-n}y\zeta}{1-y^2}\right) \right\}$$
$$= \mathcal{K}\ell\left(\frac{2\varpi^{-n}}{1-y^2}, \frac{2\varpi^{-n}\zeta^2}{1-y^2}\right) + \mathcal{K}\ell\left(\frac{2\varpi^{-n}}{1-y^2}, \frac{2\varpi^{-n}y^2\zeta^2}{1-y^2}\right)$$

since we have

$$C\left(\frac{2\varpi^{-n}\zeta}{1-y^2}\right) = \mathrm{sgn}\left(\frac{2}{1+y}\right) C\left(\frac{\varpi^{-n}\zeta}{1-y}\right) = C\left(\frac{\varpi^{-n}\zeta}{1-y}\right)$$

and

$$C\left(\frac{2\varpi^{-n}y\zeta}{1-y^2}\right) = \mathrm{sgn}\left(\frac{2}{1+y}\right) \mathrm{sgn}(y) C\left(\frac{\varpi^{-n}\zeta}{1-y}\right) = C\left(\frac{\varpi^{-n}\zeta}{1-y}\right)$$

for

$$y \in 1 + \varpi\mathcal{O}_F, \quad \text{and hence,} \quad \frac{1+y}{2} \in 1 + \varpi\mathcal{O}_F \subset (\mathcal{O}_F^\times)^2.$$

Thus we have proved (3.36).

3.3.2. The case when $|1-uu^\sigma| = 1$ and $m \geq n$. Since $A_u \in \mathrm{GL}_2(\mathcal{O}_E)$ in this case, we have

$$\mathcal{Y}_{\varpi^n A_u} = \left\{ Y = \varpi^{-n}\begin{pmatrix} \eta t & v \\ v^\sigma & -\eta t \end{pmatrix} \mid t \in \mathcal{O}_F, v \in \mathcal{O}_E, |dt^2 + vv^\sigma| = 1 \right\}.$$

Thus (3.39) becomes

$$\mathcal{K}_{\mathrm{an}}(\varpi^n A_u, \varepsilon) = q^{3n} \int_{t \in \mathcal{O}_F, v \in \mathcal{O}_E, |dt^2+vv^\sigma|=1} \psi\left[2\varpi^{-n}\left\{dt + \frac{\varepsilon(\eta uv^\sigma - \eta u^\sigma v) - d\varepsilon(1+uu^\sigma)t}{(1-uu^\sigma)^2(dt^2+vv^\sigma)}\right\}\right] dt\, dv.$$

Since $u \in \varpi^m \mathcal{O}_E \subset \varpi^n \mathcal{O}_E$, we have $u \equiv 0 \pmod{\varpi^n}$ and hence

$$\mathcal{K}_{\mathrm{an}}(\varpi^n A_u, \varepsilon) = \mathcal{K}_1 + \mathcal{K}_2$$

where

(3.44) $$\mathcal{K}_1 = q^{3n} \int_{t \in \varpi \mathcal{O}_F} \int_{v \in \mathcal{O}_E^\times} \psi \left[2\varpi^{-n} dt \left\{ 1 - \frac{\varepsilon}{(dt^2 + vv^\sigma)} \right\} \right] dt\, dv.$$

and

(3.45) $$\mathcal{K}_2 = q^{3n} \int_{t \in \mathcal{O}_F^\times} \int_{v \in \mathcal{O}_E,\ |dt^2 + vv^\sigma| = 1} \psi \left[2\varpi^{-n} dt \left\{ 1 - \frac{\varepsilon}{(dt^2 + vv^\sigma)} \right\} \right] dt\, dv.$$

Let us compute $\mathcal{K}_1$. Since $x \mapsto x^2$ is an automorphism of the multiplicative group $1 + \varpi \mathcal{O}_F$, we denote its inverse map as $y \mapsto y^{\frac{1}{2}}$. Now for $a \in \varpi \mathcal{O}_F$, we define a mapping $f_a : \mathcal{O}_E^\times \to \mathcal{O}_E^\times$ by

$$f_a(z) = z \left(1 + a (zz^\sigma)^{-1} \right)^{\frac{1}{2}}.$$

Then it is easily seen that we have

$$(f_{-a} \circ f_a)(z) = z.$$

Hence, in particular, f_a is a continuous bijection. It is also clear that

$$z_1 \equiv z_2 \pmod{\varpi^n} \implies f_a(z_1) \equiv f_a(z_2) \pmod{\varpi^n}.$$

Hence, by a change of variable $v \mapsto v \left(1 - dt^2 (vv^\sigma)^{-1} \right)^{\frac{1}{2}}$, we have

$$\mathcal{K}_1 = q^{3n} \int_{v \in \mathcal{O}_E^\times} \int_{t \in \varpi \mathcal{O}_F} \psi \left[2\varpi^{-n} dt \left(1 - \frac{\varepsilon}{vv^\sigma} \right) \right] dt\, dv.$$

Thus when $n = 1$, we have

$$\mathcal{K}_1 = q^2 \left(1 - q^{-2} \right).$$

When $n > 1$, since $\mathcal{O}_{E,n-1}^\times \ni x \mapsto xx^\sigma \in \mathcal{O}_{F,n-1}^\times$ is surjective, we have

$$\mathcal{K}_1 = q^{3n-1} \int_{\{v \in \mathcal{O}_E^\times \mid vv^\sigma \equiv \varepsilon \pmod{\varpi^{n-1}}\}} dv = q^{2n} \left(1 + q^{-1} \right)$$

by Lemma 1.12.

Now let us compute $\mathcal{K}_2$. By a change of variable $v \mapsto \eta t v$, we have

$$\mathcal{K}_2 = q^{3n} \int_{t \in \mathcal{O}_F^\times} \int_{v \in \mathcal{O}_E,\ |1 - vv^\sigma| = 1} \psi \left[2\varpi^{-n} \left\{ dt - \frac{\varepsilon}{t(1 - vv^\sigma)} \right\} \right] dt\, dv.$$

Let us move to the modulo ϖ^n setting. Then we have

$$\mathcal{K}_2 = \sum_{x \in \mathcal{O}_{F,n}^\times} \sum_{\zeta \in \mathcal{Z}} \psi^{(n)} \left[2 \left\{ dx - \frac{\varepsilon}{x(1 - \zeta \zeta^\sigma)} \right\} \right]$$

where $\mathcal{Z} = \left\{ \zeta \in \mathcal{O}_{E,n} \mid 1 - \zeta \zeta^\sigma \in \mathcal{O}_{F,n}^\times \right\}$. Let us denote by φ, the mapping

$$\varphi : \mathcal{Z} \ni \zeta \mapsto (1 - \zeta \zeta^\sigma)^{-1} \in \mathcal{O}_{F,n}^\times.$$

Let us define subgroups H_j of $\mathcal{O}_{F,n}^\times$ for $0 \leq j \leq n$ by

$$H_0 = \mathcal{O}_{F,n}^\times, \quad H_j = \left(1 + \varpi^j \mathcal{O}_F \right) / \left(1 + \varpi^n \mathcal{O}_F \right) \quad (1 \leq j \leq n)$$

and let

$$\mathcal{Z}_j = \varphi^{-1}(H_j \setminus H_{j+1}) \quad (0 \leq j \leq n-1), \quad \mathcal{Z}_n = \varphi^{-1}(H_n).$$

Then it is clear that $\mathcal{Z}_j$ is empty when j is odd and $0 < j < n$. When $0 \leq 2r < n$, the mapping $\varphi : \mathcal{Z}_{2r} \to H_{2r} \setminus H_{2r+1}$ is surjective and $q^n (1 + q^{-1})$ to one by Lemma 1.12. Also $\varphi : \mathcal{Z}_n \to H_n$ is surjective and q^n to one when n is even and q^{n-1} to one when n is odd. Thus we have

$$\mathcal{K}_2 = \sum_{x \in \mathcal{O}_{F,n}^\times} \psi^{(n)} (2dx) \left\{ q^n \left(1 + q^{-1}\right) \sum_{0 \leq 2r < n} \sum_{y \in H_{2r} \setminus H_{2r+1}} \psi^{(n)} \left(-2\varepsilon x^{-1} y\right) \right\}$$
$$+ \mathcal{K}\ell \left(2\varpi^{-n}, -2\varpi^{-n} d\varepsilon\right) \cdot \begin{cases} q^{2n}, & \text{when } n \text{ is even} \\ q^{2n-1}, & \text{when } n \text{ is odd.} \end{cases}$$

Here we note that for $0 < j < n$, we have

$$\sum_{y \in H_j} \psi^{(n)} \left(-2\varepsilon x^{-1} y\right) = \psi^{(n)} \left(-2\varepsilon x^{-1}\right) \sum_{t \in \varpi^j \mathcal{O}_F / \varpi^n \mathcal{O}_F} \psi^{(n)} \left(-2\varepsilon x^{-1} t\right) = 0$$

and

$$\sum_{y \in H_0} \psi^{(n)} \left(-2\varepsilon x^{-1} y\right) = \begin{cases} -1, & \text{when } n = 1 \\ 0, & \text{when } n > 1. \end{cases}$$

Hence by putting these all together, we have shown that

$$\mathcal{K}_2 = (-1)^n q^{2n} \cdot \mathcal{K}\ell \left(2\varpi^{-n}, -2\varpi^{-n} d\varepsilon\right) + \begin{cases} q+1, & \text{when } n = 1 \\ 0, & \text{when } n > 1. \end{cases}$$

Thus we have proved (3.35).

3.3.3. The case when $|1 - uu^\sigma| = 1$ and $0 \leq m < n$. We shall employ the $\mathrm{Mat}_{2 \times 2}(F)$ realization in this case. We note that $A_u \in \mathrm{GL}_2(\mathcal{O}_E)$ in this case and we recall that for $C = \begin{pmatrix} 1 & 1 \\ \eta & -\eta \end{pmatrix}$, we have

$$CA_u C^{-1} = \begin{pmatrix} \frac{2+u+u^\sigma}{2} & \frac{u^\sigma - u}{2\eta} \\ \frac{\eta(u-u^\sigma)}{2} & \frac{2-u-u^\sigma}{2} \end{pmatrix}, \quad C \begin{pmatrix} \eta & 0 \\ 0 & -\eta \end{pmatrix} C^{-1} = \begin{pmatrix} 0 & 1 \\ d & 0 \end{pmatrix}.$$

Thus

$$\mathcal{K}_{\mathrm{an}} \left(\varpi^n A_u, \varepsilon\right) = q^{3n} \int_{S = \left(\begin{smallmatrix} r & s \\ t & -r \end{smallmatrix}\right) \in \mathrm{GL}_2(\mathcal{O}_F)}$$
$$\psi \left[\varpi^{-n} \cdot \mathrm{tr} \left\{ C \begin{pmatrix} \eta & 0 \\ 0 & -\eta \end{pmatrix} C^{-1} S - \frac{\varepsilon}{1 - uu^\sigma} CA_u \begin{pmatrix} \eta & 0 \\ 0 & -\eta \end{pmatrix} A_u^{-1} C^{-1} S^{-1} \right\} \right] dS$$
$$= q^{3n} \int_{S = \left(\begin{smallmatrix} r & s \\ t & -r \end{smallmatrix}\right) \in \mathrm{GL}_2(\mathcal{O}_F)} \psi \left[\varpi^{-n} (ds + t) \right]$$
$$\psi \left[\frac{-\varpi^{-n} \varepsilon \left\{ -2\eta (u - u^\sigma) r + d(1-u)(1-u^\sigma) s + (1+u)(1+u^\sigma) t \right\}}{(1 - uu^\sigma)^2 (r^2 + st)} \right] dS.$$

Let us put

(3.46)
$$B_1 = \frac{\varepsilon \eta (u - u^\sigma)}{(1 - uu^\sigma)^2}, \quad B_2 = \frac{-d\varepsilon (1-u)(1-u^\sigma)}{(1 - uu^\sigma)^2}, \quad B_3 = \frac{-\varepsilon (1+u)(1+u^\sigma)}{(1 - uu^\sigma)^2}.$$

Here we note that

(3.47) $$\begin{cases} B_1 = 0, & \text{when } u \in \mathcal{O}_F \\ B_1 \in \varpi^m \mathcal{O}_F^\times, & \text{when } u \notin \mathcal{O}_F \end{cases}$$

by (3.38) and

(3.48) $$B_2 \in \mathcal{O}_F^\times, \qquad B_3 \in \mathcal{O}_F^\times$$

since $|1 - uu^\sigma| = 1$ implies that $u \notin \pm(1 + \varpi \mathcal{O}_E)$.

Then we may write
$$\mathcal{K}_{\mathrm{an}}(\varpi^n A_u, \varepsilon) = \mathcal{L}_1 + \mathcal{L}_2$$

where

(3.49)
$$\mathcal{L}_1 = q^{3n} \int_{s \in \mathcal{O}_F^\times} \int_{r,t \in \mathcal{O}_F, |r^2 + st| = 1} \psi\left[\varpi^{-n}\left(ds + t + \frac{2B_1 r + B_2 s + B_3 t}{r^2 + st}\right)\right] dr\, ds\, dt$$

and

(3.50)
$$\mathcal{L}_2 = q^{3n} \int_{s \in \varpi \mathcal{O}_F} \int_{r \in \mathcal{O}_F^\times} \int_{t \in \mathcal{O}_F} \psi\left[\varpi^{-n}\left(ds + t + \frac{2B_1 r + B_2 s + B_3 t}{r^2 + st}\right)\right] dr\, ds\, dt.$$

3.3.3.3.1. *The evaluation of $\mathcal{L}_1$.* By a change of variable $t \mapsto s^{-1} t$ in (3.49), we have

$$\mathcal{L}_1 = q^{3n} \int_{s \in \mathcal{O}_F^\times} \int_{r, t \in \mathcal{O}_F, |r^2 + t| = 1}$$
$$\psi\left[\varpi^{-n}\left(ds + s^{-1} t + \frac{2B_1 r + B_2 s + B_3 s^{-1} t}{r^2 + t}\right)\right] dr\, ds\, dt.$$

Then by a change of variable $t \mapsto t - r^2$, we have

$$\mathcal{L}_1 = q^{3n} \int_{s \in \mathcal{O}_F^\times} \int_{r \in \mathcal{O}_F} \int_{t \in \mathcal{O}_F^\times} \psi\left[\varpi^{-n}\left(ds + s^{-1} t + B_2 s t^{-1} + B_3 s^{-1}\right)\right]$$
$$\psi\left[\varpi^{-n}\left\{-s^{-1}\left(1 + B_3 t^{-1}\right) r^2 + 2B_1 r t^{-1}\right\}\right] dr\, ds\, dt.$$

By a change of variable $s \mapsto s^{-1}$, we have

$$\mathcal{L}_1 = q^{3n} \int_{s \in \mathcal{O}_F^\times} \int_{r \in \mathcal{O}_F} \int_{t \in \mathcal{O}_F^\times} \psi\left[\varpi^{-n}\left(ds^{-1} + st + B_2 s^{-1} t^{-1} + B_3 s\right)\right]$$
$$\psi\left[\varpi^{-n}\left\{-s\left(1 + B_3 t^{-1}\right) r^2 + 2B_1 r t^{-1}\right\}\right] dr\, ds\, dt.$$

Then by a change of variable $r \mapsto rt$, $s \mapsto st^{-1}$, we have

$$\mathcal{L}_1 = \sum_{j=0}^{n} \mathcal{L}_1^{(j)}$$

where

(3.51) $$\mathcal{L}_1^{(j)} = q^{3n} \int_{s \in \mathcal{O}_F^\times} \int_{t \in A_j} \psi\left[\varpi^{-n}\left\{st^{-1}(t + B_3) + s^{-1}(dt + B_2)\right\}\right]$$
$$\int_{r \in \mathcal{O}_F} \psi\left[\varpi^{-n}\left\{-s(t + B_3) r^2 + 2B_1 r\right\}\right] dr\, ds\, dt$$

and
(3.52) $$A_j = \{t \in \mathcal{O}_F^\times \mid t + B_3 \in \varpi^j \mathcal{O}_F^\times\}$$
for $0 \leq j < n$ and
(3.53) $$A_n = \{t \in \mathcal{O}_F^\times \mid t + B_3 \in \varpi^n \mathcal{O}_F\}.$$

Let us compute $\mathcal{L}_1^{(n)}$ first. We note that
(3.54) $$B_2 - dB_3 = \frac{2d\varepsilon (u + u^\sigma)}{(1 - uu^\sigma)^2} \in \varpi^m \mathcal{O}_F^\times$$
by (3.37). Therefore
(3.55)
$$\mathcal{L}_1^{(n)} = q^{3n} \int_{s \in \mathcal{O}_F^\times} \int_{t \in A_n} \psi\left[\varpi^{-n} s^{-1} (-dB_3 + B_2)\right] \int_{r \in \mathcal{O}_F} \psi\left(2\varpi^{-n} B_1 r\right) dr\, ds\, dt$$
$$= q^{2n} \left(\int_{\mathcal{O}_F^\times} \psi\left[2\varpi^{-n} d\varepsilon (u + u^\sigma)(1 - uu^\sigma)^{-2} s\right] ds\right) \left(\int_{\mathcal{O}_F} \psi\left(2\varpi^{-n} B_1 r\right) dr\right)$$
$$= \begin{cases} -q^{2n-1}, & \text{when } m = n - 1 \text{ and } u \in \mathcal{O}_F \\ 0, & \text{otherwise.} \end{cases}$$

Let us compute $\mathcal{L}_1^{(j)}$ for $0 \leq j < n$. By a change of variable $s \mapsto \varpi^j (t + B_3)^{-1} s$, we have
(3.56) $$\mathcal{L}_1^{(j)} = q^{3n} \int_{s \in \mathcal{O}_F^\times} \int_{t \in A_j} \psi\left[\varpi^{-n+j} s t^{-1} + \varpi^{-n-j} s^{-1} (dt + B_2)(t + B_3)\right]$$
$$\left(\int_{r \in \mathcal{O}_F} \psi\left(-\varpi^{-n+j} s r^2 + 2\varpi^{-n} B_1 r\right) dr\right) ds\, dt.$$

By Proposition 2.5, the inner integral is given as
(3.57) $$\int_{\mathcal{O}_F} \psi\left(-\varpi^{-n+j} s r^2 + 2\varpi^{-n} B_1 r\right) dr$$
$$= \begin{cases} 0, & \text{when } j > m \text{ and } u \notin \mathcal{O}_F \\ C_{n-j} \operatorname{sgn}(-s)^{n-j} \psi\left(\varpi^{-n-j} B_1^2 s^{-1}\right), & \text{otherwise} \end{cases}$$
by (3.47).

Now let us show that
(3.58) $$\mathcal{L}_1^{(j)} = 0 \quad \text{for} \quad j > m.$$
By (3.57), we may assume that $u \in \mathcal{O}_F$ and hence $B_1 = 0$. Then we have
$$\mathcal{L}_1^{(j)} = q^{3n} C_{n-j} \operatorname{sgn}(-1)^{n-j} \int_{t \in A_j} \int_{s \in \mathcal{O}_F^\times}$$
$$\operatorname{sgn}(s)^{n-j} \psi\left[\varpi^{-n+j} s t^{-1} + \varpi^{-n-j} s^{-1} (dt + B_2)(t + B_3)\right] ds\, dt.$$
Here we note that
$$dt + B_2 = d(t + B_3) + B_2 - dB_3 \in \varpi^m \mathcal{O}_F^\times$$
by (3.54). Thus the inner integral vanishes by Proposition 2.7 and Proposition 2.13.

3.3. EVALUATION OF THE ANISOTROPIC KLOOSTERMAN SUM

Hence now our task is to compute $\mathcal{L}_1^{(j)}$ for $0 \leq j \leq m$. By (3.57), we have

$$\mathcal{L}_1^{(j)} = q^{3n} C_{n-j} \operatorname{sgn}(-1)^{n-j} \int_{t \in A_j} \int_{s \in \mathcal{O}_F^\times}$$
$$\operatorname{sgn}(s)^{n-j} \psi \left[\varpi^{-n+j} st^{-1} + \varpi^{-n-j} s^{-1} \left\{ (dt + B_2)(t + B_3) + B_1^2 \right\} \right] ds\, dt.$$

Thus we may write
$$\mathcal{L}_1^{(j)} = \mathcal{M}_j - \mathcal{N}_j$$
where

(3.59) $\mathcal{M}_j = q^{3n} C_{n-j} \operatorname{sgn}(-1)^{n-j} \int_{t \in \mathcal{O}_F^\times \cap (-B_3(1+\varpi^j \mathcal{O}_F))} \int_{s \in \mathcal{O}_F^\times}$
$$\operatorname{sgn}(s)^{n-j} \psi \left[\varpi^{-n+j} st^{-1} + \varpi^{-n-j} s^{-1} \left\{ (dt + B_2)(t + B_3) + B_1^2 \right\} \right] ds\, dt$$

and

(3.60) $\mathcal{N}_j = q^{3n} C_{n-j} \operatorname{sgn}(-1)^{n-j} \int_{t \in -B_3(1+\varpi^{j+1}\mathcal{O}_F)} \int_{s \in \mathcal{O}_F^\times}$
$$\operatorname{sgn}(s)^{n-j} \psi \left[\varpi^{-n+j} st^{-1} + \varpi^{-n-j} s^{-1} \left\{ (dt + B_2)(t + B_3) + B_1^2 \right\} \right] ds\, dt.$$

We shall evaluate $\mathcal{M}_j$ first and then $\mathcal{N}_j$.

Let us compute $\mathcal{M}_0$. By a change of variable $t \mapsto st$, we have

$$\mathcal{M}_0 = q^{3n} C_n \operatorname{sgn}(-1)^n \int_{t \in \mathcal{O}_F^\times} \psi \left[\varpi^{-n} t^{-1} + \varpi^{-n} t (B_2 + dB_3)\right]$$
$$\left(\int_{s \in \mathcal{O}_F^\times} \operatorname{sgn}(s)^n \psi \left[\varpi^{-n} dt^2 s + \varpi^{-n} s^{-1} \left(B_1^2 + B_2 B_3 \right) \right] ds \right) dt.$$

Here we have

(3.61) $$B_1^2 + B_2 B_3 = d\varepsilon^2 (1 - uu^\sigma)^{-2}.$$

Thus by Proposition 2.11 and Proposition 2.14, we have

(3.62) $\mathcal{M}_0 = q^{2n} (-1)^n \int_{t \in \mathcal{O}_F^\times} \psi \left[\varpi^{-n} t^{-1} - 2\varpi^{-n} d\varepsilon t (1 + uu^\sigma)(1 - uu^\sigma)^{-2}\right]$
$$\left\{ \psi \left(2\varpi^{-n} d\varepsilon t (1 - uu^\sigma)^{-1} \right) + \psi \left(-2\varpi^{-n} d\varepsilon t (1 - uu^\sigma)^{-1} \right) \right\} dt$$
$$= (-1)^n q^{2n} \left\{ \mathcal{K}\ell \left(\frac{2\varpi^{-n}}{1 - uu^\sigma}, \frac{-2\varpi^{-n} d\varepsilon}{1 - uu^\sigma} \right) + \mathcal{K}\ell \left(\frac{2\varpi^{-n}}{1 - uu^\sigma}, \frac{-2\varpi^{-n} d\varepsilon uu^\sigma}{1 - uu^\sigma} \right) \right\}.$$

Here we note that the second term of (3.62) vanishes unless $m = 0$ by Proposition 2.7.

Now let us compute $\mathcal{M}_j$ for $0 < j \leq m$. By a change of variable $s \mapsto st$, we have

(3.63) $\mathcal{M}_j = q^{3n} C_{n-j} \operatorname{sgn}(B_3)^{n-j} \int_{s \in \mathcal{O}_F^\times}$
$$\operatorname{sgn}(s)^{n-j} \psi \left[\varpi^{-n+j} s - 2\varpi^{-n-j} d\varepsilon s^{-1} (1 + uu^\sigma)(1 - uu^\sigma)^{-2}\right]$$
$$\left(\int_{-B_3(1+\varpi^j \mathcal{O}_F)} \psi \left[\varpi^{-n-j} ds^{-1} t + \varpi^{-n-j} d\varepsilon^2 (1 - uu^\sigma)^{-2} s^{-1} t^{-1}\right] dt \right) ds.$$

Here by Lemma 2.10, the inner integral vanishes unless
$$-B_3 = \varepsilon\left(1+u\right)\left(1+u^\sigma\right)\left(1-uu^\sigma\right)^{-2} \equiv \pm\varepsilon\left(1-uu^\sigma\right)^{-1} \pmod{\varpi^j}.$$
Since $u \in \varpi^m \mathcal{O}_E^\times \subset \varpi^j \mathcal{O}_E$, i.e. $u \equiv 0 \pmod{\varpi^j}$, by Lemma 2.10, we have
$$\mathcal{M}_j = (-1)^{n+j} q^{2n} \int_{s \in \mathcal{O}_F^\times} \psi\left[\varpi^{-n+j} s - 4\varpi^{-n-j} d\varepsilon uu^\sigma \left(1-uu^\sigma\right)^{-2} s^{-1}\right] ds.$$

Thus when $j = m$, we have

(3.64) $$\mathcal{M}_m = (-1)^{n+m} q^{2n} \cdot \mathcal{K}\ell\left(\frac{2\varpi^{m-n}}{1-uu^\sigma}, \frac{-2\varpi^{m-n} d\varepsilon \varepsilon_u \varepsilon_u^\sigma}{1-uu^\sigma}\right).$$

When $j < m$, we have $n - j > n - m > 0$ and hence $n - j > 1$. Therefore
$$n - j \neq n + j - 2m = \operatorname{ord}\left(4\varpi^{-n-j} d\varepsilon uu^\sigma \left(1-uu^\sigma\right)^{-2}\right).$$

Thus $\mathcal{M}_j$ vanishes by Proposition 2.7.

Let us compute $\mathcal{N}_j$ for $0 \le j \le m$. When $0 \le j \le n-2$, by Proposition 2.7 and Proposition 2.13, the inner integral for $\mathcal{N}_j$ in (3.60) vanishes unless

(3.65)
$$(dt + B_2)(t + B_3) + B_1^2 = d(t + B_3)^2 + (B_2 - dB_3)(t + B_3) + B_1^2 \in \varpi^{2j} \mathcal{O}_F^\times.$$

Since $\operatorname{ord}(B_2 - dB_3) = m$ by (3.54) and $\operatorname{ord}(t + B_3) = j + 1$, we have
$$\operatorname{ord}\left\{d(t + B_3)^2 + (B_2 - dB_3)(t + B_3)\right\} \ge 2j + 1.$$

Hence (3.65) holds only when $\operatorname{ord}(B_1) = m$, i.e. $j = m$ and $u \notin \mathcal{O}_F$. Then, by a change of variable $s \mapsto st$ in (3.60), we have

$$\mathcal{N}_m = q^{3n} C_{n-m} \operatorname{sgn}(B_3)^{n-m}$$
$$\int_{\mathcal{O}_F^\times} \operatorname{sgn}(s)^{n-m} \psi\left[\varpi^{-n+m} s - 2\varpi^{-n-m} s^{-1} d\varepsilon \left(1+uu^\sigma\right)\left(1-uu^\sigma\right)^{-2}\right]$$
$$\left(\int_{-B_3(1+\varpi^{m+1}\mathcal{O}_F)} \psi\left[\varpi^{-n-m} ds^{-1} t + \varpi^{-n-m} ds^{-1} \varepsilon^2 \left(1-uu^\sigma\right)^{-2} t^{-1}\right] dt\right) ds.$$

By Lemma 2.10, the inner integral vanishes unless
$$-B_3 = \varepsilon\left(1+u\right)\left(1+u^\sigma\right)\left(1-uu^\sigma\right)^{-2} \equiv \pm\varepsilon\left(1-uu^\sigma\right)^{-1} \pmod{\varpi^{m+1}}.$$

It is readily seen that

(3.66)
$$-B_3 \equiv \varepsilon\left(1-uu^\sigma\right)^{-1} \pmod{\varpi^{m+1}} \iff m = 0 \text{ and } u + u^\sigma + 2uu^\sigma \in \varpi \mathcal{O}_F$$

and

(3.67)
$$-B_3 \equiv -\varepsilon\left(1-uu^\sigma\right)^{-1} \pmod{\varpi^{m+1}} \iff m = 0 \text{ and } u + u^\sigma + 2 \in \varpi \mathcal{O}_F.$$

3.3. EVALUATION OF THE ANISOTROPIC KLOOSTERMAN SUM

When (3.66) holds, we have

(3.68)
$$\begin{aligned}\mathcal{N}_0 =& q^{3n}C_n\,\mathrm{sgn}\,(B_3)^n \int_{\mathcal{O}_F^\times} \mathrm{sgn}\,(s)^n\,\psi\left[\varpi^{-n}s - 2\varpi^{-n}s^{-1}d\varepsilon\,(1+uu^\sigma)(1-uu^\sigma)^{-2}\right]\\
&C_n\,\mathrm{sgn}\left(ds^{-1}\varepsilon(1-uu^\sigma)^{-1}\right)^n\psi\left[2\varpi^{-n}ds^{-1}\varepsilon(1-uu^\sigma)^{-1}\right]ds\\
=&(-1)^n\,q^{2n}\cdot\mathcal{K}\ell\left(\frac{2\varpi^{-n}}{1-uu^\sigma},\frac{-2\varpi^{-n}d\varepsilon uu^\sigma}{1-uu^\sigma}\right).\end{aligned}$$

Similarly when (3.67) holds, we have

(3.69)
$$\mathcal{N}_0 = (-1)^n\,q^{2n}\cdot\mathcal{K}\ell\left(\frac{2\varpi^{-n}}{1-uu^\sigma},\frac{-2\varpi^{-n}d\varepsilon}{1-uu^\sigma}\right).$$

Finally let us compute $\mathcal{N}_j$ when $j = n-1$. We note that then $m = n-1$ since $j \le m < n$. Since $t + B_3 \in \varpi^n \mathcal{O}_F$ in (3.60), we have

$$\mathcal{N}_{n-1} = q^{3n}C_1\,\mathrm{sgn}\,(-1)\int_{s\in\mathcal{O}_F^\times}\mathrm{sgn}\,(s)\,\psi\left(\varpi^{-2n+1}B_1^2 s^{-1}\right)$$
$$\left(\int_{-B_3(1+\varpi^n\mathcal{O}_F)}\psi\left(\varpi^{-1}st^{-1}\right)dt\right)ds$$
$$= q^{2n}C_1\,\mathrm{sgn}\,(-1)\int_{\mathcal{O}_F^\times}\mathrm{sgn}\,(s)\,\psi\left(-\varpi^{-1}B_3^{-1}s + \varpi^{-2n+1}B_1^2 s^{-1}\right)ds.$$

Hence by Proposition 2.13 and Proposition 2.14, we have
$$\mathcal{N}_{n-1} = q^{2n-1}\mathrm{sgn}\,(\varepsilon)$$
when $u \in \mathcal{O}_F$, and
$$\mathcal{N}_{n-1} = q^{2n-1}\left\{\psi\left(2\varpi^{-n}B_1\sqrt{-B_3^{-1}}\right) + \psi\left(-2\varpi^{-n}B_1\sqrt{-B_3^{-1}}\right)\right\}$$
when $u \notin \mathcal{O}_F$ and $\mathrm{sgn}\,(-B_3) = 1$, and
$$\mathcal{N}_{n-1} = 0$$
when $u \notin \mathcal{O}_F$ and $\mathrm{sgn}\,(-B_3) = -1$.

Thus we have finished the computation of $\mathcal{L}_1$. Here we summarize the result of our computation so far. Let $\mathcal{R}$ be the right hand side of (3.36).
1. When $0 < m < n-1$, we have
$$\mathcal{L}_1 = \mathcal{R}.$$
2. When $m = n-1$ and $u \in \mathcal{O}_F$, we have
$$\mathcal{L}_1 = \mathcal{R} - q^{2n-1}\left\{1 + \mathrm{sgn}\,(\varepsilon)\right\}.$$
3. When $m = n-1$ and $u \notin \mathcal{O}_F$, we have
$$\mathcal{L}_1 = \mathcal{R}$$
when $\mathrm{sgn}\,(-B_3) = -1$, and
$$\mathcal{L}_1 = \mathcal{R} - q^{2n-1}\left\{\psi\left(2\varpi^{-n}B_1\sqrt{-B_3^{-1}}\right) + \psi\left(-2\varpi^{-n}B_1\sqrt{-B_3^{-1}}\right)\right\}$$
when $\mathrm{sgn}\,(-B_3) = 1$.

4. When $m = 0$, $n > 1$ and $u \in \mathcal{O}_F$, we have
$$\mathcal{L}_1 = \mathcal{R}.$$
When $m = 0$, $n > 1$ and $u \notin \mathcal{O}_F$, we have
$$\mathcal{L}_1 = \mathcal{R} + \begin{cases} (-1)^{n+1} q^{2n} \cdot \mathcal{K}\ell\left(\frac{2\varpi^{-n}}{1-uu^\sigma}, \frac{-2\varpi^{-n}d\varepsilon}{1-uu^\sigma}\right), & \text{when } u + u^\sigma + 2 \in \varpi\mathcal{O}_F \\ (-1)^{n+1} q^{2n} \cdot \mathcal{K}\ell\left(\frac{2\varpi^{-n}}{1-uu^\sigma}, \frac{-2\varpi^{-n}d\varepsilon uu^\sigma}{1-uu^\sigma}\right), & \text{when } u + u^\sigma + 2 \in \varpi\mathcal{O}_F \\ 0, & \text{otherwise.} \end{cases}$$

3.3.3.2. *The evaluation of $\mathcal{L}_2$.* Let us consider the case when $n = 1$ and $m = 0$ first. Then by a change of variable $r \mapsto r^{-1}$ in (3.50), we have

(3.70) $$\mathcal{L}_2 = q^2 \int_{\mathcal{O}_F^\times} \psi\left(2\varpi^{-1} B_1 r\right) \left(\int_{\mathcal{O}_F} \psi\left[\varpi^{-1} B_3 \left(r^2 + B_3^{-1}\right) t\right] dt\right) dr.$$

When $u \in \mathcal{O}_F$, we have $B_1 = 0$. Hence it is clear from (3.70) that we have
$$\mathcal{L}_2 = \begin{cases} 2q, & \text{when sgn}(-B_3) = 1 \\ 0, & \text{when sgn}(-B_3) = -1. \end{cases}$$
Since $B_3 = -\varepsilon (1-u)^{-2}$ in this case, we have
$$\mathcal{L}_2 = q\left\{1 + \text{sgn}(\varepsilon)\right\}.$$
Suppose that $u \notin \mathcal{O}_F$. When $\text{sgn}(-B_3) = 1$, by (3.70), we have
$$\mathcal{L}_2 = 0.$$
When $\text{sgn}(-B_3) = 1$, from (3.70), we have
$$\mathcal{L}_2 = q\left\{\psi\left(2\varpi^{-1} B_1 \sqrt{-B_3^{-1}}\right) + \psi\left(-2\varpi^{-1} B_1 \sqrt{-B_3^{-1}}\right)\right\}.$$
Hence we are done with the case when $n = 1$ and $m = 0$.

Let us consider the case when $n > 1$. By a change of variable $r \mapsto r\left(1 - r^{-2} st\right)^{\frac{1}{2}}$ in (3.50), we have

(3.71) $$\mathcal{L}_2 = q^{3n} \int_{r \in \mathcal{O}_F^\times} \int_{s \in \varpi\mathcal{O}_F} \int_{t \in \mathcal{O}_F} \psi\left[\varpi^{-n}(ds + t)\right]$$
$$\psi\left[2\varpi^{-n} B_1 r^{-1} \left(1 - r^{-2} st\right)^{\frac{1}{2}} + \varpi^{-n} r^{-2} (B_2 s + B_3 t)\right] dr\, ds\, dt.$$

We shall evaluate $\mathcal{L}_2$ by *shrinking the support* of the integral (3.71). First we observe that since $1 + \varpi^{n-1}\lambda \in \mathcal{O}_F^\times$ for $\lambda \in \mathcal{O}_F$, we have

(3.72) $$\mathcal{L}_2 = q^{3n} \int_{\lambda \in \mathcal{O}_F} \int_{r \in \mathcal{O}_F^\times} \int_{s \in \varpi\mathcal{O}_F} \int_{t \in \mathcal{O}_F} \psi\left[\varpi^{-n}(ds + t)\right]$$
$$\psi\left[2\varpi^{-n} B_1 r^{-1} \left(1 + \varpi^{n-1}\lambda\right)^{-1} \left(1 - r^{-2} \left(1 + \varpi^{n-1}\lambda\right)^{-2} st\right)^{\frac{1}{2}}\right]$$
$$\psi\left[\varpi^{-n} r^{-2} \left(1 + \varpi^{n-1}\lambda\right)^{-2} (B_2 s + B_3 t)\right] d\lambda\, dr\, ds\, dt.$$

Then since
$$\left(1 + \varpi^{n-1}\lambda\right)^{-1} \equiv 1 - \varpi^{n-1}\lambda \pmod{\varpi^n}$$
$$\left(1 + \varpi^{n-1}\lambda\right)^{-2} \equiv 1 - 2\varpi^{n-1}\lambda \pmod{\varpi^n},$$

we may write (3.72) as

$$(3.73) \quad \mathcal{L}_2 = q^{3n} \int_{r \in \mathcal{O}_F^\times} \int_{s \in \varpi \mathcal{O}_F} \int_{t \in \mathcal{O}_F} \psi\left[\varpi^{-n}(ds+t)\right]$$
$$\psi\left[2\varpi^{-n} B_1 r^{-1}\left(1 - r^{-2} st\right)^{\frac{1}{2}} + \varpi^{-n} r^{-2}(B_2 s + B_3 t)\right]$$
$$\left(\int_{\mathcal{O}_F} \psi\left[-2\varpi^{-1} r^{-2}\left(B_1 r\left(1-r^{-2}st\right)^{\frac{1}{2}} + B_3 t\right)\lambda\right] d\lambda\right) dr\, ds\, dt.$$

When $m > 0$, or, when $m = 0$ and $u \in \mathcal{O}_F$, the inner integral of (3.73) vanishes unless $t \in \varpi \mathcal{O}_F$. Hence we have

$$(3.74) \quad \mathcal{L}_2 = q^{3n} \int_{r \in \mathcal{O}_F^\times} \int_{s \in \varpi \mathcal{O}_F} \int_{t \in \varpi \mathcal{O}_F} \psi\left[\varpi^{-n}(ds+t)\right]$$
$$\psi\left[2\varpi^{-n} B_1 r^{-1}\left(1 - r^{-2} st\right)^{\frac{1}{2}} + \varpi^{-n} r^{-2}(B_2 s + B_3 t)\right] dr\, ds\, dt.$$

On the other hand when $m = 0$ and $u \notin \mathcal{O}_F$, the inner integral of (3.73) vanishes when $t \in \varpi \mathcal{O}_F$. Hence we have

$$(3.75) \quad \mathcal{L}_2 = q^{3n} \int_{r \in \mathcal{O}_F^\times} \int_{s \in \varpi \mathcal{O}_F} \int_{t \in \mathcal{O}_F^\times} \psi\left[\varpi^{-n}(ds+t)\right]$$
$$\psi\left[2\varpi^{-n} B_1 r^{-1}\left(1 - r^{-2} st\right)^{\frac{1}{2}} + \varpi^{-n} r^{-2}(B_2 s + B_3 t)\right] dr\, ds\, dt.$$

Suppose that $m > 0$. By a change of variable $s \mapsto r^2 s$ in (3.74), we have

$$(3.76) \quad \mathcal{L}_2 = q^{3n} \int_{r \in \mathcal{O}_F^\times} \int_{s \in \varpi \mathcal{O}_F} \int_{t \in \varpi \mathcal{O}_F} \psi\left[\varpi^{-n}(dr^2 s + t)\right]$$
$$\psi\left[2\varpi^{-n} B_1 r^{-1}(1 - st)^{\frac{1}{2}} + \varpi^{-n} r^{-2}(B_2 r^2 s + B_3 t)\right] dr\, ds\, dt.$$

Let us consider the case when $0 < m < n-1$ first. Then we note that for $\lambda \in \mathcal{O}_F$, we have

$$1 - \left(s + \varpi^{n-m-1}\lambda\right) t \equiv 1 - st \pmod{\varpi^{n-m}}$$

since $t \in \varpi \mathcal{O}_F$. Thus we have

$$\mathcal{L}_2 = q^{3n} \int_{r \in \mathcal{O}_F^\times} \int_{s \in \varpi \mathcal{O}_F} \int_{t \in \varpi \mathcal{O}_F} \psi\left[\varpi^{-n}(dr^2 s + t)\right]$$
$$\psi\left[2\varpi^{-n} B_1 r^{-1}(1-st)^{\frac{1}{2}} + \varpi^{-n} r^{-2}(B_2 r^2 s + B_3 t)\right]$$
$$\left(\int_{\mathcal{O}_F} \psi\left[\varpi^{-m-1} d\left(r^2 + d^{-1} B_2\right)\lambda\right] d\lambda\right) dr\, ds\, dt.$$

Hence the integral (3.76) is supported on $r \in \mathcal{O}_F^\times$ such that

$$(3.77) \qquad r^2 + d^{-1} B_2 \equiv 0 \pmod{\varpi^{m+1}}.$$

Similarly by replacing t by $t + \varpi^{n-m-1}\lambda$ where $\lambda \in \mathcal{O}_F$, we have

$$\mathcal{L}_2 = q^{3n} \int_{r \in \mathcal{O}_F^\times} \int_{s \in \varpi \mathcal{O}_F} \int_{t \in \varpi \mathcal{O}_F} \psi\left[\varpi^{-n}\left(dr^2 s + t\right)\right]$$
$$\psi\left[2\varpi^{-n} B_1 r^{-1} (1 - st)^{\frac{1}{2}} + \varpi^{-n} r^{-2} \left(B_2 r^2 s + B_3 t\right)\right]$$
$$\left(\int_{\mathcal{O}_F} \psi\left[\varpi^{-m-1} r^{-2} \left(r^2 + B_3\right) \lambda\right] d\lambda \right) dr\, ds\, dt.$$

Hence the integral (3.76) is supported on $r \in \mathcal{O}_F^\times$ such that

(3.78) $$r^2 + B_3 \equiv 0 \mod \varpi^{m+1}.$$

Then by (3.54), there does not exist $r \in \mathcal{O}_F^\times$ which satisfies both (3.77) and (3.78). Hence $\mathcal{L}_2$ vanishes in this case.

Now let us consider the case when $m = n - 1$. Then since

$$B_1 \equiv 0 \pmod{\varpi^{n-1}}, \quad (1 - st)^{\frac{1}{2}} \equiv 1 \pmod{\varpi},$$

after a change of variable $r \mapsto r^{-1}$, the integral (3.76) becomes

$$\mathcal{L}_2 = q^{3n} \int_{r \in \mathcal{O}_F^\times} \psi\left(2\varpi^{-n} B_1 r\right) \left(\int_{s \in \varpi \mathcal{O}_F} \psi\left[\varpi^{-n} B_2 r^{-2} \left(r^2 + dB_2^{-1}\right) s\right] ds \right)$$
$$\left(\int_{t \in \varpi \mathcal{O}_F} \psi\left[\varpi^{-n} B_3 \left(r^2 + B_3^{-1}\right) t\right] dt \right) dr.$$

Here we note that

$$B_2 \equiv -d\varepsilon \pmod{\varpi^{n-1}}, \quad B_3 \equiv -\varepsilon \pmod{\varpi^{n-1}}.$$

Hence

$$\mathcal{L}_2 = q^{3n} \int_{\{r \in \mathcal{O}_F^\times \mid r^2 \in \varepsilon^{-1} + \varpi^{n-1} \mathcal{O}_F\}} \psi\left(2\varpi^{-n} B_1 r\right) dr.$$

When $u \in \mathcal{O}_F$, we have $B_1 = 0$ and hence

$$\mathcal{L}_2 = q^{2n-1} \left\{1 + \operatorname{sgn}(\varepsilon)\right\}.$$

When $u \notin \mathcal{O}_F$, we have $\operatorname{ord}(B_1) = n - 1$ and hence

$$\mathcal{L}_2 = \begin{cases} q^{2n-1} \left\{\psi\left(2\varpi^{-n} B_1 \sqrt{\varepsilon^{-1}}\right) + \psi\left(-2\varpi^{-n} B_1 \sqrt{\varepsilon^{-1}}\right)\right\}, & \text{when } \operatorname{sgn}(\varepsilon) = 1 \\ 0, & \text{when } \operatorname{sgn}(\varepsilon) = -1. \end{cases}$$

Finally let us consider the case when $m = 0$. When $u \in \mathcal{O}_F$, then $B_1 = 0$ and hence (3.74) becomes

$$\mathcal{L}_2 = q^{3n} \int_{r \in \mathcal{O}_F^\times} \left(\int_{s \in \varpi \mathcal{O}_F} \psi\left[\varpi^{-n} dr^{-2} \left(r^2 + d^{-1} B_2\right) s\right] ds \right)$$
$$\left(\int_{t \in \varpi \mathcal{O}_F} \psi\left[\varpi^{-n} r^{-2} \left(r^2 + B_3\right) t\right] dt \right) dr$$

where $d^{-1} B_2 - B_3 \in \mathcal{O}_F^\times$ by (3.54). Thus $\mathcal{L}_2$ vanishes in this case.

Suppose that $u \notin \mathcal{O}_F$. By a change of variable $s \mapsto r^2 s t^{-1}$, (3.75) becomes

$$(3.79) \quad \mathcal{L}_2 = q^{3n} \int_{r \in \mathcal{O}_F^\times} \int_{s \in \varpi \mathcal{O}_F} \int_{t \in \mathcal{O}_F^\times} \psi \left[\varpi^{-n} \left(dr^2 s t^{-1} + t \right) \right]$$
$$\psi \left[2\varpi^{-n} B_1 r^{-1} (1-s)^{\frac{1}{2}} + \varpi^{-n} B_2 s t^{-1} + \varpi^{-n} B_3 r^{-2} t \right] dr\, ds\, dt.$$

For $\lambda \in \mathcal{O}_F$, we have

$$s \left(1 + \varpi^{n-1} \lambda \right)^{-1} \equiv s \pmod{\varpi^n}$$

since $s \in \varpi \mathcal{O}_F$. Thus by replacing t by $t \left(1 + \varpi^{n-1} \lambda \right)$ in (3.79), we have

$$(3.80) \quad \mathcal{L}_2 = q^{3n} \int_{r \in \mathcal{O}_F^\times} \int_{s \in \varpi \mathcal{O}_F} \int_{t \in \mathcal{O}_F^\times} \psi \left[\varpi^{-n} \left(dr^2 s t^{-1} + t \right) \right]$$
$$\psi \left[2\varpi^{-n} B_1 r^{-1} (1-s)^{\frac{1}{2}} + \varpi^{-n} B_2 s t^{-1} + \varpi^{-n} B_3 r^{-2} t \right]$$
$$\left(\int_{\lambda \in \mathcal{O}_F} \psi \left[\varpi^{-1} r^{-2} \left(r^2 + B_3 \right) \right] d\lambda \right) dr\, ds\, dt.$$

Thus $\mathcal{L}_2$ vanishes unless $\operatorname{sgn}(-B_3) = 1$. We note that

$$(3.81) \quad \operatorname{sgn}(-B_3) = \operatorname{sgn} \left\{ \varepsilon \left(1 + u \right) \left(1 + u^\sigma \right) \right\}.$$

Suppose that $\operatorname{sgn}(-B_3) = 1$. Then we have $\mathcal{L}_2 = \mathcal{L}_2^+ + \mathcal{L}_2^-$ where

$$(3.82) \quad \mathcal{L}_2^\pm = q^{3n} \int_{r \in \pm \sqrt{-B_3}(1+\varpi \mathcal{O}_F)} \int_{s \in \varpi \mathcal{O}_F} \int_{t \in \mathcal{O}_F^\times} \psi \left[\varpi^{-n} \left(dr^2 s t^{-1} + t \right) \right]$$
$$\psi \left[2\varpi^{-n} B_1 r^{-1} (1-s)^{\frac{1}{2}} + \varpi^{-n} B_2 s t^{-1} + \varpi^{-n} B_3 r^{-2} t \right] dr\, ds\, dt.$$

Since $a \mapsto a^{\frac{1}{2}}$ is an automorphism of $1 + \varpi \mathcal{O}_F$, we may have a change of variable $(1-s)^{\frac{1}{2}} = 1 + \varpi y$ in (3.82). Then we have

$$\mathcal{L}_2^\pm = q^{3n-1} \int_{r \in \pm \sqrt{-B_3}(1+\varpi \mathcal{O}_F)} \int_{t \in \mathcal{O}_F^\times} \int_{y \in \mathcal{O}_F} \psi \left[\varpi^{-n} \left(2 B_1 r^{-1} + B_3 r^{-2} t + t \right) \right]$$
$$\psi \left[-\varpi^{-n+2} d t^{-1} \left\{ \left(r^2 + d^{-1} B_2 \right) y^2 - 2 \varpi^{-1} \left(\left(r^2 + d^{-1} B_2 \right) + d^{-1} B_1 r^{-1} t \right) y \right\} \right]$$
$$dr\, dt\, dy.$$

By a change of variable $t \mapsto rt$, we have

$$(3.83)$$
$$\mathcal{L}_2^\pm = q^{3n-1} \int_{r \in \pm \sqrt{-B_3}(1+\varpi \mathcal{O}_F)} \int_{t \in \mathcal{O}_F^\times} \int_{y \in \mathcal{O}_F} \psi \left[\varpi^{-n} \left(2 B_1 r^{-1} + B_3 r^{-1} t + r t \right) \right]$$
$$\psi \left[-\varpi^{-n+2} d r^{-1} t^{-1} \left\{ \left(r^2 + d^{-1} B_2 \right) y^2 - 2 \varpi^{-1} \left(\left(r^2 + d^{-1} B_2 \right) + d^{-1} B_1 t \right) y \right\} \right]$$
$$dr\, dt\, dy.$$

In (3.83), we have

$$r^2 + d^{-1} B_2 = \left(r^2 + B_3 \right) + \left(d^{-1} B_2 - B_3 \right) \in \mathcal{O}_F^\times$$

by (3.54). Hence by Proposition 2.5, the integral (3.83) is supported on

$$t \in -B_1^{-1} \left(B_2 - d B_3 \right) + \varpi \mathcal{O}_F$$

and we have

$$(3.84) \quad \mathcal{L}_2^\pm = (-1)^n q^{3n-1} C_{n-2} \operatorname{sgn}(-1)^n \int_{\pm\sqrt{-B_3}(1+\varpi\mathcal{O}_F)} \int_{-B_1^{-1}(B_2-dB_3)+\varpi\mathcal{O}_F}$$
$$\psi\left[\varpi^{-n}\left(2B_1 r^{-1} + r^{-1}t\left(r^2+B_3\right)\right)\right]$$
$$\operatorname{sgn}\left\{rt\left(r^2+d^{-1}B_2\right)\right\}^n \psi\left[\varpi^{-n}dr^{-1}t^{-1}\cdot\frac{\left(\left(r^2+d^{-1}B_2\right)+d^{-1}B_1 t\right)^2}{r^2+d^{-1}B_2}\right] dr\, dt.$$

Since
$$r^2 + d^{-1}B_2 \equiv -B_3 + d^{-1}B_2 \pmod{\varpi}$$
in (3.84), by a change of variable $t \mapsto \left(r^2+d^{-1}B_2\right)t$, we have

$$\mathcal{L}_2^\pm = (-1)^n q^{3n-1} C_{n-2} \operatorname{sgn}(-1)^n \int_{r\in\pm\sqrt{-B_3}(1+\varpi\mathcal{O}_F)} \int_{t\in dB_1^{-1}+\varpi\mathcal{O}_F}$$
$$\psi\left[\varpi^{-n}r^{-1}t\left(r^2+d^{-1}B_2\right)\left(r^2+B_3\right)\right]$$
$$\operatorname{sgn}(rt)^n \psi\left[\varpi^{-n}dr^{-1}t^{-1}\left(1+d^{-2}B_1^2 t^2\right)\right] dr\, dt.$$

Further by a change of variable $t \mapsto r^{-1}t$, we have

$$\mathcal{L}_2^\pm = (-1)^n q^{3n-1} C_{n-2} \operatorname{sgn}(-1)^n \int_{r\in\pm\sqrt{-B_3}(1+\varpi\mathcal{O}_F)} \int_{t\in\pm dB_1^{-1}\sqrt{-B_3}+\varpi\mathcal{O}_F}$$
$$\operatorname{sgn}(t)^n \psi\left[\varpi^{-n}\left(d^{-1}B_2+B_3\right)t + \varpi^{-n}dt^{-1}\right]$$
$$\psi\left[\varpi^{-n}tr^2 + \varpi^{-n}t\left(d^{-1}B_2 B_3 - d^{-1}B_1^2\right)r^{-2}\right] dr\, dt.$$

Then by a change of variable $z = r^2$, we have

$$(3.85) \quad \mathcal{L}_2^\pm = (-1)^n q^{3n-1} C_{n-2} \operatorname{sgn}(-1)^n$$
$$\int_{t\in\mp dB_1^{-1}\sqrt{-B_3^{-1}}+\varpi\mathcal{O}_F} \operatorname{sgn}(t)^n \psi\left[\frac{-2\varpi^{-n}\varepsilon(1+uu^\sigma)}{(1-uu^\sigma)^2}t + \varpi^{-n}dt^{-1}\right]$$
$$\left(\int_{z\in -B_3(1+\varpi\mathcal{O}_F)} \psi\left[\varpi^{-n}tz + \varpi^{-n}tz^{-1}\varepsilon^2(1-uu^\sigma)^{-2}\right] dz\right) dt.$$

By Lemma 2.10, the inner integral of (3.85) vanishes unless
$$-B_3 = \varepsilon(1+u)(1+u^\sigma)(1-uu^\sigma)^{-2} \equiv \pm\varepsilon(1-uu^\sigma)^{-1} \pmod{\varpi},$$
i.e.
$$(3.86) \quad u + u^\sigma + 2uu^\sigma \in \varpi\mathcal{O}_F$$
or
$$(3.87) \quad u + u^\sigma + 2 \in \varpi\mathcal{O}_F.$$

Suppose that (3.86) holds. Then we note that
$$\operatorname{sgn}\left\{\varepsilon(1-uu^\sigma)^{-1}\right\} = 1$$
since
$$\operatorname{sgn}(-B_3) = 1, \quad -B_3 \equiv \varepsilon(1-uu^\sigma)^{-1} \pmod{\varpi}.$$

Hence by Lemma 2.10, we have

$$\mathcal{L}_2^\pm = (-1)^n q^{2n} \int_{\pm dB_1^{-1}\sqrt{-B_3}+\varpi\mathcal{O}_F} \psi\left[\frac{-4\varpi^{-n}\varepsilon uu^\sigma}{(1-uu^\sigma)^2}t + \varpi^{-n}dt^{-1}\right]dt.$$

Here we note that

(3.88) $\left(\pm dB_1^{-1}\sqrt{-B_3}\right)^2 + \dfrac{d(1-uu^\sigma)^2}{4\varepsilon uu^\sigma} = \dfrac{(1-uu^\sigma)^2(u+u^\sigma+2uu^\sigma)^2}{4\varepsilon uu^\sigma(u-u^\sigma)^2} \in \varpi\mathcal{O}_F$

by (3.86). Hence by Lemma 2.10 and Proposition 2.11, we have

$$\mathcal{L}_2 = (-1)^n q^{2n} \cdot \mathcal{K}\ell\left(\frac{2\varpi^{-n}}{1-uu^\sigma}, \frac{-2\varpi^{-n}d\varepsilon uu^\sigma}{1-uu^\sigma}\right).$$

Here we remark that (3.88) shows that under (3.86), we have

$$\operatorname{sgn}(-B_3) = 1 \iff \operatorname{sgn}(-d\varepsilon uu^\sigma) = 1.$$

Similarly when (3.87) holds, we have

$$\operatorname{sgn}(-B_3) = 1 \iff \operatorname{sgn}(-d\varepsilon) = 1$$

and

$$\mathcal{L}_2 = (-1)^n q^{2n} \cdot \mathcal{K}\ell\left(\frac{2\varpi^{-n}}{1-uu^\sigma}, \frac{-2\varpi^{-n}d\varepsilon}{1-uu^\sigma}\right).$$

Thus we have finished the proof of Theorem 3.10.

3.4. Evaluation of the quadratic Kloosterman sum

Let us denote by $\operatorname{Herm}^2(E)$ the set of 2 by 2 Hermitian matrices over E, i.e.

(3.89) $\qquad \operatorname{Herm}^2(E) = \left\{T \in \operatorname{Mat}_{2\times 2}(E) \mid {}^tT^\sigma = T\right\}$

and let $\operatorname{Herm}^2(\mathcal{O}_E) = \operatorname{Herm}^2(E) \cap \operatorname{Mat}_{2\times 2}(\mathcal{O}_E)$.

DEFINITION 3.13. For $S \in \operatorname{Sym}^2(E)$, $T \in \operatorname{Herm}^2(E)$ such that $\det(T) \neq 0$, and $\varepsilon \in \mathcal{O}_E^\times$, we define $\mathcal{H}(S,T,\varepsilon)$, the *quadratic Kloosterman sum*, by

(3.90) $\qquad \mathcal{H}(S,T,\varepsilon) = \displaystyle\int_{\mathcal{Z}_T} \psi_E\left[\varepsilon \cdot \operatorname{tr}(SZ)\right]dZ$

where
(3.91)
$$\mathcal{Z}_T = \left\{Z \in \operatorname{Sym}^2(E) \mid TZ \in \operatorname{Mat}_{2\times 2}(\mathcal{O}_E), T^{-\sigma} - Z^\sigma TZ \in \operatorname{Mat}_{2\times 2}(\mathcal{O}_E)\right\}.$$

First we observe some elementary properties of $\mathcal{H}(S,T,\varepsilon)$.

PROPOSITION 3.14. *1. For $\varepsilon' \in \mathcal{O}_E^\times$, we have*

$$\mathcal{H}(\varepsilon'S,T,\varepsilon) = \mathcal{H}(S,T,\varepsilon\varepsilon').$$

For $\mu \in \mathcal{O}_F^\times$, we have

$$\mathcal{H}(S,\mu T,\varepsilon) = \mathcal{H}\left(S,T,\mu^{-1}\varepsilon\right).$$

2. For $k \in \operatorname{GL}_2(\mathcal{O}_E)$, we have

$$\mathcal{H}\left(S,{}^tk^\sigma Tk,\varepsilon\right) = \mathcal{H}\left({}^tk^{-1}Sk^{-1},T,\varepsilon\right).$$

3. For $h \in \operatorname{GL}_2(\mathcal{O}_E)$ such that ${}^thSh = \lambda S$ where $\lambda \in \mathcal{O}_E^\times$, we have

$$\mathcal{H}\left(S,{}^th^\sigma Th,\varepsilon\right) = \mathcal{H}\left(S,T,\varepsilon\lambda^{-1}\right).$$

4. For $\zeta \in \mathcal{O}_E^\times$ such that $\zeta \zeta^\sigma = 1$, we have
$$\mathcal{H}(S, T, \varepsilon \zeta) = \mathcal{H}(S, T, \varepsilon).$$
5. $\mathcal{H}(S, T^\sigma, \varepsilon) = \mathcal{H}(S^\sigma, T, \varepsilon)$.

PROOF. 1. The first assertion is trivial. The second one is also clear since $\mathcal{Z}_{\mu T} = \mu^{-1} \mathcal{Z}_T$.

2. Let $T_1 = {}^t k^\sigma T k$. Then for $Z \in \mathrm{Sym}^2(E)$, we have $Z \in \mathcal{Z}_{T_1}$ if and only if
$$T_1 Z = {}^t k^\sigma T k Z = {}^t k^\sigma T \left(k Z {}^t k \right) {}^t k^{-1} \in \mathrm{Mat}_{2 \times 2}(\mathcal{O}_E)$$
and
$$T_1^{-\sigma} - Z^\sigma T_1 Z = k^{-\sigma} T^{-\sigma t} k^{-1} - Z^\sigma {}^t k^\sigma T k Z$$
$$= k^{-\sigma} \left\{ T^{-\sigma} - \left(k Z {}^t k \right)^\sigma T \left(k Z {}^t k \right) \right\} {}^t k^{-1} \in \mathrm{Mat}_{2 \times 2}(\mathcal{O}_E).$$
Hence we have
$$(3.92) \qquad \mathcal{Z}_{T_1} = \mathcal{Z}_{{}^t k^\sigma T k} = k^{-1} \mathcal{Z}_T {}^t k^{-1}.$$
Thus
$$\mathcal{H}(S, T_1, \varepsilon) = \int_{\mathcal{Z}_T} \psi_E \left[\varepsilon \cdot \mathrm{tr}\left(S k^{-1} Z {}^t k^{-1} \right) \right] dZ$$
$$= \int_{\mathcal{Z}_T} \psi_E \left[\varepsilon \cdot \mathrm{tr}\left({}^t k^{-1} S k^{-1} Z \right) \right] dZ = \mathcal{H}\left({}^t k^{-1} S k^{-1}, T, \varepsilon \right).$$

3. This follows from the previous two assertions.

4. First we note that $\zeta^{-1} \mathcal{Z}_T = \mathcal{Z}_T$. Hence we have
$$\mathcal{H}(S, T, \varepsilon \zeta) = \int_{\mathcal{Z}_T} \psi_E \left[\varepsilon \zeta \cdot \mathrm{tr}\left(S \left(\zeta^{-1} Z \right) \right) \right] dZ = \mathcal{H}(S, T, \varepsilon).$$

5. It is clear that $\mathcal{Z}_{T^\sigma} = (\mathcal{Z}_T)^\sigma$. Hence
$$\mathcal{H}(S, T^\sigma, \varepsilon) = \int_{\mathcal{Z}_T} \psi_E \left[\varepsilon \cdot \mathrm{tr}(S Z^\sigma) \right] dZ$$
$$= \int_{\mathcal{Z}_T} \psi_E \left[\varepsilon^\sigma \cdot \mathrm{tr}(S^\sigma Z) \right] dZ = \mathcal{H}(S^\sigma, T, \varepsilon^\sigma).$$
Then by the previous assertion, we have
$$\mathcal{H}(S^\sigma, T, \varepsilon^\sigma) = \mathcal{H}\left(S^\sigma, T, \varepsilon^\sigma \left(\varepsilon \varepsilon^{-\sigma} \right) \right) = \mathcal{H}(S^\sigma, T, \varepsilon).$$
□

From now on we specify S to be
$$S_1 = \begin{pmatrix} 0 & 1 \\ 1 & 0 \end{pmatrix}.$$

DEFINITION 3.15. For $T \in \mathrm{Herm}^2(E)$ such that $\det(T) \neq 0$ and $\varepsilon \in \mathcal{O}_E^\times$, we define $\mathcal{H}(T, \varepsilon)$ by
$$(3.93) \qquad \mathcal{H}(T, \varepsilon) = \mathcal{H}(S_1, T, \varepsilon) = \int_{\mathcal{Z}_T} \psi_E \left[\varepsilon \cdot \mathrm{tr}\left(\begin{pmatrix} 0 & 1 \\ 1 & 0 \end{pmatrix} Z \right) \right] dZ.$$

Now the goal of this section is to prove the following theorem, which may be regarded as a generalization of the Davenport-Hasse relation, Proposition 2.16.

3.4. EVALUATION OF THE QUADRATIC KLOOSTERMAN SUM

THEOREM 3.16. *Let $T \in \mathrm{Herm}^2(\mathcal{O}_E)$ such that $\det T \neq 0$. Let us write*

$$T = \begin{pmatrix} a & y \\ y^\sigma & b \end{pmatrix}, \qquad \Delta = \det T,$$

and, $y = \varpi^{\mathrm{ord}(y)} \varepsilon_y$ when $y \neq 0$. Let us define a non-negative integer m by

$$m = \min\{\mathrm{ord}(a), \mathrm{ord}(b)\}.$$

1. *Suppose that*

(3.94) $$m \leq \mathrm{ord}(y).$$

Then the quadratic Kloosterman sum $\mathcal{H}(T, \varepsilon)$ is evaluated as follows.

(a) *When $m = 0$ and $\mathrm{ord}(\Delta) = 0$, we have*

$$\mathcal{H}(T, \varepsilon) = 1.$$

(b) *When $m = 0$ and $0 < \mathrm{ord}(\Delta) \leq \mathrm{ord}(y)$, we have*

$$\mathcal{H}(T, \varepsilon) = |\Delta|^{-1}(1 + q^{-1}).$$

(c) *When $m = 0$ and $0 \leq \mathrm{ord}(y) < \mathrm{ord}(\Delta)$, we have*

$$\mathcal{H}(T, \varepsilon) = (-1)^{\mathrm{ord}(\Delta) - \mathrm{ord}(y)} |\Delta|^{-1} \cdot \mathcal{K}\ell\left(2\varpi^{\mathrm{ord}(y)}\Delta^{-1}, 2\varpi^{\mathrm{ord}(y)} \varepsilon \varepsilon^\sigma \varepsilon_y \varepsilon_y^\sigma \Delta^{-1}\right).$$

(d) *When $m > 0$ and $\mathrm{ord}(\Delta) \leq \mathrm{ord}(y)$, we have*

$$\mathcal{H}(T, \varepsilon) = (-1)^{\mathrm{ord}(\Delta)} |\Delta|^{-1} \cdot \mathcal{K}\ell\left(2a\Delta^{-1}, 2\varepsilon\varepsilon^\sigma b \Delta^{-1}\right) + |\Delta|^{-1}(1 + q^{-1}).$$

(e) *When $m > 0$ and $\mathrm{ord}(y) < \mathrm{ord}(\Delta)$, we have*

$$\mathcal{H}(T, \varepsilon) = (-1)^{\mathrm{ord}(\Delta)} |\Delta|^{-1}$$
$$\left\{\mathcal{K}\ell\left(2a\Delta^{-1}, 2\varepsilon\varepsilon^\sigma b \Delta^{-1}\right) + (-1)^{\mathrm{ord}(y)} \mathcal{K}\ell\left(2\varpi^{\mathrm{ord}(y)}\Delta^{-1}, 2\varpi^{\mathrm{ord}(y)} \varepsilon \varepsilon^\sigma \varepsilon_y \varepsilon_y^\sigma \Delta^{-1}\right)\right\}.$$

2. *Suppose that*

(3.95) $$m > \mathrm{ord}(y).$$

Then $\mathcal{H}(T, \varepsilon)$ is evaluated as follows.

(a) *When $\mathrm{ord}(y) = 0$, we have*

$$\mathcal{H}(T, \varepsilon) = 1.$$

(b) *When $m \geq 2\,\mathrm{ord}(y)$, we have*

$$\mathcal{H}(T, \varepsilon)$$
$$= |\Delta|^{-1}\left\{(-1)^{\mathrm{ord}(y)} \mathcal{K}\ell\left(2\varpi^{\mathrm{ord}(y)}\Delta^{-1}, 2\varpi^{\mathrm{ord}(y)} \varepsilon \varepsilon^\sigma \varepsilon_y \varepsilon_y^\sigma \Delta^{-1}\right) + (1 - q^{-1})\right\}.$$

(c) *When $\mathrm{ord}(y) < m < 2\,\mathrm{ord}(y)$, we have*

$$\mathcal{H}(T, \varepsilon) = |\Delta|^{-1} (-1)^{\mathrm{ord}(y)} \cdot \mathcal{K}\ell\left(2\varpi^{\mathrm{ord}(y)}\Delta^{-1}, 2\varpi^{\mathrm{ord}(y)} \varepsilon \varepsilon^\sigma \varepsilon_y \varepsilon_y^\sigma \Delta^{-1}\right)$$
$$+ |\Delta|^{-1} \mathrm{sgn}\left(\varepsilon_y \varepsilon_y^\sigma\right)^m \cdot \mathcal{K}\ell\left(2a\Delta^{-1}, 2b\Delta^{-1}\right).$$

3.4.1. Proof of Theorem 3.16: When (3.94) holds.

3.4.1.1. *Simplification of the integral.* First we remark that by Proposition 3.14, we have
$$\mathcal{H}(T,\varepsilon) = \mathcal{H}\left(\begin{pmatrix} 0 & 1 \\ 1 & 0 \end{pmatrix} T \begin{pmatrix} 0 & 1 \\ 1 & 0 \end{pmatrix}, \varepsilon\right).$$
Hence we may assume that

(3.96) $$\operatorname{ord}(a) \leq \operatorname{ord}(b).$$

Then we have
$$T = \begin{pmatrix} a & y \\ y^\sigma & b \end{pmatrix} = \begin{pmatrix} 1 & 0 \\ a^{-1}y^\sigma & 1 \end{pmatrix} \begin{pmatrix} a & 0 \\ 0 & a^{-1}\Delta \end{pmatrix} \begin{pmatrix} 1 & a^{-1}y \\ 0 & 1 \end{pmatrix}$$
where $\begin{pmatrix} 1 & a^{-1}y \\ 0 & 1 \end{pmatrix} \in \operatorname{GL}_2(\mathcal{O}_E)$. Hence by Proposition 3.14, we have
$$\mathcal{H}(T,\varepsilon) = \mathcal{H}\left(\begin{pmatrix} 0 & 1 \\ 1 & -2a^{-1}y \end{pmatrix}, \begin{pmatrix} a & 0 \\ 0 & a^{-1}\Delta \end{pmatrix}, \varepsilon\right).$$
Since the mapping $N_{E/F} : \mathcal{O}_E^\times \to \mathcal{O}_F^\times$ is surjective, there exist $\zeta, \mu \in \mathcal{O}_E^\times$ such that $a = \varpi^m \zeta \zeta^\sigma$ and $a^{-1}\Delta = -\varpi^n \mu \mu^\sigma$ where $m + n = \operatorname{ord}(\Delta)$. Then
$$\begin{pmatrix} a & 0 \\ 0 & a^{-1}\Delta \end{pmatrix} = \begin{pmatrix} \zeta^\sigma & 0 \\ 0 & \mu^\sigma \end{pmatrix} \begin{pmatrix} \varpi^m & 0 \\ 0 & -\varpi^n \end{pmatrix} \begin{pmatrix} \zeta & 0 \\ 0 & \mu \end{pmatrix}.$$
Hence, by Proposition 3.14, we have

(3.97) $$\mathcal{H}(T,\varepsilon) = \mathcal{H}\left(\begin{pmatrix} 0 & \zeta^{-1}\mu^{-1} \\ \zeta^{-1}\mu^{-1} & -2a^{-1}\mu^{-2}y \end{pmatrix}, \begin{pmatrix} \varpi^m & 0 \\ 0 & -\varpi^n \end{pmatrix}, \varepsilon\right).$$

DEFINITION 3.17. Let m and n be integers such that $0 \leq m \leq n$. Then we define a matrix $A_{m,n}$ by
$$A_{m,n} = \begin{pmatrix} \varpi^m & 0 \\ 0 & -\varpi^n \end{pmatrix}$$
and let us denote $\mathcal{Z}_{A_{m,n}}$ by $\mathcal{Z}_{m,n}$. We recall that
$$\mathcal{Z}_{m,n} = \left\{ Z \in \operatorname{Sym}^2(E) \mid A_{m,n}Z,\ Z^\sigma A_{m,n} Z - A_{m,n}^{-\sigma} \in \operatorname{Mat}_{2\times 2}(\mathcal{O}_E) \right\}.$$
Then for $\nu, \varepsilon \in \mathcal{O}_E^\times$ and $\alpha \in \mathcal{O}_E$, let us define $\mathcal{H}_{m,n}(\nu, \alpha, \varepsilon)$ by
$$\mathcal{H}_{m,n}(\nu, \alpha, \varepsilon) = \mathcal{H}\left(\begin{pmatrix} 0 & \nu \\ \nu & 2\alpha \end{pmatrix}, A_{m,n}, \varepsilon\right).$$
We recall that

(3.98) $$\mathcal{H}_{m,n}(\nu, \alpha, \varepsilon) = \int_{\mathcal{Z}_{m,n}} \psi_E\left[\varepsilon \cdot \operatorname{tr}\left(\begin{pmatrix} 0 & \nu \\ \nu & 2\alpha \end{pmatrix} Z\right)\right] dZ.$$

As for $\mathcal{Z}_{m,n}$, we note that for $Z = \begin{pmatrix} r & s \\ s & t \end{pmatrix} \in \operatorname{Sym}^2(E)$, we have
$$A_{m,n} Z \in \operatorname{Mat}^2(\mathcal{O}_E) \iff \begin{cases} r &= \varpi^{-m} u \\ s &= \varpi^{-m} v \\ t &= \varpi^{-n} w \end{cases}$$

3.4. EVALUATION OF THE QUADRATIC KLOOSTERMAN SUM

for some $u, v, w \in \mathcal{O}_E$. Then for $Z = \begin{pmatrix} \varpi^{-m}u & \varpi^{-m}v \\ \varpi^{-m}v & \varpi^{-n}w \end{pmatrix}$, we have

$$Z^\sigma A_{m,n} Z - A_{m,n}^{-\sigma}$$
$$= \begin{pmatrix} \varpi^{-m}\left(uu^\sigma - \varpi^{n-m}vv^\sigma - 1\right) & \varpi^{-m}\left(u^\sigma v - v^\sigma w\right) \\ \varpi^{-m}\left(uv^\sigma - vw^\sigma\right) & \varpi^{-n}\left(\varpi^{n-m}vv^\sigma - ww^\sigma + 1\right) \end{pmatrix}.$$

Hence let $\mathcal{Y}_{m,n}$ be the set of $(u, v, w) \in \mathcal{O}_E^3$ satisfying

(3.99) $$uu^\sigma - \varpi^{n-m}vv^\sigma \equiv 1 \pmod{\varpi^m}$$

(3.100) $$u^\sigma v - v^\sigma w \equiv 0 \pmod{\varpi^m}$$

(3.101) $$ww^\sigma - \varpi^{n-m}vv^\sigma \equiv 1 \pmod{\varpi^n}.$$

Then we have

$$\mathcal{Z}_{m,n} = \left\{ \begin{pmatrix} \varpi^{-m}u & \varpi^{-m}v \\ \varpi^{-m}v & \varpi^{-n}w \end{pmatrix} \mid (u, v, w) \in \mathcal{Y}_{m,n} \right\}.$$

It is clear that

(3.102) $$\mathcal{Y}_{0,0} = \mathcal{O}_E^3$$

and

(3.103) $$\mathcal{Y}_{0,n} = \left\{ (u, v, w) \in \mathcal{O}_E^3 \mid ww^\sigma \equiv 1 \pmod{\varpi^n} \right\}$$

for $n > 0$.

Let us consider the parameterization of $\mathcal{Y}_{m,n}$ for $m > 0$.

DEFINITION 3.18. Let us denote by $\mathcal{S}_{m,n}$, the set of triples (u, v, w), where $u \in \mathcal{O}_{E,m}$, $v \in \mathcal{O}_{E,m}$, $w \in \mathcal{O}_{E,n}$ satisfying (3.99), (3.100) and (3.101).

PROPOSITION 3.19. Let m, n be integers such that $n \geq m > 0$.
1. Suppose that $m = n$. Let us denote by $\mathcal{T}_{m,m}$, the set of triples (u, a, z) where $u \in \mathcal{O}_{E,m}$, $a \in \mathcal{O}_{F,m}$ and $z \in \mathcal{O}_{E,m}^\times$, which satisfy

(3.104) $$uu^\sigma - a^2 zz^\sigma \equiv 1 \pmod{\varpi^m}.$$

Then the mapping $\varphi_{m,m}$ defined by

(3.105) $$\varphi_{m,m} : (u, a, z) \mapsto \begin{pmatrix} u & az \\ az & zz^{-\sigma}u^\sigma \end{pmatrix}$$

is a surjective mapping from $\mathcal{T}_{m,m}$ to $\mathcal{S}_{m,n}$.

For (u, a, z), $(u', a', z') \in \mathcal{T}_{m,m}$, we have $\varphi_{m,m}(u, a, z) = \varphi_{m,m}(u', a', z')$ if and only if $u = u'$ and there exists $\zeta \in \mathcal{O}_{F,m}^\times$ such that $a' = a\zeta^{-1}$ and $z' = z\zeta$.

In particular, for each $(u, v, w) \in \mathcal{S}_{m,m}$, we have

(3.106) $$\#\varphi_{m,m}^{-1}(u, v, w) = \#\mathcal{O}_{F,m}^\times = q^m \left(1 - q^{-1}\right).$$

2. Suppose that $n > m$. Let $\mathcal{T}_{m,n} = \mathcal{O}_{F,m} \times \mathcal{O}_{E,m}^\times \times \mathcal{O}_{E,n}^1$.
Then the mapping $\varphi_{m,n}$ defined by

(3.107) $$\varphi_{m,n} : (a, z, \xi) \mapsto \begin{pmatrix} zz^{-\sigma}\xi^\sigma \left(1 + \varpi^{n-m}a^2 zz^\sigma\right)^{\frac{1}{2}} & az \\ az & \xi \left(1 + \varpi^{n-m}a^2 zz^\sigma\right)^{\frac{1}{2}} \end{pmatrix}$$

is a surjective mapping from $\mathcal{T}_{m,n}$ to $\mathcal{S}_{m,n}$.

For (a, z, ξ), $(a', z', \xi') \in \mathcal{T}_{m,n}$, we have $\varphi_{m,n}(a, z, \xi) = \varphi_{m,n}(a', z', \xi')$ if and only if $\xi = \xi'$ and there exists $\zeta \in \mathcal{O}_{F,m}^\times$ such that $a' = a\zeta^{-1}$ and $z' = z\zeta$.

In particular, for each $(u,v,w) \in \mathcal{S}_{m,n}$, we have

(3.108) $$\varphi_{m,n}^{-1}(u,v,w) = \#\mathcal{O}_{F,m}^{\times} = q^m(1-q^{-1}).$$

PROOF. 1. First we note that when $m = n$, the conditions (3.99), (3.100) and (3.101) are equivalent to

(3.109) $$\begin{pmatrix} u^\sigma & -v^\sigma \\ -v^\sigma & w^\sigma \end{pmatrix} \begin{pmatrix} u & v \\ v & w \end{pmatrix} \equiv \begin{pmatrix} 1 & 0 \\ 0 & 1 \end{pmatrix} \pmod{\varpi^m}.$$

Then we indeed have

$$\begin{pmatrix} u^\sigma & -az^\sigma \\ -az^\sigma & z^\sigma z^{-1}u \end{pmatrix} \begin{pmatrix} u & az \\ az & zz^{-\sigma}u^\sigma \end{pmatrix} \equiv \begin{pmatrix} 1 & 0 \\ 0 & 1 \end{pmatrix} \pmod{\varpi^m}$$

and hence $\varphi_{m,m}$ is a mapping from $\mathcal{T}_{m,m}$ to $\mathcal{S}_{m,m}$.

Let us prove the surjectivity of $\varphi_{m,m}$. For $(u,v,w) \in \mathcal{S}_{m,m}$, let $M = \begin{pmatrix} u & v \\ v & w \end{pmatrix}$ and $\mu = \det M$. Then by taking the determinants of the both sides of (3.109), we have $\mu \in \mathcal{O}_{E,m}^1$. Hence by Lemma 1.12, there exists $z \in \mathcal{O}_{E,m}^\times$ such that $\mu = zz^{-\sigma}$. Then we have $\mu \in \mathcal{O}_{E,m}^\times$ and

$$M^{-1} = \mu^{-1}\begin{pmatrix} w & -v \\ -v & u \end{pmatrix} = \begin{pmatrix} u^\sigma & -v^\sigma \\ -v^\sigma & w^\sigma \end{pmatrix}.$$

Hence we have $w = \mu u^\sigma = zz^{-\sigma}u^\sigma$. We also have $\mu^{-1}v = v^\sigma$, i.e. $vz^{-1} = v^\sigma z^{-\sigma}$. Let $a = vz^{-1} \in \mathcal{O}_{F,m}$. Then we have

$$M = \begin{pmatrix} u & az \\ az & zz^{-\sigma}u^\sigma \end{pmatrix} = \varphi_{m,m}(u,a,z)$$

and hence $\varphi_{m,m} : \mathcal{T}_{m,m} \to \mathcal{S}_{m,m}$ is surjective.

For $(u,a,z), (u',a',z') \in \mathcal{T}_{m,m}$, suppose that $\varphi_{m,m}(u,a,z) = \varphi_{m,m}(u',a',z')$, i.e.

$$\begin{pmatrix} u & az \\ az & zz^{-\sigma}u^\sigma \end{pmatrix} = \begin{pmatrix} u' & a'z' \\ a'z' & z'(z')^{-\sigma}u^\sigma \end{pmatrix}.$$

Then we immediately have $u = u'$. When $u \in \mathcal{O}_{E,m}^\times$, we have $zz^{-\sigma} = z'(z')^{-\sigma}$, i.e. $z'z^{-1} = (z'z^{-1})^\sigma$. Then $z' = z\zeta$ where $\zeta = z'z^{-1} \in \mathcal{O}_{F,m}^\times$ and $az = a'z'$ implies that $a' = a\zeta^{-1}$. When $u \notin \mathcal{O}_{E,m}$, we have $a \in \mathcal{O}_{F,m}^\times$ by (3.109). Hence for $\zeta = a(a')^{-1}$, we have $a' = a\zeta^{-1}$ and $z' = z\zeta$.

2. It is easily seen that $\varphi_{m,n}$ is indeed a mapping from $\mathcal{T}_{m,n}$ to $\mathcal{S}_{m,n}$. Let us show the surjectivity of $\varphi_{m,n}$. Suppose that $(u,v,w) \in \mathcal{S}_{m,n}$. Then since $n - m > 0$, the condition (3.99) implies that $u \in \mathcal{O}_{E,m}^\times$. Similarly the condition (3.101) implies that $w \in \mathcal{O}_{E,n}^\times$. Also (3.99) and (3.101) together imply that $uu^\sigma \equiv ww^\sigma \pmod{\varpi^m}$. Thus we have $u^{-1}w^\sigma \in \mathcal{O}_{E,m}^1$ and there exists $z \in \mathcal{O}_{E,m}^\times$ such that $u^{-1}w^\sigma = z^{-1}z^\sigma$ by Lemma 1.12. Then we have

$$u^\sigma v - v^\sigma w \equiv u^\sigma(v - z^{-\sigma}zv^\sigma) \equiv u^\sigma z^{-1}(vz^{-1} - v^\sigma z^{-\sigma}) \pmod{\varpi^m}.$$

Hence the condition (3.100) implies that $vz^{-1} \in \mathcal{O}_{F,m}$. Let $a = vz^{-1} \in \mathcal{O}_{F,m}$. Then we have $v = az$. The condition (3.101) implies that

$$ww^\sigma \equiv 1 + \varpi^{n-m}vv^\sigma \equiv 1 + \varpi^{n-m}a^2zz^\sigma \pmod{\varpi^n}.$$

3.4. EVALUATION OF THE QUADRATIC KLOOSTERMAN SUM

Hence by Lemma 1.12, we have
$$w = \xi \left(1 + \varpi^{n-m}a^2 zz^\sigma\right)^{\frac{1}{2}}$$

for some $\xi \in \mathcal{O}_{E,}^1$. Then we have
$$u = \left(uw^{-\sigma}\right) w^\sigma = zz^{-\sigma}w = zz^{-\sigma}\xi^\sigma \left(1 + \varpi^{n-m}a^2 zz^\sigma\right)^{\frac{1}{2}}$$

and we have shown the surjectivity of $\varphi_{m,n}$.

For (a, z, ξ), $(a', z', \xi') \in \mathcal{T}_{m,n}$, suppose that $\varphi_{m,n}(a, z, \xi) = \varphi_{m,n}(a', z', \xi')$. Then since
$$zz^{-\sigma} = \left\{zz^{-\sigma}\xi^\sigma \left(1 + \varpi^{n-m}a^2 zz^\sigma\right)^{\frac{1}{2}}\right\} \left\{\xi \left(1 + \varpi^{n-m}a^2 zz^\sigma\right)^{\frac{1}{2}}\right\}^{-\sigma},$$

we have $zz^{-\sigma} = z'(z')^{-\sigma}$. Thus $z'z^{-1} \in \mathcal{O}_{F,m}^\times$. Let $\zeta = z'z^{-1} \in \mathcal{O}_{F,m}^\times$. Then $az = a'z'$ implies that $a' = a\zeta^{-1}$. Finally we have $\xi = \xi'$, since
$$\xi \left(1 + \varpi^{n-m}a^2 zz^\sigma\right)^{\frac{1}{2}} = \xi' \left(1 + \varpi^{n-m} (a')^2 z'(z')^\sigma\right)^{\frac{1}{2}}.$$

$\square$

Now we are ready to evaluate $\mathcal{H}_{m,n}(\nu, \alpha, \varepsilon)$. We recall that

(3.110)
$$\mathcal{H}_{m,n}(\nu, \alpha, \varepsilon) = q^{4m+2n} \int_{\mathcal{Y}_{m,n}} \psi_E \left[\varepsilon \cdot \mathrm{tr} \left\{\begin{pmatrix} 0 & \nu \\ \nu & 2\alpha \end{pmatrix} \begin{pmatrix} \varpi^{-m}u & \varpi^{-m}v \\ \varpi^{-m}v & \varpi^{-n}w \end{pmatrix}\right\}\right] du\, dv\, dw$$
$$= q^{4m+2n} \int_{\mathcal{Y}_{m,n}} \psi_E \left[2\varepsilon \left(\varpi^{-m}\nu v + \varpi^{-n}\alpha w\right)\right] du\, dv\, dw.$$

First we consider the case when $m = 0$.

PROPOSITION 3.20. 1. $\mathcal{H}_{0,0}(\nu, \alpha, \varepsilon) = 1$.
2. Suppose that $n > 0$. When $\alpha \in \varpi^n \mathcal{O}_E$, we have
$$\mathcal{H}_{0,n}(\nu, \alpha, \varepsilon) = q^n \left(1 + q^{-1}\right).$$

When $\alpha = \varpi^k \varepsilon_\alpha$ where $0 \leq k < n$ and $\varepsilon_\alpha \in \mathcal{O}_E^\times$, we have
$$\mathcal{H}_{0,n}(\nu, \alpha, \varepsilon) = (-1)^{n-k} q^n \cdot \mathcal{K}\ell \left(2\varpi^{k-n}, 2\varpi^{k-n}\varepsilon\varepsilon^\sigma \varepsilon_\alpha \varepsilon_\alpha^\sigma\right).$$

PROOF. 1. This is clear from (3.102).
2. By (3.103), we have
$$\mathcal{H}_{0,n}(\nu, \alpha, \varepsilon) = q^{2n} \int_{\{w \in \mathcal{O}_E^\times \mid ww^\sigma \equiv 1 \pmod{\varpi^n}\}} \psi_E \left(2\varpi^{-n}\varepsilon\alpha w\right) dw$$
$$= \sum_{x \in \mathcal{O}_{E,n}^1} \psi_E \left(2\varpi^{-n}\varepsilon\alpha x\right).$$

Thus when $\alpha \in \varpi^n \mathcal{O}_E$, we have
$$\mathcal{H}_{0,n}(\nu, \alpha, \varepsilon) = \#\mathcal{O}_{E,n}^1 = q^n \left(1 + q^{-1}\right)$$

by Lemma 1.12. When $\alpha \in \varpi^k \mathcal{O}_E^\times$ $(0 \leq k < n)$, by Lemma 1.12 and Proposition 2.16, we have

$$\mathcal{H}_{0,n}(\nu, \alpha, \varepsilon) = q^k \sum_{y \in \mathcal{O}_{E,n-k}^1} \psi_E \left(2\varpi^{k-n} \varepsilon \varepsilon_\alpha y\right)$$

$$= q^{2n-k} \mathcal{H}_{n-k} \left(\varepsilon \varepsilon^\sigma \varepsilon_\alpha \varepsilon_\alpha^\sigma\right)$$

$$= (-1)^{n-k} q^n \cdot \mathcal{K}\ell \left(2\varpi^{k-n}, 2\varpi^{k-n} \varepsilon \varepsilon^\sigma \varepsilon_\alpha \varepsilon_\alpha^\sigma\right).$$

□

PROPOSITION 3.21. *Suppose that $m = n > 0$.*
1. *When $\alpha \in \varpi^m \mathcal{O}_E$, we have*

$$\mathcal{H}_{m,m}(\nu, \alpha, \varepsilon) = q^{2m} \cdot \mathcal{K}\ell \left(2\varpi^{-m}, 2\varpi^{-m} \varepsilon \varepsilon^\sigma (\alpha \alpha^\sigma - \nu \nu^\sigma)\right) + q^{2m} \left(1 + q^{-1}\right).$$

2. *When $\alpha = \varpi^k \varepsilon_\alpha$ where $0 \leq k < m$ and $\varepsilon_\alpha \in \mathcal{O}_E^\times$, we have*

$$\mathcal{H}_{m,m}(\nu, \alpha, \varepsilon) = q^{2m} \cdot \mathcal{K}\ell \left(2\varpi^{-m}, 2\varpi^{-m} \varepsilon \varepsilon^\sigma (\alpha \alpha^\sigma - \nu \nu^\sigma)\right)$$

$$+ (-1)^{m-k} q^{2m} \cdot \mathcal{K}\ell \left(2\varpi^{k-m}, 2\varpi^{k-m} \varepsilon \varepsilon^\sigma \varepsilon_\alpha \varepsilon_\alpha^\sigma\right).$$

PROOF. We may write (3.110) as

$$\mathcal{H}_{m,m}(\nu, \alpha, \varepsilon) = \sum_{(u,v,w) \in \mathcal{S}_{m,m}} \psi_E \left[2\varpi^{-m} \varepsilon (\nu v + \alpha w)\right].$$

Then by Proposition 3.19, we have

$$\mathcal{H}_{m,m}(\nu, \alpha, \varepsilon) = \frac{q^{-m}}{1 - q^{-1}} \sum_{(u,a,z) \in \mathcal{T}_{m,m}} \psi_E \left[2\varpi^{-m} \varepsilon \left(\nu a z + \alpha z z^{-\sigma} u^\sigma\right)\right]$$

where we recall that

$$\mathcal{T}_{m,m} = \left\{ (u, a, z) \in \mathcal{O}_{E,m} \times \mathcal{O}_{F,m} \times \mathcal{O}_{E,m}^\times \mid uu^\sigma - a^2 zz^\sigma \equiv 1 \pmod{\varpi^m} \right\}.$$

Since $(u, a, z) \mapsto (zz^{-\sigma} u^\sigma, a, z)$ is a permutation of $\mathcal{T}_{m,m}$, we have

$$(3.111) \qquad \mathcal{H}_{m,m}(\nu, \alpha, \varepsilon) = \frac{q^{-m}}{1 - q^{-1}} \sum_{(u,a,z) \in \mathcal{T}_{m,m}} \psi_E \left[2\varpi^{-m} \varepsilon (\nu a z + \alpha u)\right].$$

Let

$$\mathcal{U} = \{(u, v) \in \mathcal{O}_{E,m} \times \mathcal{O}_{E,m} \mid uu^\sigma - vv^\sigma \equiv 1 \pmod{\varpi^m}\}.$$

Then it is clear that the mapping

$$\phi : \mathcal{T}_{m,m} \ni (u, a, z) \mapsto (u, az) \in \mathcal{U}$$

is surjective. For each integer j $(0 \leq j \leq m)$, let

$$\mathcal{U}_j = \left\{ (u, v) \in \mathcal{U} \mid v \in \varpi^j \mathcal{O}_{E,m} \right\}.$$

Let us count $\#\phi^{-1}(u, v)$ for each $(u, v) \in \mathcal{U}$. It is clear that when $(u, v) \in \mathcal{U}_m$,

$$\#\phi^{-1}(u, v) = \#\mathcal{O}_{E,m}^\times = q^{2m} \left(1 - q^{-2}\right).$$

Suppose that $(u, v) \in \mathcal{U}_j \setminus \mathcal{U}_{j+1}$ where $0 \leq j < m$. We observe that for $\beta_1, \beta_2 \in \mathcal{O}_F^\times$ and $z_1, z_2 \in \mathcal{O}_E^\times$, we have

$$\varpi^j \beta_1 z_1 \equiv \varpi^j \beta_2 z_2 \pmod{\varpi^m} \iff \beta_1 z_1 \equiv \beta_2 z_2 \pmod{\varpi^{m-j}}.$$

Hence in this case, $\#\phi^{-1}(u,v)$ is equal to the number of elements in the kernel of the homomorphism
$$\mathcal{O}_{F,m-j}^\times \times \mathcal{O}_{E,m}^\times \ni (\beta, z) \mapsto \beta z \in \mathcal{O}_{E,m-j}^\times.$$

Therefore
$$\#\phi^{-1}(u,v) = \frac{\#\left(\mathcal{O}_{F,m-j}^\times \times \mathcal{O}_{E,m}^\times\right)}{\#\mathcal{O}_{E,m-j}^\times} = q^{m+j}\left(1 - q^{-1}\right).$$

Thus for an integer j such that $0 \leq j \leq m$, let
$$A_j = \sum_{(u,v) \in \mathcal{U}_j} \psi_E\left[2\varpi^{-m}\varepsilon(\alpha u + \nu v)\right].$$

Then we have
(3.112)
$$\mathcal{H}_{m,m}(\nu, \alpha, \varepsilon) = \frac{q^{-m}}{1-q^{-1}}\left\{q^{2m}\left(1-q^{-2}\right)A_m + \sum_{j=1}^{m-1} q^{m+j}\left(1-q^{-1}\right)(A_j - A_{j+1})\right\}$$
$$= A_0 + q^m A_m + \sum_{0 < j < m} q^m \left(1 - q^{-1}\right) A_j.$$

As for A_m, we have
$$A_m = \sum_{u \in \mathcal{O}_{E,m}^1} \psi_E\left(2\varpi^{-m}\varepsilon\alpha u\right).$$

Then as we have shown in the proof of Proposition 3.20, we have
$$A_m = \begin{cases} q^m\left(1+q^{-1}\right), & \text{when } \alpha \in \varpi^m \mathcal{O}_E \\ (-1)^{m-k} q^m \mathcal{K}\ell\left(2\varpi^{k-m}, 2\varpi^{k-m}\varepsilon\varepsilon^\sigma \varepsilon_\alpha \varepsilon_\alpha^\sigma\right), & \text{when } \alpha \in \varpi^k \mathcal{O}_E^\times, 0 \leq k < m. \end{cases}$$

Now let us compute A_j for $0 < j < m$. First suppose that $\alpha \in \varpi^m \mathcal{O}_E$. Then
$$A_j = \sum_{(u,v) \in \mathcal{U}_j} \psi_E\left(2\varpi^{-m}\varepsilon\nu v\right).$$

Since $v \in \varpi^j \mathcal{O}_E$ where $j > 0$, we have $1 + vv^\sigma \in \mathcal{O}_{F,m}^\times$. Hence for given $v \in \varpi^j \mathcal{O}_E$,
$$\#\{u \in \mathcal{O}_{E,m} \mid uu^\sigma \equiv 1 + vv^\sigma \pmod{\varpi^m}\} = \#\mathcal{O}_{E,m}^1 = q^m\left(1+q^{-1}\right)$$
by Lemma 1.12. Thus
$$A_j = q^m\left(1+q^{-1}\right) \sum_{v \in \varpi^j \mathcal{O}_{E,m}} \psi_E\left(2\varpi^{-m}\varepsilon\nu v\right) = 0$$
since $j < m$.

Suppose that $\alpha \in \varpi^k \mathcal{O}_E^\times$ where $0 \leq k < m$. When $j \geq \frac{m}{2}$, we have
$$vv^\sigma \equiv 0 \pmod{\varpi^m}$$
for $v \in \varpi^j \mathcal{O}_E$. Hence
$$\mathcal{U}_j = \left\{(u,v) \mid u \in \mathcal{O}_{E,m}^1, v \in \varpi^j \mathcal{O}_{E,m}\right\}.$$

Thus
$$A_j = \sum_{(u,v)\in \mathcal{U}_j} \psi_E\left[2\varpi^{-m}\varepsilon\left(\alpha u + \nu v\right)\right]$$
$$= \left(\sum_{u\in\mathcal{O}_{E,m}^1} \psi_E\left(2\varpi^{-m}\varepsilon\alpha u\right)\right)\left(\sum_{v\in\varpi^j\mathcal{O}_{E,m}} \psi_E\left(2\varpi^{-m}\varepsilon\nu v\right)\right) = 0.$$

When $0 < j < \frac{m}{2}$, let $h = m - j$. Then for $v \in \varpi^j \mathcal{O}_E$ and $\xi \in \varpi^h \mathcal{O}_E$, we have
$$(v+\xi)(v+\xi)^\sigma \equiv vv^\sigma \pmod{\varpi^m}.$$
Hence $(u,v) \mapsto (u, v+\xi)$ is a permutation of $\mathcal{U}_j$. Therefore
$$A_j = q^{2(h-m)} \sum_{\xi\in\varpi^h\mathcal{O}_{E,m}} \sum_{(u,v)\in\mathcal{U}_j} \psi_E\left[2\varpi^{-m}\varepsilon\left(\alpha u + \nu(v+\xi)\right)\right]$$
$$= q^{2(h-m)} A_j \sum_{\xi\in\varpi^h\mathcal{O}_{E,m}} \psi_E\left(2\varpi^{-m}\varepsilon\nu\xi\right) = 0.$$

Finally let us compute A_0. First we note that we may identify $\mathcal{U}_0$ with
$$\mathrm{SU}_{1,1}(\mathcal{O}_{E,m}) = \left\{\begin{pmatrix} u & v \\ v^\sigma & u^\sigma \end{pmatrix} \in \mathrm{Mat}_{2\times 2}(\mathcal{O}_{E,m}) \mid uu^\sigma - vv^\sigma \equiv 1 \pmod{\varpi^m}\right\}.$$
Then
$$A_0 = \sum_{(u,v)\in\mathcal{U}_0} \psi_E\left[2\varpi^{-m}\varepsilon(\alpha u + \nu v)\right]$$
$$= \sum_{g\in\mathrm{SU}_{1,1}(\mathcal{O}_{E,m})} \psi\left[2\varpi^{-m}\cdot\mathrm{tr}\left(\begin{pmatrix} \varepsilon\alpha & \varepsilon^\sigma\nu^\sigma \\ \varepsilon\nu & \varepsilon^\sigma\alpha^\sigma \end{pmatrix}g\right)\right].$$

By the Cayley transform
$$g \mapsto \begin{pmatrix} 1 & 1 \\ \eta & -\eta \end{pmatrix} g \begin{pmatrix} 1 & 1 \\ \eta & -\eta \end{pmatrix}^{-1},$$
we have
$$\mathrm{SU}_{1,1}(\mathcal{O}_{E,m}) \simeq \mathrm{SL}_2(\mathcal{O}_{F,m}).$$
Also by the theory of elementary divisors, there exist $P, Q \in \mathrm{GL}_2(\mathcal{O}_{F,m})$ such that
$$\mathrm{Mat}_{2\times 2}(\mathcal{O}_{F,m}) \ni \begin{pmatrix} 1 & 1 \\ \eta & -\eta \end{pmatrix} \begin{pmatrix} \varepsilon\alpha & \varepsilon^\sigma\nu^\sigma \\ \varepsilon\nu & \varepsilon^\sigma\alpha^\sigma \end{pmatrix} \begin{pmatrix} 1 & 1 \\ \eta & -\eta \end{pmatrix}^{-1}$$
$$= P \begin{pmatrix} 1 & 0 \\ 0 & \varepsilon\varepsilon^\sigma(\alpha\alpha^\sigma - \nu\nu^\sigma) \end{pmatrix} Q$$
and
(3.113) $$\det(PQ) \equiv 1 \pmod{\varpi^m}.$$

Let $\delta = \varepsilon\varepsilon^\sigma(\alpha\alpha^\sigma - \nu\nu^\sigma)$. Then we have
$$A_0 = \sum_{g\in\mathrm{SL}_2(\mathcal{O}_{F,m})} \psi\left[2\varpi^{-m}\cdot\mathrm{tr}\left(P\begin{pmatrix} 1 & 0 \\ 0 & \delta \end{pmatrix}Qg\right)\right]$$
$$= \sum_{g\in\mathrm{SL}_2(\mathcal{O}_{F,m})} \psi\left[2\varpi^{-m}\cdot\mathrm{tr}\left(\begin{pmatrix} 1 & 0 \\ 0 & \delta \end{pmatrix}g\right)\right],$$

since $g \mapsto QgP$ is a permutation of $\mathrm{SL}_2(\mathcal{O}_{F,m})$ by (3.113).

Let us write $A_0 = A_0^{(1)} + A_0^{(2)}$, where

(3.114) $$A_0^{(1)} = \sum_{\left(\begin{smallmatrix} w & x \\ y & z \end{smallmatrix}\right) \in \mathrm{SL}_2(\mathcal{O}_{F,m}), x \in \mathcal{O}_{F,m}^\times} \psi\left[2\varpi^{-m}(w + \delta z)\right]$$

and

(3.115) $$A_0^{(2)} = \sum_{\left(\begin{smallmatrix} w & x \\ y & z \end{smallmatrix}\right) \in \mathrm{SL}_2(\mathcal{O}_{F,m}), x \in \varpi \mathcal{O}_{F,m}} \psi\left[2\varpi^{-m}(w + \delta z)\right].$$

In (3.114), $y \in \mathcal{O}_{F,m}$ is uniquely determined by $y = x^{-1}(wz - 1)$ for given $x \in \mathcal{O}_{F,m}^\times$, $w \in \mathcal{O}_{F,m}$ and $z \in \mathcal{O}_{F,m}$. Hence

$$A_0^{(1)} = q^m(1 - q^{-1}) \sum_{z \in \mathcal{O}_{F,m}} \psi(2\varpi^{-m}\delta z) \sum_{w \in \mathcal{O}_{F,m}} \psi(2\varpi^{-m}w) = 0.$$

In (3.115), we have $w \in \mathcal{O}_{F,m}^\times$. Hence $z \in \mathcal{O}_{F,m}^\times$ is uniquely determined by $z = w^{-1}(1 + xy)$ for given $w \in \mathcal{O}_{F,m}^\times$, $x \in \varpi\mathcal{O}_{F,m}$ and $y \in \mathcal{O}_{F,m}$. Thus

(3.116)
$$A_0^{(2)} = \sum_{w \in \mathcal{O}_{F,m}^\times} \psi\left[2\varpi^{-m}(w + \delta w^{-1})\right] \sum_{x \in \varpi\mathcal{O}_{F,m}} \left(\sum_{y \in \mathcal{O}_{F,m}} \psi(2\varpi^{-m}\delta w^{-1}xy)\right).$$

In (3.116), the inner sum vanishes unless $x\delta \equiv 0 \pmod{\varpi^m}$. Let
$$\ell = \mathrm{ord}(\delta) = \mathrm{ord}(\alpha\alpha^\sigma - \nu\nu^\sigma).$$

Then we have
$$\#\{x \in \varpi\mathcal{O}_{F,m} \mid x\delta \equiv 0 \pmod{\varpi^m}\} = \begin{cases} q^{m-1}, & \text{when } \ell \geq m-1 \\ q^\ell, & \text{when } 0 \leq \ell < m-1 \end{cases}$$
$$= q^{\min\{\ell, m-1\}}.$$

Therefore

(3.117) $$A_0 = A_0^{(2)} = q^{m + \min\{\ell, m-1\}} \sum_{w \in \mathcal{O}_{F,m}} \psi\left[2\varpi^{-n}(w + \delta w^{-1})\right]$$
$$= q^{2m + \min\{\ell, m-1\}} \mathcal{K}\ell\left(2\varpi^{-m}, 2\varpi^{-m}\varepsilon\varepsilon^\sigma(\alpha\alpha^\sigma - \nu\nu^\sigma)\right).$$

Here suppose that $\min\{\ell, m-1\} > 0$. Then we have $m - 1 > 0$, i.e. $m \geq 2$ and $\ell > 0$. Hence by Proposition 2.7
$$\mathcal{K}\ell\left(2\varpi^{-m}, 2\varpi^{-m}\varepsilon\varepsilon^\sigma(\alpha\alpha^\sigma - \nu\nu^\sigma)\right) = 0.$$

Therefore we may write (3.117) as
$$A_0 = q^{2m} \cdot \mathcal{K}\ell\left(2\varpi^{-m}, 2\varpi^{-m}\varepsilon\varepsilon^\sigma(\alpha\alpha^\sigma - \nu\nu^\sigma)\right)$$
and we have proved the proposition. $\square$

PROPOSITION 3.22. *Suppose that $n > m > 0$.*
1. *When $\alpha \in \varpi^n \mathcal{O}_E$, we have*
$$\mathcal{H}_{m,n}(\nu, \alpha, \varepsilon) = q^{m+n}(1 + q^{-1}).$$

2. When $\alpha = \varpi^k \varepsilon_\alpha$ where $0 \leq k < n$ and $\varepsilon_\alpha \in \mathcal{O}_E^\times$, we have

$$\mathcal{H}_{m,n}(\nu, \alpha, \varepsilon) = (-1)^{n-k} q^{m+n} \cdot \mathcal{K}\ell\left(2\varpi^{k-n}, 2\varpi^{k-n} \varepsilon \varepsilon^\sigma \varepsilon_\alpha \varepsilon_\alpha^\sigma\right)$$
$$+ (-1)^{m+n} q^{m+n} \cdot \mathcal{K}\ell\left(2\varpi^{-n}, 2\varpi^{-n} \varepsilon \varepsilon^\sigma \left(\alpha \alpha^\sigma - \varpi^{n-m} \nu \nu^\sigma\right)\right).$$

PROOF. By Proposition 3.19, we may rewrite (3.110) as

(3.118) $\mathcal{H}_{m,n}(\nu, \alpha, \varepsilon)$
$$= \frac{q^{-m}}{1-q^{-1}} \sum_{(a,z,\xi) \in \mathcal{T}_{m,n}} \psi_E\left[2\varepsilon\left\{\varpi^{-m}\nu a z + \varpi^{-n}\alpha\xi\left(1 + \varpi^{n-m}a^2 z z^\sigma\right)^{\frac{1}{2}}\right\}\right]$$

where

$$\mathcal{T}_{m,n} = \mathcal{O}_{F,m} \times \mathcal{O}_{E,m}^\times \times \mathcal{O}_{E,n}^1.$$

Suppose that $\alpha \in \varpi^n \mathcal{O}_E$. Then we have

$$\mathcal{H}_{m,n}(\nu, \alpha, \varepsilon) = \frac{q^{-m}}{1-q^{-1}} \cdot q^n \left(1+q^{-1}\right) \sum_{z \in \mathcal{O}_{E,m}^\times} \left(\sum_{a \in \mathcal{O}_{F,m}} \psi\left[2\varpi^{-m} a \cdot \operatorname{tr}_{E/F}(\varepsilon \nu z)\right]\right).$$

Here the inner sum vanishes unless

$$\operatorname{tr}_{E/F}(\varepsilon \nu z) \equiv 0 \pmod{\varpi^m}, \text{ i.e. } z \in \varepsilon^{-1} \nu^{-1} \eta \mathcal{O}_{F,m}^\times.$$

Thus

$$\mathcal{H}_{m,n}(\nu, \alpha, \varepsilon) = \frac{q^{n-m}\left(1+q^{-1}\right)}{1-q^{-1}} \cdot q^m\left(1-q^{-1}\right) \cdot q^m = q^{m+n}\left(1+q^{-1}\right).$$

Now suppose that $\alpha = \varpi^k \varepsilon_\alpha$ where $0 \leq k < n$ and $\varepsilon_\alpha \in \mathcal{O}_E^\times$. Let us denote $\mathcal{O}_{E,m} \times \mathcal{O}_{E,n}^1$ by $\mathcal{V}$ and let

$$\mathcal{V}_j = \left\{(f,\xi) \mid f \in \varpi^j \mathcal{O}_{E,m}, \xi \in \mathcal{O}_{E,n}^1\right\}$$

for $0 \leq j \leq m$. Then the mapping

$$\varrho : \mathcal{T}_{m,n} \ni (a,z,\xi) \mapsto (az,\xi) \in \mathcal{V}$$

is surjective and for each $(f,\xi) \in \mathcal{V}$,

$$\#\varrho^{-1}(f,\xi) = \begin{cases} q^{2m}\left(1-q^{-2}\right), & \text{when } (f,\xi) \in \mathcal{V}_m \\ q^{m+j}\left(1-q^{-1}\right), & \text{when } (f,\xi) \in \mathcal{V}_j \setminus \mathcal{V}_{j+1} \text{ for } 0 \leq j < m. \end{cases}$$

Hence let

$$B_j = \sum_{(f,\xi) \in \mathcal{V}_j} \psi_E\left[2\varepsilon\left\{\varpi^{-m}\nu f + \varpi^{k-n}\varepsilon_\alpha \xi\left(1 + \varpi^{n-m} f f^\sigma\right)^{\frac{1}{2}}\right\}\right].$$

Then we have

$$\mathcal{H}_{m,n}(\nu, \alpha, \varepsilon) = B_0 + q^m B_m + \sum_{0 < j < m} q^m\left(1-q^{-1}\right) B_j.$$

First suppose that j satisfies $2j + n - m \geq n - k$, i.e. $j \geq \frac{m-k}{2}$. Then for $f \in \varpi^j \mathcal{O}_{E,m}$, we have

$$1 + \varpi^{n-m} f f^\sigma \equiv 1 \pmod{\varpi^{n-k}}.$$

3.4. EVALUATION OF THE QUADRATIC KLOOSTERMAN SUM

Hence

$$B_j = \sum_{\xi \in \mathcal{O}^1_{E,n}} \psi_E \left(2\varpi^{k-n}\varepsilon\varepsilon_\alpha\xi\right) \left(\sum_{f \in \varpi^j \mathcal{O}_{E,m}} \psi_E \left(2\varpi^{-m}\varepsilon\nu f\right)\right).$$

Here the inner sum vanishes unless $j = m$. When $j = m$, we have

$$B_m = \sum_{\xi \in \mathcal{O}^1_{E,n}} \psi_E \left(2\varpi^{k-n}\varepsilon\varepsilon_\alpha\xi\right) = (-1)^{n-k} q^n \cdot \mathcal{K}\ell \left(2\varpi^{k-n}, 2\varpi^{k-n}\varepsilon\varepsilon^\sigma \varepsilon_\alpha \varepsilon_\alpha^\sigma\right)$$

by Proposition 2.16.

Now suppose that $0 \leq j < \frac{m-k}{2}$. Let $h = m - k - j$. Then we note that

$$\frac{m-k}{2} < h \leq m - k.$$

Thus for $f \in \varpi^j \mathcal{O}_{E,m}$ and $\rho \in \varpi^h \mathcal{O}_{E,m}$, we have

$$1 + (f + \rho)(f + \rho)^\sigma \equiv 1 + f f^\sigma \pmod{\varpi^{n-k}}.$$

Then since $(f, \xi) \mapsto (f + \rho, \xi)$ is a permutation of $\mathcal{V}_j$, we have

$$B_j = \sum_{(f,\xi) \in \mathcal{V}_j} \psi_E \left[2\varepsilon \left(\varpi^{-m}\nu(f+\rho) + \varpi^{k-n}\varepsilon_\alpha \xi \left(1 + \varpi^{n-m}(f+\rho)(f+\rho)^\sigma\right)^{\frac{1}{2}}\right)\right]$$
$$=\psi_E \left(2\varpi^{-m}\varepsilon\nu\rho\right) B_j.$$

Thus B_j vanishes unless $h = m$, i.e. $j = k = 0$.

Suppose that $k = 0$. Then we have

(3.119)

$$B_0 = \sum_{f \in \mathcal{O}_{E,m}} \psi_E \left(2\varpi^{-m}\varepsilon\nu f\right) \left(\sum_{\xi \in \mathcal{O}^1_{E,n}} \psi_E \left[2\varpi^{-n}\varepsilon\alpha \left(1 + \varpi^{n-m} f f^\sigma\right)^{\frac{1}{2}} \xi\right]\right)$$
$$=q^{2n} \sum_{f \in \mathcal{O}_{E,m}} \psi_E \left(2\varpi^{-m}\varepsilon\nu f\right) \mathcal{H}_n \left[\varepsilon\varepsilon^\sigma \alpha\alpha^\sigma \left(1 + \varpi^{n-m} f f^\sigma\right)\right]$$
$$= (-1)^n q^n \sum_{f \in \mathcal{O}_{E,m}} \psi_E \left(2\varpi^{-m}\varepsilon\nu f\right) \mathcal{K}\ell \left(2\varpi^{-n}, 2\varpi^{-n}\varepsilon\varepsilon^\sigma \alpha\alpha^\sigma \left(1 + \varpi^{n-m} f f^\sigma\right)\right)$$

by Proposition 2.16. Let $\gamma = \varepsilon\varepsilon^\sigma \alpha\alpha^\sigma$. Then by rewriting (3.119) in terms of the integrals, we have

$$B_0 = (-1)^n q^{n+2m} \int_{z \in \mathcal{O}_E} \int_{t \in \mathcal{O}_F^\times} \psi \left[2\varpi^{-m}\left(\varepsilon\nu z + \varepsilon^\sigma \nu^\sigma z^\sigma\right)\right]$$
$$\psi \left[2\varpi^{-n} t + 2\varpi^{-n} t^{-1}\gamma \left(1 + \varpi^{n-m} z z^\sigma\right)\right] dt\, dz$$
$$= (-1)^n q^{n+2m} \int_{t \in \mathcal{O}_F^\times} \psi \left[2\varpi^{-n}\left\{t\left(1 - \varpi^{n-m}\alpha^{-1}\alpha^{-\sigma}\nu\nu^\sigma\right) + t^{-1}\gamma\right\}\right]$$
$$\left(\int_{z \in \mathcal{O}_E} \psi \left[2\varpi^{-m} t^{-1}\gamma \left(z + t\gamma^{-1}\varepsilon^\sigma \nu^\sigma\right)\left(z^\sigma + t\gamma^{-1}\varepsilon\nu\right)\right] dz\right) dt.$$

Let us compute the inner integral. Let us express $z + t\gamma^{-1}\varepsilon^\sigma = x + \eta y$ where $x, y \in \mathcal{O}_F$. Then by Proposition 2.5, the inner integral is given by

$$\left(\int_{x \in \mathcal{O}_F} \psi\left(2\varpi^{-m}t^{-1}\gamma x^2\right) dx\right) \left(\int_{y \in \mathcal{O}_F} \psi\left(-2\varpi^{-m}t^{-1}\gamma dy^2\right) dy\right)$$
$$= C\left(2\varpi^{-m}t^{-1}\gamma\right) C\left(-2\varpi^{-m}t^{-1}\gamma d\right) = (-1)^m q^{-m}.$$

Therefore

$$B_0 = (-1)^{n+m} q^{n+m} \cdot \mathcal{K}\ell\left(2\varpi^{-n}\left(1 - \varpi^{n-m}\alpha^{-1}\alpha^{-\sigma}\nu\nu^\sigma\right), 2\varpi^{-n}\gamma\right)$$
$$= (-1)^{n+m} q^{n+m} \cdot \mathcal{K}\ell\left(2\varpi^{-n}, 2\varpi^{-n}\varepsilon\varepsilon^\sigma\left(\alpha\alpha^\sigma - \varpi^{n-m}\nu\nu^\sigma\right)\right).$$

Here we remark that when $0 < k < n$, by Proposition 2.7, we have

$$\mathcal{K}\ell\left(2\varpi^{-n}, 2\varpi^{-n}\varepsilon\varepsilon^\sigma\left(\alpha\alpha^\sigma - \varpi^{n-m}\nu\nu^\sigma\right)\right) = 0$$

since $n \geq 2$ and $\alpha\alpha^\sigma - \varpi^{n-m}\nu\nu^\sigma \in \varpi\mathcal{O}_F$. Hence we obtain the uniform expression as in the statement of the proposition. $\square$

3.4.1.2. *Verification of Theorem 3.16 when* (3.94) *holds.* We recall that

$$\mathcal{H}\left(\begin{pmatrix} a & y \\ y^\sigma & b \end{pmatrix}, \varepsilon\right) = \mathcal{H}_{m,n}(\nu, \alpha, \varepsilon)$$

where the dictionary between the parameters are given by

$$m = \mathrm{ord}(a), \quad m + n = \mathrm{ord}(\Delta), \quad \nu = \zeta^{-1}\mu^{-1}, \quad \alpha = -a^{-1}y, \quad k = \mathrm{ord}(\alpha),$$

where $\zeta, \mu \in \mathcal{O}_E^\times$ such that

$$a = \varpi^m \zeta\zeta^\sigma, \quad a^{-1}\Delta = -\varpi^n \mu\mu^\sigma.$$

Thus we have

$$\begin{cases} n - k = \mathrm{ord}(\Delta) - \mathrm{ord}(y) \\ \varepsilon_\alpha \varepsilon_\alpha^\sigma = \varpi^{2\mathrm{ord}(\Delta) - 2\mathrm{ord}(y)} \Delta^{-2} yy^\sigma \\ \nu\nu^\sigma = -\varpi^{\mathrm{ord}(\Delta)} \Delta^{-1} \\ \alpha\alpha^\sigma - \varpi^{n-m}\nu\nu^\sigma = \varpi^{2\mathrm{ord}(\Delta) - 2m} \Delta^{-2} ab. \end{cases}$$

Hence

$$\mathcal{K}\ell\left(2\varpi^{k-n}, 2\varpi^{k-n}\varepsilon\varepsilon^\sigma \varepsilon_\alpha \varepsilon_\alpha^\sigma\right) = \mathcal{K}\ell\left(2\varpi^{\mathrm{ord}(y) - \mathrm{ord}(\Delta)}, 2\varpi^{\mathrm{ord}(\Delta) - \mathrm{ord}(y)} \varepsilon\varepsilon^\sigma \Delta^{-2} yy^\sigma\right)$$
$$= \mathcal{K}\ell\left(2\varpi^{\mathrm{ord}(y)}\Delta^{-1}, 2\varpi^{\mathrm{ord}(y)}\varepsilon\varepsilon^\sigma \Delta^{-1} \varepsilon_y \varepsilon_y^\sigma\right)$$

since $\varpi^{\mathrm{ord}(\Delta)}\Delta^{-1} \in \mathcal{O}_F^\times$, and, similarly we have

$$\mathcal{K}\ell\left(2\varpi^{-n}, 2\varpi^{-n}\varepsilon\varepsilon^\sigma\left(\alpha\alpha^\sigma - \varpi^{n-m}\nu\nu^\sigma\right)\right)$$
$$= \mathcal{K}\ell\left(2\varpi^{m - \mathrm{ord}(\Delta)}, 2\varpi^{\mathrm{ord}(\Delta) - m}\varepsilon\varepsilon^\sigma \Delta^{-2} ab\right) = \mathcal{K}\ell\left(2a\Delta^{-1}, 2\varepsilon\varepsilon^\sigma b\Delta^{-1}\right)$$

since $\varpi^{\mathrm{ord}(\Delta) - m}\Delta^{-1}a \in \mathcal{O}_F^\times$.

We remark that when $n > m > 0$, i.e. $\mathrm{ord}(\Delta) > 2m$, and $\mathrm{ord}(\Delta) \leq \mathrm{ord}(y)$, we have

$$\mathrm{ord}(ab) = \mathrm{ord}(\Delta + yy^\sigma) = \mathrm{ord}(\Delta) > 2m.$$

Hence $\mathrm{ord}(b) > m$ and thus by Proposition 2.7, we have

$$\mathcal{K}\ell\left(2a\Delta^{-1}, 2\varepsilon\varepsilon^\sigma b\Delta^{-1}\right) = 0$$

since $|a\Delta^{-1}| = q^n > q$.

Therefore Theorem 3.16 holds under the condition (3.94) by Proposition 3.20, Proposition 3.21 and Proposition 3.22.

3.4.2. Proof of Theorem 3.16: When (3.95) holds. Let us denote $\operatorname{ord}(y)$ by ℓ. Then we recall that we write $y = \varpi^\ell \varepsilon_y$.

When $\ell = 0$, we have $T \in \operatorname{GL}_2(\mathcal{O}_E)$ and hence $\mathcal{H}(T, \varepsilon) = 1$.

Suppose that $\ell > 0$. First we note that by Proposition 3.14, we have
$$\mathcal{H}(T,\varepsilon) = \mathcal{H}\left(\begin{pmatrix} 1 & 0 \\ 0 & \varepsilon_y^{-\sigma} \end{pmatrix} T \begin{pmatrix} 1 & 0 \\ 0 & \varepsilon_y^{-1} \end{pmatrix}, \varepsilon \varepsilon_y^{-1}\right)$$
where
$$\begin{pmatrix} 1 & 0 \\ 0 & \varepsilon_y^{-\sigma} \end{pmatrix} T \begin{pmatrix} 1 & 0 \\ 0 & \varepsilon_y^{-1} \end{pmatrix} = \begin{pmatrix} 1 & 0 \\ 0 & \varepsilon_y^{-\sigma} \end{pmatrix} \begin{pmatrix} a & y \\ y^\sigma & b \end{pmatrix} \begin{pmatrix} 1 & 0 \\ 0 & \varepsilon_y^{-1} \end{pmatrix} = \begin{pmatrix} a & \varpi^\ell \\ \varpi^\ell & b(\varepsilon_y \varepsilon_y^\sigma)^{-1} \end{pmatrix}.$$

Now let us define $k = \begin{pmatrix} 1 & k_1 \\ 1 & k_2 \end{pmatrix} \in \operatorname{GL}_2(\mathcal{O}_F)$ by

(3.120)
$$\begin{cases} k_1 = -\left\{1 + \varpi^{-\ell} b (\varepsilon_y \varepsilon_y^\sigma)^{-1}\right\} \left\{1 - ab(yy^\sigma)^{-1}\right\}^{-\frac{1}{2}} \\ k_2 = \left\{1 + \varpi^{-\ell} a\right\} \left\{1 - ab(yy^\sigma)^{-1}\right\}^{-\frac{1}{2}}. \end{cases}$$

Then we have
$${}^t k \begin{pmatrix} a & \varpi^\ell \\ \varpi^\ell & b(\varepsilon_y \varepsilon_y^\sigma)^{-1} \end{pmatrix} k = \left\{2 + \varpi^{-\ell} a + \varpi^{-\ell} b (\varepsilon_y \varepsilon_y^\sigma)^{-1}\right\} \begin{pmatrix} \varpi^\ell & 0 \\ 0 & -\varpi^\ell \end{pmatrix}.$$

Further
$${}^t C^{-\sigma} \begin{pmatrix} \varpi^\ell & 0 \\ 0 & -\varpi^\ell \end{pmatrix} C^{-1} = \frac{1}{2d} \begin{pmatrix} 0 & \eta \varpi^\ell \\ -\eta \varpi^\ell & 0 \end{pmatrix}$$

where $C = \begin{pmatrix} 1 & 1 \\ \eta & -\eta \end{pmatrix}$. Thus by Proposition 3.14, we have

(3.121)
$$\mathcal{H}(T,\varepsilon) = \mathcal{H}\left(\begin{pmatrix} d(1+k_1)(1+k_2) & \eta(1-k_1 k_2) \\ \eta(1-k_1 k_2) & (1-k_1)(1-k_2) \end{pmatrix}, \begin{pmatrix} 0 & \eta \varpi^\ell \\ -\eta \varpi^\ell & 0 \end{pmatrix}, \nu\right)$$

where

(3.122)
$$\nu = \frac{\varepsilon \varepsilon_y^{-1}}{2 + \varpi^{-\ell} a + \varpi^{-\ell} b (\varepsilon_y \varepsilon_y^\sigma)^{-1}}.$$

Here let us write
$$B_\ell = \begin{pmatrix} 0 & \eta \varpi^\ell \\ -\eta \varpi^\ell & 0 \end{pmatrix}$$
and
$$\mathcal{Z}_\ell = \left\{Z \in \operatorname{Sym}^2(E) \mid B_\ell Z \in \operatorname{Mat}_{2 \times 2}(\mathcal{O}_E), B_\ell^{-\sigma} - Z^\sigma B_\ell Z \in \operatorname{Mat}_{2 \times 2}(\mathcal{O}_E)\right\}.$$

Since $\varpi^{-\ell} B_\ell \in \operatorname{GL}_2(\mathcal{O}_E)$, we have
$$B_\ell Z \in \operatorname{Mat}_{2 \times 2}(\mathcal{O}_E) \iff \varpi^\ell Z \in \operatorname{Sym}^2(\mathcal{O}_E)$$
for $Z \in \operatorname{Sym}^2(E)$. Also we note that

(3.123) $\quad B_\ell^{-\sigma} - Z^\sigma B_\ell Z = \varpi^{-\ell} \eta \left\{d^{-1} \begin{pmatrix} 0 & 1 \\ -1 & 0 \end{pmatrix} - (\varpi^\ell Z)^\sigma \begin{pmatrix} 0 & 1 \\ -1 & 0 \end{pmatrix} (\varpi^\ell Z)\right\}.$

Let us write $Z = \varpi^{-\ell} M$ where $M = \begin{pmatrix} u & v \\ v & w \end{pmatrix} \in \mathrm{Sym}^2(\mathcal{O}_E)$. Then by (3.123), we have $B_\ell^{-\sigma} - Z^\sigma B_\ell Z \in \mathrm{Mat}_{2\times 2}(\mathcal{O}_E)$ if and only if

(3.124) $$M^\sigma \begin{pmatrix} 0 & 1 \\ -1 & 0 \end{pmatrix} M \equiv d^{-1} \begin{pmatrix} 0 & 1 \\ -1 & 0 \end{pmatrix} \pmod{\varpi^\ell}.$$

Let $\mathcal{S}_\ell$ denote the set of $M \in \mathrm{Sym}^2(\mathcal{O}_{E,\ell})$ satisfying (3.124). Then we have the following proposition.

PROPOSITION 3.23. *Let*
$$\mathcal{T}_\ell = \left\{ (z, S) \in \mathcal{O}_{E,\ell}^\times \times \left(\mathrm{Sym}^2(\mathcal{O}_{F,\ell}) \cap \mathrm{GL}_2(\mathcal{O}_{F,\ell}) \right) \mid zz^\sigma \cdot \det S = d^{-1} \right\}.$$

Then the mapping $\varphi_\ell : \mathcal{T}_\ell \to \mathcal{S}_\ell$ *defined by*
$$\varphi_\ell(z, S) = zS$$

is surjective and we have
$$\#\varphi_\ell^{-1}(M) = q^\ell \left(1 - q^{-1}\right)$$

for each $M \in \mathcal{S}_\ell$.

PROOF. Since
$${}^t g \begin{pmatrix} 0 & 1 \\ -1 & 0 \end{pmatrix} g = \det g \cdot \begin{pmatrix} 0 & 1 \\ -1 & 0 \end{pmatrix}$$

for $g \in \mathrm{GL}_2(\mathcal{O}_{F,\ell})$, it is clear that $\varphi_\ell(\mathcal{T}_\ell) \subset \mathcal{S}_\ell$.

Let us show that φ_ℓ is surjective. Let $M \in \mathcal{S}_\ell$ and let $\mu = \det M$. Then by taking the determinants of the both sides of (3.124), we have $\mu \mu^\sigma = d^{-2}$. Hence by Lemma 1.12 there exists $z \in \mathcal{O}_{E,\ell}^\times$ such that $\mu = d^{-1} z z^{-\sigma}$. Then (3.124) implies that
$$M^{-1} = d \begin{pmatrix} 0 & -1 \\ 1 & 0 \end{pmatrix} M^\sigma \begin{pmatrix} 0 & 1 \\ -1 & 0 \end{pmatrix},$$
i.e.
$$dz^{-1} z^\sigma \begin{pmatrix} w & -v \\ -v & u \end{pmatrix} = d \begin{pmatrix} w^\sigma & -v^\sigma \\ -v^\sigma & u^\sigma \end{pmatrix}.$$

Thus
$$z^{-1} \begin{pmatrix} w & -v \\ -v & u \end{pmatrix} = \left\{ z^{-1} \begin{pmatrix} w & -v \\ -v & u \end{pmatrix} \right\}^\sigma$$

and hence we have $M = zS$ where $S \in \mathrm{Sym}^2(\mathcal{O}_{F,\ell}) \cap \mathrm{GL}_2(\mathcal{O}_{F,\ell})$. Then we have $zz^\sigma \cdot \det S = d^{-1}$ by (3.124). Hence $\varphi_\ell : \mathcal{T}_\ell \to \mathcal{S}_\ell$ is surjective.

Finally we note that for $z, z' \in \mathcal{O}_{E,\ell}^\times$ and $S, S' \in \mathrm{Sym}^2(\mathcal{O}_{F,\ell}) \cap \mathrm{GL}_2(\mathcal{O}_{F,\ell})$, we have
$$zS = z'S' \iff z = az', \ S = a^{-1} S' \text{ for some } a \in \mathcal{O}_{F,\ell}^\times.$$

Hence for each $M \in \mathcal{S}_\ell$,
$$\#\varphi_\ell^{-1}(M) = \#\mathcal{O}_{F,\ell}^\times = q^\ell \left(1 - q^{-1}\right).$$

□

3.4. EVALUATION OF THE QUADRATIC KLOOSTERMAN SUM

Thus we have

$$(3.125) \quad \mathcal{H}(T,\varepsilon) = \frac{q^{-\ell}}{1-q^{-1}} \sum_{(z,(\begin{smallmatrix}r & s \\ s & t\end{smallmatrix}))\in\mathcal{T}_\ell}$$

$$\psi_E \left[\varpi^{-\ell}\nu z \cdot \operatorname{tr} \left\{ \begin{pmatrix} d(1+k_1)(1+k_2) & \eta(1-k_1k_2) \\ \eta(1-k_1k_2) & (1-k_1)(1-k_2) \end{pmatrix} \begin{pmatrix} r & s \\ s & t \end{pmatrix} \right\} \right].$$

3.4.2.1. *When $m \geq 2\ell$.* Then we have

$$k_1 \equiv -1 \pmod{\varpi^\ell}, \qquad k_2 \equiv 1 \pmod{\varpi^\ell}.$$

Hence

$$(3.126) \quad \mathcal{H}(T,\varepsilon) = \frac{q^{-\ell}}{1-q^{-1}} \sum_{(z,(\begin{smallmatrix}r & s \\ s & t\end{smallmatrix}))\in\mathcal{T}_\ell} \psi_E\left(4\varpi^{-\ell}\eta\nu zs\right).$$

Let

$$(3.127) \quad \mathcal{T}^{(1)} = \left\{ \left(z, \begin{pmatrix} r & s \\ s & t \end{pmatrix}\right) \in \mathcal{T}_\ell \mid t \in \mathcal{O}_{F,\ell}^\times \right\},$$

$$(3.128) \quad \mathcal{T}^{(2)} = \left\{ \left(z, \begin{pmatrix} r & s \\ s & t \end{pmatrix}\right) \in \mathcal{T}_\ell \mid t \in \varpi\mathcal{O}_{F,\ell} \right\}$$

and

$$(3.129) \quad \mathcal{H}^{(i)} = \frac{q^{-\ell}}{1-q^{-1}} \sum_{(z,(\begin{smallmatrix}r & s \\ s & t\end{smallmatrix}))\in\mathcal{T}^{(i)}} \psi_E\left(4\varpi^{-\ell}\eta\nu zs\right) \quad (i=1,2).$$

Let us compute $\mathcal{H}^{(1)}$ first. We note that $r \in \mathcal{O}_{F,\ell}$ such that

$$\left(z, \begin{pmatrix} r & s \\ s & t \end{pmatrix}\right) \in \mathcal{T}^{(1)},$$

for given $t \in \mathcal{O}_{F,\ell}^\times$, $s \in \mathcal{O}_{F,\ell}$ and $z \in \mathcal{O}_{E,\ell}^\times$, is uniquely determined by the condition $r = t^{-1}\left\{s^2 + d^{-1}(zz^\sigma)^{-1}\right\}$. Thus

$$\mathcal{H}^{(1)} = \frac{q^{-\ell}}{1-q^{-1}} \sum_{z\in\mathcal{O}_{E,\ell}^\times, t\in\mathcal{O}_{F,\ell}^\times} \left(\sum_{s\in\mathcal{O}_{F,\ell}} \psi_E\left(4\varpi^{-\ell}\eta\nu zs\right) \right).$$

Here the inner sum vanishes unless

$$z = \nu^{-1}a, \quad a \in \mathcal{O}_{F,\ell}^\times.$$

Hence

$$\mathcal{H}^{(1)} = \frac{q^{-\ell}}{1-q^{-1}} \sum_{a\in\mathcal{O}_{F,\ell}^\times, t\in\mathcal{O}_{F,\ell}^\times} q^\ell = q^{2\ell}\left(1-q^{-1}\right).$$

As for $\mathcal{H}^{(2)}$, the condition $zz^\sigma\left(rt - s^2\right) \equiv d^{-1} \pmod{\varpi^\ell}$ implies that $s \in \mathcal{O}_{F,\ell}^\times$. Hence by a change of variable $r \mapsto rs$, $t \mapsto st$, we have

$$\mathcal{H}^{(2)} = \frac{q^{-\ell}}{1-q^{-1}} \sum_{\{(z,r,s,t)\in\mathcal{O}_{E,\ell}^\times\times\mathcal{O}_{F,\ell}\times\mathcal{O}_{F,\ell}^\times\times\varpi\mathcal{O}_{F,\ell}|zz^\sigma s^2(rt-1)=d^{-1}\}} \psi_E\left(4\varpi^{-\ell}\eta\nu zs\right).$$

Let $\zeta = zs$. Then we have

$$\mathcal{H}^{(2)} = \sum_{r \in \mathcal{O}_{F,\ell},\, t \in \varpi\mathcal{O}_{F,\ell}} \left(\sum_{\{\zeta \in \mathcal{O}_{E,\ell}^\times \mid \zeta\zeta^\sigma = d^{-1}(rt-1)^{-1}\}} \psi_E\left(4\varpi^{-\ell}\eta\nu\zeta\right) \right).$$

By Proposition 2.16, we have

$$\mathcal{H}^{(2)} = (-1)^\ell \sum_{r \subset \mathcal{O}_{F,\ell},\, t \subset \varpi\mathcal{O}_{F,\ell}} \sum_{x \in \mathcal{O}_{F,\ell}^\times} \psi\left[2\varpi^{-\ell}\left(x^{-1} + \frac{4\nu\nu^\sigma}{1-rt}x\right)\right].$$

Here we note that

$$\nu = \frac{\varepsilon\varepsilon_y^{-1}}{2 + \varpi^{-\ell}a + \varpi^{-\ell}b\left(\varepsilon_y\varepsilon_y^\sigma\right)^{-1}} \equiv \frac{\varepsilon}{2\varepsilon_y} \pmod{\varpi^\ell}.$$

Thus

$$\mathcal{H}^{(2)} = (-1)^\ell \sum_{r \in \mathcal{O}_{F,\ell},\, t \in \varpi\mathcal{O}_{F,\ell}} \sum_{x \in \mathcal{O}_{F,\ell}^\times} \psi\left[2\varpi^{-\ell}\left\{x^{-1} + \frac{\varepsilon\varepsilon^\sigma}{\varepsilon_y\varepsilon_y^\sigma(1-rt)}x\right\}\right].$$

Hence when $\ell = 1$, we have

$$\mathcal{H}^{(2)} = -q \sum_{x \in \mathcal{O}_{F,1}^\times} \psi\left[2\varpi^{-1}\left\{x^{-1} + \varepsilon\varepsilon^\sigma\left(\varepsilon_y\varepsilon_y^\sigma\right)^{-1}x\right\}\right]$$

$$= -q^2 \cdot \mathcal{K}\ell\left(2\varpi^{-1}, 2\varpi^{-1}\varepsilon\varepsilon^\sigma\left(\varepsilon_y\varepsilon_y^\sigma\right)^{-1}\right).$$

Suppose that $\ell \geq 2$. Let i be an integer such that $1 \leq i < \ell$. Then for $t \in \varpi^i \mathcal{O}_{F,\ell}^\times$, the mapping $\mathcal{O}_{F,\ell} \ni r \mapsto 1 - rt \in 1 + \varpi^i \mathcal{O}_{F,\ell}$ is surjective and for $r, r' \in \mathcal{O}_{F,\ell}$,

$$1 - rt = 1 - r't \iff r - r' \in \varpi^{\ell-i}\mathcal{O}_{F,\ell}.$$

We also note that the mapping

$$1 + \varpi^i \mathcal{O}_{F,\ell} \ni u \mapsto u^{-1} \in 1 + \varpi^i \mathcal{O}_{F,\ell}$$

is an automorphism of the multiplicative group $1 + \varpi^i \mathcal{O}_{F,\ell}$. Hence when $\rho \in \mathcal{O}_{F,\ell}^\times$ and $1 \leq i < \ell$, we have

(3.130)
$$\sum_{r \in \mathcal{O}_{F,\ell},\, t \in \varpi^i \mathcal{O}_{F,\ell}^\times} \psi\left(\varpi^{-\ell}\rho(1-rt)^{-1}\right) = q^i \sum_{t \in \varpi^i \mathcal{O}_{F,\ell}^\times} \sum_{u \in 1+\varpi^i\mathcal{O}_{F,\ell}} \psi\left(\varpi^{-\ell}\rho u\right) = 0.$$

Thus we have

$$\mathcal{H}^{(2)} = (-1)^\ell q^\ell \sum_{x \in \mathcal{O}_{F,\ell}^\times} \psi\left(2\varpi^{-\ell}x^{-1} + 2\varpi^{-\ell}\varepsilon\varepsilon^\sigma\left(\varepsilon_y\varepsilon_y^\sigma\right)^{-1}x\right)$$

$$= (-1)^\ell q^{2\ell} \cdot \mathcal{K}\ell\left(2\varpi^{-\ell}, 2\varpi^{-\ell}\varepsilon\varepsilon^\sigma\left(\varepsilon_y\varepsilon_y^\sigma\right)^{-1}\right).$$

Hence we have shown that

$$\mathcal{H}(T,\varepsilon) = (-1)^\ell q^{2\ell} \cdot \mathcal{K}\ell\left(2\varpi^{-\ell}, 2\varpi^{-\ell}\varepsilon\varepsilon^\sigma\left(\varepsilon_y\varepsilon_y^\sigma\right)^{-1}\right) + q^{2\ell}\left(1 - q^{-1}\right).$$

Here we note that $\varpi^{-2\ell}\Delta \equiv -\varepsilon_y\varepsilon_y^\sigma \pmod{\varpi^\ell}$ since $m \geq 2\ell$. Thus

$$\mathcal{K}\ell\left(2\varpi^{-\ell}, 2\varpi^{-\ell}\varepsilon\varepsilon^\sigma\left(\varepsilon_y\varepsilon_y^\sigma\right)^{-1}\right) = \mathcal{K}\ell\left(2\varpi^{-\ell}, 2\varpi^{-\ell}\varepsilon\varepsilon^\sigma\varepsilon_y\varepsilon_y^\sigma\left(\varpi^{-2\ell}\Delta\right)^{-2}\right)$$
$$= \mathcal{K}\ell\left(2\varpi^\ell\Delta^{-1}, 2\varpi^\ell\varepsilon\varepsilon^\sigma\varepsilon_y\varepsilon_y^\sigma\Delta^{-1}\right).$$

We have proved our assertion when $m \geq 2\ell$.

3.4.2.2. *When $\ell < m < 2\ell$*. As in the previous case, let $\mathcal{T}^{(1)}$ and $\mathcal{T}^{(2)}$ be the subset of $\mathcal{T}_\ell$ defined by (3.127) and (3.128), respectively. Then for $i = 1, 2$, let

$$\mathcal{H}^{(i)} = \frac{q^{-\ell}}{1 - q^{-1}} \sum_{(z, (\begin{smallmatrix}r & s \\ s & t\end{smallmatrix})) \in \mathcal{T}^{(i)}} \psi_E\left[\varpi^{-\ell}\nu z \cdot \operatorname{tr}\left\{\begin{pmatrix} d(1+k_1)(1+k_2) & \eta(1-k_1k_2) \\ \eta(1-k_1k_2) & (1-k_1)(1-k_2) \end{pmatrix}\begin{pmatrix} r & s \\ s & t \end{pmatrix}\right\}\right].$$

We note that

$$1 - k_1 k_2 = \frac{2 + \varpi^{-\ell}a + \varpi^{-\ell}b\left(\varepsilon_y\varepsilon_y^\sigma\right)^{-1}}{1 - ab(yy^\sigma)^{-1}}.$$

Let $\mu = \nu\eta(1 - k_1k_2)$. Then we have

$$(3.131) \qquad \mu = -\frac{\eta\varepsilon\varepsilon_y^\sigma}{\varpi^{-2\ell}\Delta}$$

where $\Delta = ab - yy^\sigma = \det T$. Also let

$$(3.132) \qquad A_1 = \frac{(1+k_1)(1+k_2)}{1 - k_1k_2}, \qquad A_2 = \frac{(1-k_1)(1-k_2)}{1 - k_1k_2}.$$

Then we have

$$\mathcal{H}^{(i)} = \frac{q^{-\ell}}{1-q^{-1}} \sum_{(z, (\begin{smallmatrix}r & s \\ s & t\end{smallmatrix})) \in \mathcal{T}^{(i)}} \psi_E\left[\varpi^{-\ell}\mu z\left(\eta A_1 r + 2s + \eta^{-1} A_2 t\right)\right] \quad (i=1, 2).$$

Let us evaluate $\mathcal{H}^{(1)}$ first. Since r is determined as $r = t^{-1}\left\{s^2 + d^{-1}(zz^\sigma)^{-1}\right\}$ for given $t \in \mathcal{O}_{F,\ell}^\times$, $s \in \mathcal{O}_{F,\ell}$ and $z \in \mathcal{O}_{E,\ell}^\times$, we have

$$(3.133) \quad \mathcal{H}^{(1)} = \frac{q^{-\ell}}{1-q^{-1}} \sum_{z \in \mathcal{O}_{E,\ell}^\times, t \in \mathcal{O}_{F,\ell}^\times} \psi_E\left[\varpi^{-\ell}\eta^{-1}\mu z\left\{A_2 t + A_1(zz^\sigma)^{-1}t^{-1}\right\}\right]$$

$$\sum_{s \in \mathcal{O}_{F,\ell}} \psi_E\left[\varpi^{-\ell}\mu z\left(\eta A_1 t^{-1} s^2 + 2s\right)\right].$$

Here we note that

$$A_1 = \frac{(1+k_1)(1+k_2)}{1-k_1k_2} \equiv -\varpi^{-\ell}b\left(\varepsilon_y\varepsilon_y^\sigma\right)^{-1}\left(2 + \varpi^{-\ell}a\right) \pmod{\varpi^{m-\ell+1}}$$
$$\equiv -2\varpi^{-\ell}b\left(\varepsilon_y\varepsilon_y^\sigma\right)^{-1} \pmod{\varpi^{m-\ell+1}}.$$

By Proposition 3.14, we have

$$\mathcal{H}(T, \varepsilon) = \mathcal{H}\left(\begin{pmatrix} 0 & 1 \\ 1 & 0 \end{pmatrix} T \begin{pmatrix} 0 & 1 \\ 1 & 0 \end{pmatrix}, \varepsilon\right).$$

Hence we may assume that

$$(3.134) \qquad \operatorname{ord}(b) = m = \min\{\operatorname{ord}(a), \operatorname{ord}(b)\}$$

and then $A_1 \in \varpi^{m-\ell}\mathcal{O}_F^\times$. Let us write

(3.135) $$A_1 = \varpi^{m-\ell}\varepsilon_1, \quad \varepsilon_1 \in \mathcal{O}_F^\times.$$

Then the inner sum of (3.133) is written as

(3.136) $$\sum_{s \in \mathcal{O}_{F,\ell}} \psi\left[\varpi^{m-2\ell}\varepsilon_1 t^{-1} s^2 \cdot \mathrm{tr}_{E/F}(\eta\mu z) + 2\varpi^{-\ell} s \cdot \mathrm{tr}_{E/F}(\mu z)\right].$$

Since $\mu z \in \mathcal{O}_E^\times$, we have either

$$\mathrm{tr}_{E/F}(\eta\mu z) \in \mathcal{O}_F^\times \quad \text{or} \quad \mathrm{tr}_{E/F}(\mu z)\,\mathcal{O}_F^\times.$$

Suppose that $\mathrm{tr}_{E/F}(\eta\mu z) \notin \mathcal{O}_F^\times$. Then we have $\mathrm{tr}_{E/F}(\mu z) \in \mathcal{O}_F^\times$. Since $\ell > 2\ell - m$, the sum (3.136) vanishes by Proposition 2.5.

Suppose that $\mathrm{tr}_{E/F}(\eta\mu z) \in \mathcal{O}_F^\times$. Then by Proposition 2.5, the sum (3.136) vanishes unless $\mathrm{tr}_{E/F}(\mu z) \in \varpi^{m-\ell}\mathcal{O}_F$ and conversely when $\mu z = \varpi^{m-\ell} x + \eta u$, where $x \in \mathcal{O}_{F,\ell}$ and $u \in \mathcal{O}_{F,\ell}^\times$, we have

$$\sum_{s \in \mathcal{O}_{F,\ell}} \psi\left[\varpi^{m-2\ell}\varepsilon_1 t^{-1} s^2 \cdot \mathrm{tr}_{E/F}(\eta\mu z) + 2\varpi^{-\ell} s \cdot \mathrm{tr}_{E/F}(\mu z)\right]$$
$$= q^\ell\, C\left(2d\varpi^{m-2\ell}\varepsilon_1 tu\right) \psi\left(-\frac{2\varpi^{m-2\ell} x^2}{d\varepsilon_1 u} t\right).$$

Hence

$$\mathcal{H}^{(1)} = \frac{q^{\ell-m}}{1-q^{-1}} \sum_{x \in \mathcal{O}_{F,\ell}} \sum_{y \in \mathcal{O}_{F,\ell}^\times} \sum_{t \in \mathcal{O}_{F,\ell}^\times} C\left(2d\varpi^{m-2\ell}\varepsilon_1 tu\right)$$
$$\psi\left[2\varpi^{m-2\ell}\left\{\left(\varpi^{\ell-m} A_2 u - \frac{x^2}{d\varepsilon_1 u}\right) t + \frac{\varepsilon_1 u \mu \mu^\sigma}{\varpi^{2(m-\ell)} x^2 - du^2} t^{-1}\right\}\right].$$

Here we note that $A_2 \equiv 0 \pmod{\varpi^{m-\ell}}$. Hence let us write

(3.137) $$A_2 = \varpi^{m-\ell}\xi_2, \quad \xi_2 \in \mathcal{O}_F.$$

Then by a change of variable $x \mapsto xu$, we have

$$\mathcal{H}^{(1)} = \frac{q^{\ell-m}}{1-q^{-1}} \sum_{x \in \mathcal{O}_{F,\ell}} \sum_{y \in \mathcal{O}_{F,\ell}^\times} \sum_{t \in \mathcal{O}_{F,\ell}^\times} C\left(2d\varpi^{m-2\ell}\varepsilon_1 tu\right)$$
$$\psi\left[2\varpi^{m-2\ell}\left\{\left(\xi_2 - d^{-1}\varepsilon_1^{-1} x^2\right) tu + \frac{\varepsilon_1 \mu \mu^\sigma}{\varpi^{2(m-\ell)} x^2 - d} t^{-1} u^{-1}\right\}\right].$$

By a change of variable $t \mapsto \varepsilon_1 tu^{-1}$, we have

$$\mathcal{H}^{(1)} = q^{2\ell-m} \sum_{x \in \mathcal{O}_{F,\ell}} \sum_{t \in \mathcal{O}_{F,\ell}^\times} C\left(2d\varpi^{m-2\ell} t\right)$$
$$\psi\left[2\varpi^{m-2\ell}\left\{\left(\varepsilon_1 \xi_2 - d^{-1} x^2\right) t + \frac{\mu\mu^\sigma}{\varpi^{2(m-\ell)} x^2 - d} t^{-1}\right\}\right].$$

3.4. EVALUATION OF THE QUADRATIC KLOOSTERMAN SUM

Further, by a change of variable $t \mapsto \left(1 - d^{-1}\varpi^{2(m-\ell)}x^2\right)^{-1} t$, we have

$$\mathcal{H}^{(1)} = q^m \sum_{x \in \mathcal{O}_{F,2\ell-m}} \sum_{t \in \mathcal{O}^\times_{F,2\ell-m}} C\left(2d\varpi^{m-2\ell}t\right) \psi\left[2\varpi^{m-2\ell}\left(\frac{\varepsilon_1\xi_2 - d^{-1}x^2}{1 - d^{-1}\varpi^{2(m-\ell)}x^2}t - d^{-1}\mu\mu^\sigma t^{-1}\right)\right].$$

Here we note that

(3.138) $$\varepsilon_1\xi_2 = \varpi^{2(\ell-m)} A_1 A_2 = \varpi^{-2m} ab \left(\varepsilon_y \varepsilon_y^\sigma\right)^{-1}$$

and from (3.132) we have

(3.139) $$\mu\mu^\sigma = \frac{-d\varepsilon\varepsilon^\sigma \varepsilon_y \varepsilon_y^\sigma}{\left(\varpi^{-2\ell}\Delta\right)^2}.$$

By a change of variable $t \mapsto \left(\varepsilon_y \varepsilon_y^\sigma\right) t$, we have

(3.140) $$\mathcal{H}^{(1)} = q^m \sum_{x \in \mathcal{O}_{F,2\ell-m}} \sum_{t \in \mathcal{O}^\times_{F,2\ell-m}}$$
$$C\left(2d\varpi^{m-2\ell}t\right) \psi\left[2\varpi^{m-2\ell}\left\{\frac{\varpi^{-2m}ab - d^{-1}\left(\varepsilon_y\varepsilon_y^\sigma\right)x^2}{1 - d^{-1}\varpi^{2(m-\ell)}x^2}t + \frac{\varepsilon\varepsilon^\sigma}{\left(\varpi^{-2\ell}\Delta\right)^2}t^{-1}\right\}\right].$$

Here we note that

$$\frac{\varpi^{-2m}ab - d^{-1}\left(\varepsilon_y\varepsilon_y^\sigma\right)x^2}{1 - d^{-1}\varpi^{2(m-\ell)}x^2} = \varpi^{-2m}ab + \frac{-d^{-1}\varepsilon_y\varepsilon_y^\sigma\left(1 - \varpi^{-2\ell}ab\left(\varepsilon_y\varepsilon_y^\sigma\right)^{-1}\right)x^2}{1 - d^{-1}\varpi^{2(m-\ell)}x^2}$$

and the mapping

$$\mathcal{O}_{F,2\ell-m} \ni x \mapsto \frac{\left\{1 - \varpi^{-2\ell}ab\left(\varepsilon_y\varepsilon_y^\sigma\right)^{-1}\right\}^{\frac{1}{2}} x}{\left\{1 - d^{-1}\varpi^{2(m-\ell)}x^2\right\}^{\frac{1}{2}}} \in \mathcal{O}_{F,2\ell-m}$$

is a bijection. Thus (3.140) becomes

(3.141)
$$\mathcal{H}^{(1)} = q^m \sum_{t \in \mathcal{O}_{F,2\ell-m}} C\left(2d\varpi^{m-2\ell}t\right) \psi\left[2\varpi^{m-2\ell}\left(\varpi^{-2m}abt + \frac{\varepsilon\varepsilon^\sigma}{\left(\varpi^{-2\ell}\Delta\right)^2}t^{-1}\right)\right]$$
$$\left(\sum_{v \in \mathcal{O}_{F,2\ell-m}} \psi\left(-2\varpi^{m-2\ell}d^{-1}\varepsilon_y\varepsilon_y^\sigma tv^2\right)\right).$$

By Proposition 2.5, we have

$$\sum_{v \in \mathcal{O}_{F,2\ell-m}} \psi\left(-2\varpi^{m-2\ell}d^{-1}\varepsilon_y\varepsilon_y^\sigma tv^2\right) = q^{2\ell-m} C\left(-2\varpi^{m-2\ell}d^{-1}\varepsilon_y\varepsilon_y^\sigma t\right).$$

Hence we have
(3.142)
$$\mathcal{H}^{(1)} = q^m \operatorname{sgn}\left(\varepsilon_y \varepsilon_y^\sigma\right)^m \sum_{t \in \mathcal{O}_{F,2\ell-m}} \psi\left[2\varpi^{m-2\ell}\left(\varpi^{-2m}ab\,t + \frac{\varepsilon\varepsilon^\sigma}{(\varpi^{-2\ell}\Delta)^2}t^{-1}\right)\right]$$
$$= q^{2\ell} \operatorname{sgn}\left(\varepsilon_y \varepsilon_y^\sigma\right)^m \cdot \mathcal{K}\ell\left(2\varpi^{-m}ab\Delta^{-1}, 2\varpi^m \varepsilon\varepsilon^\sigma \Delta^{-1}\right).$$

Finally let us compute $\mathcal{H}^{(2)}$. Since $zz^\sigma\left(rt-s^2\right) = d^{-1}$ and $t \in \varpi\mathcal{O}_{F,\ell}$, we have $s \in \mathcal{O}_{F,\ell}^\times$. Thus by a change of variable $r \mapsto rs$, $t \mapsto st$, we have

$$\mathcal{H}^{(2)} = \frac{q^{-\ell}}{1-q^{-1}} \sum_{\{(z,r,s,t) \in \mathcal{O}_{E,\ell}^\times \times \mathcal{O}_{F,\ell} \times \mathcal{O}_{F,\ell}^\times \times \varpi\mathcal{O}_{F,\ell} \mid zz^\sigma s^2(rt-1)=d^{-1}\}} \psi_E\left[\varpi^{-\ell}\mu zs\left\{2 + \eta\varpi^{m-\ell}\left(\varepsilon_1 r + d^{-1}\xi_2 t\right)\right\}\right].$$

Let $\zeta = zs$. Then we have

$$\mathcal{H}^{(2)} = \sum_{r \in \mathcal{O}_{F,\ell},\, t \in \varpi\mathcal{O}_{F,\ell}} \sum_{\{\zeta \in \mathcal{O}_{E,\ell}^\times \mid \zeta\zeta^\sigma = d^{-1}(rt-1)^{-1}\}} \psi_E\left[2\varpi^{-\ell}\mu\zeta\left\{1 + \frac{\eta\varpi^{m-\ell}\left(\varepsilon_1 r + d^{-1}\xi_2 t\right)}{2}\right\}\right].$$

By Proposition 2.16, we have

(3.143) $\mathcal{H}^{(2)} = (-1)^\ell \sum_{r \in \mathcal{O}_{F,\ell},\, t \in \varpi\mathcal{O}_{F,\ell}}$
$$\sum_{x \in \mathcal{O}_{F,\ell}^\times} \psi\left[2\varpi^{-\ell}x^{-1} + \frac{2\varpi^{-\ell}\left(\varepsilon\varepsilon^\sigma \varepsilon_y \varepsilon_y^\sigma\right)x}{(\varpi^{-2\ell}\Delta)^2(1-rt)}\left\{1 - \frac{d\varpi^{2(m-\ell)}\left(\varepsilon_1 r + d^{-1}\xi_2 t\right)^2}{4}\right\}\right].$$

When $2(m-\ell) \geq \ell$, we have

(3.144) $\mathcal{H}^{(2)} = (-1)^\ell \sum_{r \in \mathcal{O}_{F,\ell},\, t \in \varpi\mathcal{O}_{F,\ell}} \sum_{x \in \mathcal{O}_{F,\ell}^\times} \psi\left[2\varpi^{-\ell}x^{-1} + \frac{2\varpi^{-\ell}\left(\varepsilon\varepsilon^\sigma \varepsilon_y \varepsilon_y^\sigma\right)x}{(\varpi^{-2\ell}\Delta)^2(1-rt)}\right].$

Here, by (3.130), the sum (3.144) is supported on $r \in \varpi^\ell \mathcal{O}_{F,\ell} = \{0\}$ and hence we have

$$\mathcal{H}^{(2)} = (-1)^\ell q^{2\ell} \cdot \mathcal{K}\ell\left(2\varpi^{-\ell}, 2\varpi^{-\ell}\varepsilon\varepsilon^\sigma \varepsilon_y \varepsilon_y^\sigma (\varpi^{-2\ell}\Delta)^{-2}\right)$$
$$= (-1)^\ell q^{2\ell} \cdot \mathcal{K}\ell\left(2\varpi^\ell \Delta^{-1}, 2\varpi^\ell \varepsilon\varepsilon^\sigma \varepsilon_y \varepsilon_y^\sigma \Delta^{-1}\right).$$

Suppose that $\ell > 2(m-\ell)$. We rewrite (3.143) as

(3.145) $\mathcal{H}^{(2)} = (-1)^\ell \sum_{x \in \mathcal{O}_{F,\ell}^\times} \psi\left[2\varpi^{-\ell}x^{-1} + \frac{2\varpi^{-\ell}\left(\varepsilon\varepsilon^\sigma \varepsilon_y \varepsilon_y^\sigma\right)x}{(\varpi^{-2\ell}\Delta)^2}\right]$
$$\sum_{r \in \mathcal{O}_{F,\ell},\, t \in \varpi\mathcal{O}_{F,\ell}} \psi\left[\frac{2\varpi^{-\ell}\left(\varepsilon\varepsilon^\sigma \varepsilon_y \varepsilon_y^\sigma\right)x}{(\varpi^{-2\ell}\Delta)^2(1-rt)}\left\{rt - \frac{d\varpi^{2(m-\ell)}\left(\varepsilon_1 r + d^{-1}\xi_2 t\right)^2}{4}\right\}\right].$$

3.4. EVALUATION OF THE QUADRATIC KLOOSTERMAN SUM

Thus it is enough for us to show that

$$(3.146) \quad \mathcal{I}(u) = \sum_{r \in \mathcal{O}_{F,\ell}, t \in \varpi \mathcal{O}_{F,\ell}} \psi\left[\varpi^{-\ell} u \frac{4rt - d\varpi^{2(m-\ell)}\left(\varepsilon_1 r + d^{-1}\xi_2 t\right)^2}{1 - rt}\right] = q^\ell$$

for $u \in \mathcal{O}_F^\times$. First we note that we have

$$\mathcal{I}(u) = q^{2\ell} \sum_{i=1}^\infty \mathcal{I}_i$$

where

$$(3.147) \quad \mathcal{I}_i = \int_{t \in \varpi^i \mathcal{O}_F^\times} \psi\left[2\varpi^{-\ell} u \cdot \mathcal{A}(t)\right]$$
$$\int_{r \in \mathcal{O}_F} \psi\left[\varpi^{-\ell} u \left\{\mathcal{B}(t)(1-rt) + \mathcal{C}(t)(1-rt)^{-1}\right\}\right] dr\, dt$$

and

$$(3.148) \quad \mathcal{A}(t) = -2 + d\varpi^{2(m-\ell)} t^{-2} \varepsilon_1^2 \left(1 + d^{-1}\varepsilon_1^{-1}\xi_2 t^2\right)$$
$$(3.149) \quad \mathcal{B}(t) = -d\varpi^{2(m-\ell)} t^{-2} \varepsilon_1^2$$
$$(3.150) \quad \mathcal{C}(t) = 4 - d\varpi^{2(m-\ell)} t^{-2} \varepsilon_1^2 \left(1 + d^{-1}\varepsilon_1^{-1}\xi_2 t^2\right)^2.$$

Put $1 - rt = \varepsilon$ in (3.147). Then we have

$$(3.151) \quad \mathcal{I}_i = q^i \int_{t \in \varpi^i \mathcal{O}_F^\times} \psi\left[2\varpi^{-\ell} u \cdot \mathcal{A}(t)\right]$$
$$\int_{\varepsilon \in 1 + \varpi^i \mathcal{O}_F} \psi\left[\varpi^{-\ell} u \left\{\mathcal{B}(t)\varepsilon + \mathcal{C}(t)\varepsilon^{-1}\right\}\right] d\varepsilon\, dt$$

Suppose that $1 \leq i < m - \ell$. Then we have

$$1 = |\mathcal{C}(t)| > |\mathcal{B}(t)|.$$

Hence by Lemma 2.8, we have $\mathcal{I}_i = 0$.

Suppose that $i \geq m - \ell$. Let us write $t = \varpi^i \varepsilon_t$. Then

$$\mathcal{B}(t) = -d\varpi^{2(m-\ell)-2i} \varepsilon_t^{-2} \varepsilon_1^2,$$
$$\mathcal{C}(t) = -d\varpi^{2(m-\ell)-2i} \varepsilon_t^{-2} \varepsilon_1^2 \left\{\left(1 + d^{-1}\varepsilon_1^{-1}\xi_2 t^2\right)^2 - 4d^{-1}\varpi^{2i-2(m-\ell)}\varepsilon_t^2 \varepsilon_1^{-2}\right\}.$$

Hence

$$\mathcal{B}(t)^{-1} \mathcal{C}(t) = \left(1 + d^{-1}\varepsilon_1^{-1}\xi_2 t^2\right)^2 - 4d^{-1}\varpi^{2i-2(m-\ell)}\varepsilon_t^2 \varepsilon_1^{-2}$$
$$\equiv 1 - 4d^{-1}\varpi^{2i-2(m-\ell)}\varepsilon_t^2 \varepsilon_1^{-2} \pmod{\varpi^i}.$$

Thus by Lemma 2.10 the inner integral of (3.151) vanishes unless $2i - 2(m - \ell) \geq i$, i.e. $i \geq 2(m - \ell)$. By the same lemma, when $i \geq 2(m - \ell)$, we have

$$\mathcal{I}_i = q^i C\left(-d\varpi^{2m-3\ell-2i} u\right) \int_{t \in \varpi^i \mathcal{O}_F^\times} \psi\left[2\varpi^{-\ell} u (\mathcal{A}(t) + \mathcal{B}(t) Z(t))\right] dt$$

where

$$Z(t) = \left(1 + d^{-1}\varepsilon_1^{-1}\xi_2 t^2\right) \left\{1 - \frac{4d^{-1}\varpi^{-2(m-\ell)} t^2 \varepsilon_1^{-2}}{1 + d^{-1}\varepsilon_1^{-1}\xi_2 t^2}\right\}^{\frac{1}{2}}.$$

Here
$$\mathcal{A}(t) + \mathcal{B}(t) Z(t)$$
$$= -2 + d\varpi^{2(m-\ell)} t^{-2} \varepsilon_1^2 \left(1 + d^{-1}\varepsilon_1^{-1}\xi_2 t^2\right) \left\{ 1 - \left\{ 1 - \frac{4d^{-1}\varpi^{-2(m-\ell)} t^2 \varepsilon_1^{-2}}{1 + d^{-1}\varepsilon_1^{-1}\xi_2 t^2} \right\}^{\frac{1}{2}} \right\}$$
$$= -2 + 4 \left\{ 1 + \left\{ 1 - \frac{4d^{-1}\varpi^{-2(m-\ell)} t^2 \varepsilon_1^{-2}}{1 + d^{-1}\varepsilon_1^{-1}\xi_2 t^2} \right\}^{\frac{1}{2}} \right\}^{-1}.$$

Since
$$\varpi^i \mathcal{O}_F^\times \ni t \mapsto 2d^{-1}\varepsilon_1^{-1} t \left(1 + d^{-1}\varepsilon_1^{-1}\xi_2 t^2\right)^{-\frac{1}{2}} \in \varpi^i \mathcal{O}_F^\times$$
is a bijection, put $x = 2d^{-1}\varepsilon_1^{-1} t \left(1 + d^{-1}\varepsilon_1^{-1}\xi_2 t^2\right)^{-\frac{1}{2}}$. Then we have
$$\mathcal{I}_i = C\left(-d\varpi^{2m-3\ell} u\right) \int_{x \in \varpi^i \mathcal{O}_F^\times} \psi \left[2\varpi^{-\ell} u \left\{ -2 + \frac{4}{1 + \sqrt{1 - d\varpi^{-2(m-\ell)} x^2}} \right\} \right] dx$$
$$= q^{-m+\ell} C\left(-d\varpi^{2m-3\ell} u\right) \int_{v \in \varpi^{i-(m-\ell)} \mathcal{O}_F^\times} \psi \left[4\varpi^{-\ell} u \cdot \frac{1 - \sqrt{1 - dv^2}}{1 + \sqrt{1 - dv^2}} \right] dv$$
where $v = \varpi^{-(m-\ell)} x$. Here
$$\frac{1 - \sqrt{1 - dv^2}}{1 + \sqrt{1 - dv^2}} = d^{-1} \left\{ \frac{1 - \sqrt{1 - dv^2}}{v} \right\}^2$$
and
$$\varpi^{i-(m-\ell)} \mathcal{O}_F^\times \ni v \mapsto \frac{1 - \sqrt{1 - dv^2}}{v} \in \varpi^{i-(m-\ell)} \mathcal{O}_F^\times$$
is a bijection. Hence
$$\mathcal{I}_i = q^{-m+\ell} C\left(-d\varpi^{2m-3\ell} u\right) \int_{w \in \varpi^{i-(m-\ell)} \mathcal{O}_F^\times} \psi \left(4\varpi^{-\ell} d^{-1} u w^2\right) dw.$$

Thus
$$\mathcal{I}(u) = q^{2\ell} \sum_{i=2(m-\ell)}^{\infty} \mathcal{I}_i$$
$$= q^{-m+3\ell} C\left(-d\varpi^{2m-3\ell} u\right) \int_{w \in \varpi^{m-\ell} \mathcal{O}_F} \psi \left(4\varpi^{-\ell} d^{-1} u w^2\right) dw$$
$$= q^{-2m+4\ell} C\left(-d\varpi^{2m-3\ell} u\right) \int_{w \in \mathcal{O}_F} \psi \left(4\varpi^{2m-3\ell} d^{-1} u w^2\right) dw.$$

Hence by Proposition 2.5, we have
$$\mathcal{I}(u) = q^{-2m+4\ell} C\left(-d\varpi^{2m-3\ell} u\right) C\left(4\varpi^{2m-3\ell} d^{-1} u\right) = q^\ell.$$

Thus we have shown (3.146) and we have finished the proof of Theorem 3.16.

CHAPTER 4

Evaluation of the Novodvorsky Orbital Integral

In this chapter we shall evaluate the Novodvorsky orbital integral explicitly for the main relevant double cosets.

4.1. Double coset decomposition

4.1.1. $\bar{H}\backslash G/H$ double coset decomposition. We recall that G denotes $\mathrm{GSp}(4, F)$, the group of 4 by 4 symplectic similitude matrices,

$$G = \left\{ g \in \mathrm{GL}(4, F) \mid {}^t g \begin{pmatrix} 0 & 1_2 \\ -1_2 & 0 \end{pmatrix} g = \lambda(g) \begin{pmatrix} 0 & 1_2 \\ -1_2 & 0 \end{pmatrix}, \lambda(g) \in \mathrm{GL}(1, F) \right\}.$$

We also recall that H denotes the upper Novodvorsky subgroup of G defined by

$$(4.1) \qquad H = \left\{ \begin{pmatrix} a & 0 & 0 & 0 \\ 0 & b & 0 & 0 \\ 0 & 0 & b & 0 \\ 0 & 0 & 0 & a \end{pmatrix} \begin{pmatrix} 1_2 & X \\ 0 & 1_2 \end{pmatrix} \mid a, b \in \mathrm{GL}_1(F), X \in \mathrm{Sym}^2(F) \right\}$$

and $\bar{H}$ denotes the lower Novodvorsky subgroup defined by

$$(4.2) \qquad \bar{H} = \{ {}^t h \mid h \in H \}.$$

Then our goal in this section is to describe the double coset decomposition $\bar{H}\backslash G/H$.

We recall that the upper Siegel parabolic subgroup P of G is given by the Levi decomposition $P = MU$ where

$$M = \left\{ \begin{pmatrix} h & 0 \\ 0 & \lambda \cdot {}^t h^{-1} \end{pmatrix} \mid h \in \mathrm{GL}_2(F), \lambda \in \mathrm{GL}_1(F) \right\},$$

$$U = \left\{ \begin{pmatrix} 1_2 & X \\ 0 & 1_2 \end{pmatrix} \mid X \in \mathrm{Sym}^2(F) \right\}.$$

Then the lower Siegel parabolic subgroup $\bar{P}$ of G is given by the Levi decomposition $\bar{P} = M\bar{U}$ where

$$\bar{U} = \left\{ \begin{pmatrix} 1_2 & 0 \\ Y & 1_2 \end{pmatrix} \mid Y \in \mathrm{Sym}^2(F) \right\}.$$

First we recall that, by the Bruhat decomposition, we have

$$(4.3) \qquad G = \bar{P}P \cup \bar{P}w_1 P \cup \bar{P}w_2 P$$

where

$$(4.4) \qquad w_1 = \begin{pmatrix} 0 & 0 & 1 & 0 \\ 0 & 0 & 0 & 1 \\ -1 & 0 & 0 & 0 \\ 0 & -1 & 0 & 0 \end{pmatrix}, \quad w_2 = \begin{pmatrix} 1 & 0 & 0 & 0 \\ 0 & 0 & 0 & 1 \\ 0 & 0 & 1 & 0 \\ 0 & -1 & 0 & 0 \end{pmatrix}.$$

Here we note that $\bar{P}P$ is the unique Zariski open double coset in G.

For $h \in \mathrm{GL}_2(F)$ and $\lambda \in \mathrm{GL}_1(F)$, we denote by (h, λ) an element of M given by

(4.5) $$(h, \lambda) = \begin{pmatrix} h & 0 \\ 0 & \lambda \cdot {}^t h^{-1} \end{pmatrix} \in M.$$

We shall identify $\mathrm{GL}_2(F)$ with
$$\{(h, 1) \in M \mid h \in \mathrm{GL}_2(F)\},$$
and, $\mathrm{GL}_1(F)$ with
$$\{(1, \lambda) \in M \mid \lambda \in \mathrm{GL}_1(F)\},$$
respectively. Finally let us denote by A the subgroup of M defined by

(4.6) $$A = \left\{ \begin{pmatrix} a & 0 & 0 & 0 \\ 0 & b & 0 & 0 \\ 0 & 0 & b & 0 \\ 0 & 0 & 0 & a \end{pmatrix} \mid a, b \in \mathrm{GL}_1(F) \right\}.$$

We shall often identify A with the subgroup of $\mathrm{GL}_2(F)$ as

(4.7) $$A = \left\{ \begin{pmatrix} a & 0 \\ 0 & b \end{pmatrix} \mid a, b \in \mathrm{GL}_1(F) \right\}.$$

We remark that then we have

(4.8) $$A = \left\{ g \in \mathrm{GL}_2(F) \mid {}^t g \begin{pmatrix} 0 & 1 \\ 1 & 0 \end{pmatrix} g = \det g \cdot \begin{pmatrix} 0 & 1 \\ 1 & 0 \end{pmatrix} \right\}.$$

PROPOSITION 4.1. *1. We have the disjoint decomposition*
$$\bar{P} P = \bigcup_{h \in A \backslash \mathrm{GL}_2(F)/A,\, \lambda \in \mathrm{GL}_1(F)} \bar{H}(h, \lambda) H.$$

2. We have the disjoint decomposition
$$\bar{P} w_1 P = \bigcup_{h \in A \backslash \mathrm{GL}_2(F)/A,\, \lambda \in \mathrm{GL}_1(F)} \bar{H} w_1 (h, \lambda) H.$$

3. As the representatives of the $(\bar{H}, H)$-double cosets in $\bar{P} w_2 P$, we may take elements of the form
$$m_1 w_2 m_2$$
where $m_2 \in M$ and
$$m_1 = 1_4,\ \begin{pmatrix} 0 & 1 & 0 & 0 \\ -1 & 0 & 0 & 0 \\ 0 & 0 & 0 & 1 \\ 0 & 0 & -1 & 0 \end{pmatrix},\ \begin{pmatrix} 1 & 1 & 0 & 0 \\ 0 & 1 & 0 & 0 \\ 0 & 0 & 1 & 0 \\ 0 & 0 & -1 & 1 \end{pmatrix} \begin{pmatrix} 0 & 1 & 0 & 0 \\ -1 & 0 & 0 & 0 \\ 0 & 0 & 0 & 1 \\ 0 & 0 & -1 & 0 \end{pmatrix}.$$

PROOF. Since the proofs are similar for the first two assertions, here we prove only the second one. We have
$$\bar{P} w_1 P = \bar{U} w_1 \left(w_1^{-1} M w_1 \right) M U = \bar{U} w_1 M U,$$
since $w_1^{-1} M w_1 = M$. We also have $w_1^{-1} A w_1 = A$. Hence we may take elements of the form (h, λ) as the representatives of the double cosets.

Let us prove the disjointness. Let us suppose that for $h, h_1 \in \mathrm{GL}_2(F)$ and $\lambda, \lambda_1 \in \mathrm{GL}_1(F)$, we have
$$\bar{H} w_1 (h, \lambda) H = \bar{H} w_1 (h_1, \lambda_1) H.$$

Then there exist $a, a_1 \in A$, $\bar{u} \in \bar{U}$ and $u \in U$ such that
$$\bar{u} a w_1 (h, \lambda) a_1 u = w_1 (h_1, \lambda_1).$$
Hence
$$(h_1, \lambda_1) = w_1^{-1} \bar{u} w_1 \cdot w_1^{-1} a w_1 (h, \lambda) a_1 \cdot u$$
where $w_1^{-1} \bar{u} w_1 \in U$. Thus by considering the reductive part of the Levi decomposition in $P = MU$, we have
$$(h_1, \lambda_1) = w_1^{-1} a w_1 \cdot (h, \lambda) \cdot a_1.$$
Hence h and h_1 belong to the same (A, A)-double coset in $\mathrm{GL}_2(F)$ and thus $h = h_1$. Then we have

(4.9) $\begin{pmatrix} h & 0 \\ 0 & \lambda \cdot {}^t h^{-1} \end{pmatrix} = \begin{pmatrix} \alpha & 0 & 0 & 0 \\ 0 & \beta & 0 & 0 \\ 0 & 0 & \beta & 0 \\ 0 & 0 & 0 & \alpha \end{pmatrix} \begin{pmatrix} h & 0 \\ 0 & \lambda \cdot {}^t h^{-1} \end{pmatrix} \begin{pmatrix} \alpha_1 & 0 & 0 & 0 \\ 0 & \beta_1 & 0 & 0 \\ 0 & 0 & \beta_1 & 0 \\ 0 & 0 & 0 & \alpha_1 \end{pmatrix}$

for some $\alpha, \beta, \alpha_1, \beta_1 \in F^\times$. By looking at the GL_2 part of (4.9), we have
$$h = \begin{pmatrix} \alpha & 0 \\ 0 & \beta \end{pmatrix} h \begin{pmatrix} \alpha_1 & 0 \\ 0 & \beta_1 \end{pmatrix}.$$

Hence by taking the determinants, we have $(\alpha \beta)(\alpha_1 \beta_1) = 1$. Then by taking the similitude of the both sides of (4.9), we have $\lambda = \lambda_1$.

Finally let us prove the third assertion. Since $\bar{P} w_2 P = \bar{U} M w_2 M U$, it is clear that we may take elements of the form $m_1 w_2 m_2$ where $m_1, m_2 \in M$ as the representatives. Let us denote by M_B the intersection of M and the upper Borel subgroup of G, i.e.

$$M_B = \left\{ \begin{pmatrix} \begin{pmatrix} a & x \\ 0 & b \end{pmatrix} & 0 \\ 0 & \lambda \cdot {}^t \begin{pmatrix} a & x \\ 0 & b \end{pmatrix}^{-1} \end{pmatrix} \in M \right\}.$$

Then since $w_2^{-1} M_B w_2 \subset P$, we may assume m_1 to be of the form
$$\begin{pmatrix} h & 0 \\ 0 & {}^t h^{-1} \end{pmatrix}$$
where h is taken from the representatives of $A \backslash \mathrm{GL}_2(F) / B_0$. Here B_0 denotes the upper Borel subgroup of $\mathrm{GL}_2(F)$. By the Bruhat decomposition, we have
$$\mathrm{GL}_2(F) = B_0 \cup N_0 \begin{pmatrix} 0 & 1 \\ -1 & 0 \end{pmatrix} B_0$$
where N_0 denotes the unipotent radical of B_0. Then for $x \in F^\times$, we observe that
$$\begin{pmatrix} 1 & x \\ 0 & 1 \end{pmatrix} \begin{pmatrix} 0 & 1 \\ -1 & 0 \end{pmatrix} = \begin{pmatrix} x & 0 \\ 0 & 1 \end{pmatrix} \begin{pmatrix} 1 & 1 \\ 0 & 1 \end{pmatrix} \begin{pmatrix} 0 & 1 \\ -1 & 0 \end{pmatrix} \begin{pmatrix} 1 & 0 \\ 0 & x^{-1} \end{pmatrix}.$$
Hence we have
$$\mathrm{GL}_2(F) = A B_0 \cup A \begin{pmatrix} 0 & 1 \\ -1 & 0 \end{pmatrix} B_0 \cup A \begin{pmatrix} 1 & 1 \\ 0 & 1 \end{pmatrix} \begin{pmatrix} 0 & 1 \\ -1 & 0 \end{pmatrix} B_0$$

and this proves the third assertion of the proposition. □

4.1.2. $A\backslash\mathrm{GL}_2(F)/A$ double coset decomposition. For the sake of the reader, here we recollect some facts on the $A\backslash\mathrm{GL}_2(F)/A$ double coset decomposition following Jacquet [**J1**], mutatis mutandis.

Let us define a mapping $X : \mathrm{GL}_2(F) \to F \cup \{\infty\}$ by

$$(4.10) \qquad X\begin{pmatrix} a & b \\ c & d \end{pmatrix} = \frac{bc}{ad}.$$

Then for $a_1, a_2, b_1, b_2 \in F^\times$, we have

$$X\left[\begin{pmatrix} a_1 & 0 \\ 0 & a_2 \end{pmatrix}\begin{pmatrix} a & b \\ c & d \end{pmatrix}\begin{pmatrix} b_1 & 0 \\ 0 & b_2 \end{pmatrix}\right] = X\begin{pmatrix} a_1 b_1 a & a_1 b_2 b \\ a_2 b_1 c & a_2 b_2 d \end{pmatrix} = X\begin{pmatrix} a & b \\ c & d \end{pmatrix}.$$

Hence we may regard X as a mapping from $A\backslash\mathrm{GL}_2(F)/A$ to $F \cup \{\infty\}$.

DEFINITION 4.2. We say that $g \in \mathrm{GL}_2(F)$ and also its double coset AgA is A-*regular* when $X(g) \neq 0, \infty$.

We say that $g \in \mathrm{GL}_2(F)$ and also its double coset AgA is A-*singular* when $X(g) = 0$ or ∞.

PROPOSITION 4.3. 1. *There exist six A-singular double cosets in $\mathrm{GL}_2(F)$ and they are represented by*

$$1_2, \quad n_+, \quad n_-, \quad w, \quad wn_+, \quad wn_-,$$

where $w = \begin{pmatrix} 0 & 1 \\ 1 & 0 \end{pmatrix}$, $n_+ = \begin{pmatrix} 1 & 1 \\ 0 & 1 \end{pmatrix}$ *and* $n_- = \begin{pmatrix} 1 & 0 \\ 1 & 1 \end{pmatrix}$.

2. *The mapping $X : A\backslash\mathrm{GL}_2(F)/A \to F \cup \{\infty\}$ induces a bijection between the A-regular double cosets in $\mathrm{GL}_2(F)$ and $F \setminus \{0, 1\}$. In particular the A-regular double cosets in $\mathrm{GL}_2(F)$ are represented by matrices*

$$\begin{pmatrix} 1 & x \\ 1 & 1 \end{pmatrix}$$

where $x \in F$ and $x \neq 0, 1$.

PROOF. 1. Suppose that $g = \begin{pmatrix} a & b \\ c & d \end{pmatrix} \in \mathrm{GL}_2(F)$ is A-singular.

Suppose that $X(g) = 0$, then we have $bc = 0$.
If $b = c = 0$, then $g \in A$.
If $b = 0$ but $c \neq 0$, then

$$g = \begin{pmatrix} a & 0 \\ c & d \end{pmatrix} = \begin{pmatrix} a & 0 \\ 0 & d \end{pmatrix}\begin{pmatrix} d & 0 \\ 0 & c \end{pmatrix}\begin{pmatrix} 1 & 0 \\ 1 & 1 \end{pmatrix}\begin{pmatrix} d & 0 \\ 0 & c \end{pmatrix}^{-1}.$$

If $c = 0$ but $b \neq 0$, then

$$g = \begin{pmatrix} a & b \\ 0 & d \end{pmatrix} = \begin{pmatrix} a & 0 \\ 0 & d \end{pmatrix}\begin{pmatrix} b & 0 \\ 0 & a \end{pmatrix}\begin{pmatrix} 1 & 1 \\ 0 & 1 \end{pmatrix}\begin{pmatrix} b & 0 \\ 0 & a \end{pmatrix}^{-1}.$$

Suppose that $X(g) = \infty$. Then we have $ad = 0$.
If $a = d = 0$, then

$$g = \begin{pmatrix} 0 & b \\ c & 0 \end{pmatrix} = \begin{pmatrix} 0 & 1 \\ 1 & 0 \end{pmatrix}\begin{pmatrix} c & 0 \\ 0 & b \end{pmatrix}.$$

If $a = 0$ but $d \neq 0$, then

$$g = \begin{pmatrix} 0 & b \\ c & d \end{pmatrix} = \begin{pmatrix} bc & 0 \\ 0 & cd \end{pmatrix}\begin{pmatrix} 0 & 1 \\ 1 & 0 \end{pmatrix}\begin{pmatrix} 1 & 1 \\ 0 & 1 \end{pmatrix}\begin{pmatrix} d & 0 \\ 0 & c \end{pmatrix}^{-1}.$$

If $d = 0$ but $a \neq 0$, then
$$g = \begin{pmatrix} a & b \\ c & 0 \end{pmatrix} = \begin{pmatrix} ab & 0 \\ 0 & bc \end{pmatrix} \begin{pmatrix} 0 & 1 \\ 1 & 0 \end{pmatrix} \begin{pmatrix} 1 & 0 \\ 1 & 1 \end{pmatrix} \begin{pmatrix} b & 0 \\ 0 & a \end{pmatrix}^{-1}.$$

2. Suppose that $g = \begin{pmatrix} a & b \\ c & d \end{pmatrix} \in \mathrm{GL}_2(F)$ is A-regular. Then by definition we have $X(g) \in F^\times$. Since $\det(g) = ad - bc \neq 0$, we also have $X(g) \neq 1$. Conversely let $x \in F \setminus \{0, 1\}$. Then we have $h_x = \begin{pmatrix} 1 & x \\ 1 & 1 \end{pmatrix} \in \mathrm{GL}_2(F)$ and $X(h_x) = x$. Finally suppose that $X\begin{pmatrix} a & b \\ c & d \end{pmatrix} = x$ for $x \in F \setminus \{0, 1\}$. Then we have
$$\begin{pmatrix} a & b \\ c & d \end{pmatrix} = \begin{pmatrix} a & 0 \\ 0 & c \end{pmatrix} \begin{pmatrix} 1 & x \\ 1 & 1 \end{pmatrix} \begin{pmatrix} 1 & 0 \\ 0 & c^{-1}d \end{pmatrix}.$$

Thus we have proved the assertion. □

We note the following lemma concerning the stabilizer of the A-regular double coset.

LEMMA 4.4. *Suppose that* $g = \begin{pmatrix} a & b \\ c & d \end{pmatrix} \in \mathrm{GL}_2(F)$ *is A-regular. Then for* $a_1, a_2, b_1, b_2 \in F^\times$,
$$\begin{pmatrix} a_1 & 0 \\ 0 & a_2 \end{pmatrix} \begin{pmatrix} a & b \\ c & d \end{pmatrix} \begin{pmatrix} b_1 & 0 \\ 0 & b_2 \end{pmatrix} = \begin{pmatrix} a & b \\ c & d \end{pmatrix} \begin{pmatrix} \lambda & 0 \\ 0 & \lambda \end{pmatrix}$$
for some $\lambda \in F^\times$ implies that $a_1 = b_1$ and $a_2 = b_2$.

PROOF. We have
$$\begin{pmatrix} a_1 & 0 \\ 0 & a_2 \end{pmatrix} \begin{pmatrix} a & b \\ c & d \end{pmatrix} \begin{pmatrix} b_1 & 0 \\ 0 & b_2 \end{pmatrix} = \begin{pmatrix} a_1 b_1 a & a_1 b_2 b \\ a_2 b_1 c & a_2 b_2 d \end{pmatrix}.$$
Since g is A-regular, we have $abcd \neq 0$. Hence $a_1 b_1 = a_1 b_2 = a_2 b_1 = a_2 b_2 = \lambda$ and thus $a_1 = a_2$ and $b_1 = b_2$. □

4.2. Relevant double cosets

We recall that $\psi : F \to \mathbb{C}^\times$ denotes an additive character of exponent zero. Then we denote by θ a character of H defined by

(4.11) $\quad \theta\left[\begin{pmatrix} a & 0 & 0 & 0 \\ 0 & b & 0 & 0 \\ 0 & 0 & b & 0 \\ 0 & 0 & 0 & a \end{pmatrix} \begin{pmatrix} 1_2 & X \\ 0 & 1_2 \end{pmatrix}\right] = \chi_E(ab) \cdot \psi\left[\mathrm{tr}\left(\begin{pmatrix} 0 & 1 \\ 1 & 0 \end{pmatrix} X\right)\right]$

where χ_E denotes the quadratic character of $F^\times$ attached to the unique unramified quadratic extension E of F. By abuse of notation we denote by ψ a character of $\bar{H}$ defined by

(4.12) $\quad \psi\left[\begin{pmatrix} a & 0 & 0 & 0 \\ 0 & b & 0 & 0 \\ 0 & 0 & b & 0 \\ 0 & 0 & 0 & a \end{pmatrix} \begin{pmatrix} 1_2 & 0 \\ Y & 1_2 \end{pmatrix}\right] = \psi\left[\mathrm{tr}\left(\begin{pmatrix} 0 & 1 \\ 1 & 0 \end{pmatrix} Y\right)\right].$

DEFINITION 4.5. We say that a double coset $\bar{H}rH$ in G is *relevant* if the mapping
$$\bar{H}rH \ni \bar{h}rh \mapsto \psi(\bar{h})\theta(h) \in \mathbb{C}^\times$$
is well defined, i.e.

(4.13) $$\psi(\bar{h}) = \theta(r^{-1}\bar{h}r), \quad \forall\, \bar{h} \in \bar{H} \cap rHr^{-1}.$$

It is precisely these double cosets which enter into the relative trace formula.

PROPOSITION 4.6. *The relevant $(\bar{H}, H)$-double cosets in G are represented by the following elements.*

1. (h, λ) where $\lambda \in F^\times$, $h = \begin{pmatrix} 1 & x \\ 1 & 1 \end{pmatrix}$ for $x \in F \setminus \{0, 1\}$.

2. (h, λ) where $\lambda \in F^\times$ and
$$h = \begin{pmatrix} 1 & 1 \\ 0 & 1 \end{pmatrix},\ \begin{pmatrix} 1 & 0 \\ 1 & 1 \end{pmatrix},\ \begin{pmatrix} 0 & 1 \\ 1 & 0 \end{pmatrix}\begin{pmatrix} 1 & 1 \\ 0 & 1 \end{pmatrix},\ \begin{pmatrix} 0 & 1 \\ 1 & 0 \end{pmatrix}\begin{pmatrix} 1 & 0 \\ 1 & 1 \end{pmatrix}.$$

3. $n_+ww_2 \begin{pmatrix} a & 0 & 0 & 0 \\ 0 & 1 & 0 & 0 \\ 0 & 0 & a^{-1} & 0 \\ 0 & 0 & 0 & 1 \end{pmatrix} n_+w$ where $a \in F^\times$.

4. w_2, w_2w, ww_2, ww_2w.

Here we note that
$$(h, \lambda) = \begin{pmatrix} h & 0 \\ 0 & \lambda \cdot {}^t h^{-1} \end{pmatrix}$$
for $\lambda \in \mathrm{GL}_1(F)$ and $h \in \mathrm{GL}_2(F)$, and,
$$n_+ = \begin{pmatrix} 1 & 1 & 0 & 0 \\ 0 & 1 & 0 & 0 \\ 0 & 0 & 1 & 0 \\ 0 & 0 & -1 & 1 \end{pmatrix},\ w = \begin{pmatrix} 0 & 1 & 0 & 0 \\ -1 & 0 & 0 & 0 \\ 0 & 0 & 0 & 1 \\ 0 & 0 & -1 & 0 \end{pmatrix},\ w_2 = \begin{pmatrix} 1 & 0 & 0 & 0 \\ 0 & 0 & 0 & 1 \\ 0 & 0 & 1 & 0 \\ 0 & -1 & 0 & 0 \end{pmatrix}.$$

PROOF. First let us consider the double cosets in $\bar{P}P$. Let $r = (h, \lambda)$. Then we note that
$$r^{-1} \begin{pmatrix} a & 0 & 0 & 0 \\ 0 & b & 0 & 0 \\ 0 & 0 & b & 0 \\ 0 & 0 & 0 & a \end{pmatrix} \begin{pmatrix} 1_2 & 0 \\ Y & 1_2 \end{pmatrix} r$$
$$= \begin{pmatrix} h^{-1} \begin{pmatrix} a & 0 \\ 0 & b \end{pmatrix} h & 0 \\ 0 & {}^th \begin{pmatrix} b & 0 \\ 0 & a \end{pmatrix} {}^th^{-1} \end{pmatrix} \begin{pmatrix} 1_2 & 0 \\ \lambda^{-1} \cdot {}^thYh & 1_2 \end{pmatrix}$$

and hence $\bar{H} \cap rHr^{-1} \subset A$.

Suppose that h is A-regular. Then by Lemma 4.4 we have $\bar{H} \cap rHr^{-1} = Z$ where Z denotes the center of G. Hence the double coset $\bar{H}rH$ is relevant.

4.2. RELEVANT DOUBLE COSETS

Suppose that h is A-singular. Then h is one of the six elements given in Proposition 4.3. Then since

$$\begin{pmatrix} 1 & 1 \\ 0 & 1 \end{pmatrix}^{-1} \begin{pmatrix} a & 0 \\ 0 & b \end{pmatrix} \begin{pmatrix} 1 & 1 \\ 0 & 1 \end{pmatrix} = \begin{pmatrix} a & a-b \\ 0 & b \end{pmatrix},$$

$$\begin{pmatrix} 1 & 0 \\ 1 & 1 \end{pmatrix}^{-1} \begin{pmatrix} a & 0 \\ 0 & b \end{pmatrix} \begin{pmatrix} 1 & 0 \\ 1 & 1 \end{pmatrix} = \begin{pmatrix} a & 0 \\ b-a & b \end{pmatrix},$$

we have

$$\bar{H} \cap rHr^{-1} = \begin{cases} A, & \text{when } h = 1_2, \begin{pmatrix} 0 & 1 \\ 1 & 0 \end{pmatrix} \\ Z, & \text{otherwise.} \end{cases}$$

Hence we obtain the relevant double cosets in the second case.

Now let us consider the double cosets in $\bar{P}w_1 P$ where

$$w_1 = \begin{pmatrix} 0 & 1_2 \\ -1_2 & 0 \end{pmatrix}.$$

Let $r = (h, \lambda)$. Then we note that

$$r^{-1} w_1^{-1} \begin{pmatrix} 1_2 & 0 \\ Y & 1_2 \end{pmatrix} w_1 r = \begin{pmatrix} h & 0 \\ 0 & \lambda \cdot {}^t h^{-1} \end{pmatrix}^{-1} \begin{pmatrix} 1_2 & -Y \\ 0 & 1_2 \end{pmatrix} \begin{pmatrix} h & 0 \\ 0 & \lambda \cdot {}^t h^{-1} \end{pmatrix} \in U.$$

Hence if $\bar{H}rH$ is relevant, then we have

$$\psi \left[\operatorname{tr} \left(\begin{pmatrix} 0 & 1 \\ 1 & 0 \end{pmatrix} Y \right) \right] = \psi \left[-\operatorname{tr} \left(\begin{pmatrix} 0 & 1 \\ 1 & 0 \end{pmatrix} \lambda \cdot h^{-1} Y {}^t h^{-1} \right) \right] \text{ for any } Y \in \operatorname{Sym}^2(F),$$

which is equivalent to

(4.14) $$\quad {}^t h \begin{pmatrix} 0 & 1 \\ 1 & 0 \end{pmatrix} h = -\lambda \begin{pmatrix} 0 & 1 \\ 1 & 0 \end{pmatrix}.$$

By taking the determinants of the both sides of (4.14), we have $\det h = \pm \lambda$. When $\det h = -\lambda$, we have

$$ {}^t h \begin{pmatrix} 0 & 1 \\ 1 & 0 \end{pmatrix} h = \det h \cdot \begin{pmatrix} 0 & 1 \\ 1 & 0 \end{pmatrix}.$$

Hence $h \in A$ by (4.8). Similarly when $\det h = \lambda$, we have

$${}^t \left(h \begin{pmatrix} 0 & 1 \\ -1 & 0 \end{pmatrix} \right) \begin{pmatrix} 0 & 1 \\ 1 & 0 \end{pmatrix} \left(h \begin{pmatrix} 0 & 1 \\ -1 & 0 \end{pmatrix} \right) = \det \left(h \begin{pmatrix} 0 & 1 \\ -1 & 0 \end{pmatrix} \right) \cdot \begin{pmatrix} 0 & 1 \\ 1 & 0 \end{pmatrix}$$

and hence $h \begin{pmatrix} 0 & 1 \\ -1 & 0 \end{pmatrix} \in A$.

Thus the double cosets in $\bar{P}w_1 P$ which are possibly relevant are

$$\bar{H} w_1 \begin{pmatrix} 1_2 & 0 \\ 0 & -1_2 \end{pmatrix} H, \quad \bar{H} w_1 \begin{pmatrix} 0 & 1 & 0 & 0 \\ -1 & 0 & 0 & 0 \\ 0 & 0 & 0 & 1 \\ 0 & 0 & -1 & 0 \end{pmatrix} H.$$

Here we note that

$$w_1 \begin{pmatrix} 1_2 & 0 \\ 0 & -1_2 \end{pmatrix} \begin{pmatrix} a & 0 & 0 & 0 \\ 0 & b & 0 & 0 \\ 0 & 0 & b & 0 \\ 0 & 0 & 0 & a \end{pmatrix} \begin{pmatrix} 1_2 & 0 \\ 0 & -1_2 \end{pmatrix}^{-1} w_1^{-1} = \begin{pmatrix} b & 0 & 0 & 0 \\ 0 & a & 0 & 0 \\ 0 & 0 & a & 0 \\ 0 & 0 & 0 & b \end{pmatrix}.$$

Since χ_E is non-trivial, the double coset $\bar{H} w_1 \begin{pmatrix} 1_2 & 0 \\ 0 & -1_2 \end{pmatrix} H$ is not relevant. Similarly the other double coset is not relevant either. Hence there is no relevant double coset in $\bar{P} w_1 P$.

Finally let us consider the double cosets in $\bar{P} w_2 P$. Suppose that $\bar{H} r H$ is a relevant double coset in $\bar{P} w_2 P$. By Proposition 4.1, we may assume that r is of the form $r = m_1 w_2 m_2$ where $m_2 \in M$ and $m_1 = 1_4, w, n_+ w$.

Suppose that $m_1 = 1_4$. Let us write $m_2 = \begin{pmatrix} h & 0 \\ 0 & \lambda \cdot {}^t h^{-1} \end{pmatrix}$. Then we have

$$r^{-1} \begin{pmatrix} 1 & 0 & 0 & 0 \\ 0 & 1 & 0 & 0 \\ 0 & 0 & 1 & 0 \\ 0 & x & 0 & 1 \end{pmatrix} r = \begin{pmatrix} 1_2 & \lambda \cdot h^{-1} \begin{pmatrix} 0 & 0 \\ 0 & -s \end{pmatrix} {}^t h^{-1} \\ 0 & 1_2 \end{pmatrix}$$

for any $s \in F$. Since $\bar{H} r H$ is relevant, $h^{-1} \begin{pmatrix} 0 & 0 \\ 0 & 1 \end{pmatrix} {}^t h^{-1}$ must be diagonal. Let us write $h = \begin{pmatrix} \alpha & \beta \\ \gamma & \delta \end{pmatrix}$. Then we have

$$h^{-1} \begin{pmatrix} 0 & 0 \\ 0 & 1 \end{pmatrix} {}^t h^{-1} = (\det h)^{-2} \begin{pmatrix} \beta^2 & -\alpha\beta \\ -\alpha\beta & \alpha^2 \end{pmatrix}.$$

Hence $\alpha = 0$ or $\beta = 0$ and then

$$h = \begin{cases} \begin{pmatrix} 1 & 0 \\ \alpha^{-1}\gamma & 1 \end{pmatrix} \begin{pmatrix} \alpha & 0 \\ 0 & \delta \end{pmatrix}, & \text{when } \beta = 0 \\ \begin{pmatrix} 0 & 1 \\ -1 & 0 \end{pmatrix} \begin{pmatrix} 1 & -\beta^{-1}\delta \\ 0 & 1 \end{pmatrix} \begin{pmatrix} -\gamma & 0 \\ 0 & \beta \end{pmatrix}, & \text{when } \alpha = 0. \end{cases}$$

Thus we may assume that m_2 is of the form

(4.15) $$m_2 = \begin{pmatrix} \begin{pmatrix} 1 & 0 \\ x & 1 \end{pmatrix} & 0 \\ 0 & \lambda \cdot \begin{pmatrix} 1 & -x \\ 0 & 1 \end{pmatrix} \end{pmatrix}$$

for some $x \in F$, or,

(4.16) $$m_2 = w \begin{pmatrix} \begin{pmatrix} 1 & y \\ 0 & 1 \end{pmatrix} & 0 \\ 0 & \lambda \cdot \begin{pmatrix} 1 & 0 \\ -y & 1 \end{pmatrix} \end{pmatrix}$$

for some $y \in F$.

When (4.15) holds, we have $\bar{H}w_2m_2H = \bar{H}w_2H$ since

$$w_2m_2 = \begin{pmatrix} 1 & 0 & 0 & 0 \\ 0 & \lambda & 0 & 0 \\ 0 & 0 & \lambda & 0 \\ 0 & 0 & 0 & 1 \end{pmatrix} \begin{pmatrix} 1 & 0 & 0 & 0 \\ 0 & 1 & 0 & 0 \\ 0 & -x & 1 & 0 \\ -x & 0 & 0 & 1 \end{pmatrix} w_2 \in \bar{H}w_2.$$

Similarly when (4.16) holds, we have $\bar{H}w_2wm_2H = \bar{H}w_2wH$ since

$$w_2wm_2 = \begin{pmatrix} 1 & 0 & 0 & 0 \\ 0 & \lambda & 0 & 0 \\ 0 & 0 & \lambda & 0 \\ 0 & 0 & 0 & 1 \end{pmatrix} \begin{pmatrix} 1 & 0 & 0 & 0 \\ 0 & 1 & 0 & 0 \\ 0 & y & 1 & 0 \\ y & 0 & 0 & 1 \end{pmatrix} w_2w \in \bar{H}w_2w.$$

Then since we have

$$w_2^{-1} \begin{pmatrix} a & 0 & 0 & 0 \\ 0 & b & 0 & 0 \\ 0 & 0 & b & 0 \\ 0 & 0 & 0 & a \end{pmatrix} \begin{pmatrix} 1 & 0 & 0 & 0 \\ 0 & 1 & 0 & 0 \\ x & y & 1 & 0 \\ y & z & 0 & 1 \end{pmatrix} w_2 = \begin{pmatrix} a & 0 & 0 & 0 \\ 0 & a & 0 & 0 \\ 0 & 0 & b & 0 \\ 0 & 0 & 0 & b \end{pmatrix} \begin{pmatrix} 1 & 0 & 0 & 0 \\ -y & 1 & 0 & -z \\ x & 0 & 1 & y \\ 0 & 0 & 0 & 1 \end{pmatrix}$$

and

$$(w_2w)^{-1} \begin{pmatrix} a & 0 & 0 & 0 \\ 0 & b & 0 & 0 \\ 0 & 0 & b & 0 \\ 0 & 0 & 0 & a \end{pmatrix} \begin{pmatrix} 1 & 0 & 0 & 0 \\ 0 & 1 & 0 & 0 \\ x & y & 1 & 0 \\ y & z & 0 & 1 \end{pmatrix} w_2w$$

$$= \begin{pmatrix} a & 0 & 0 & 0 \\ 0 & a & 0 & 0 \\ 0 & 0 & b & 0 \\ 0 & 0 & 0 & b \end{pmatrix} \begin{pmatrix} 1 & y & -z & 0 \\ 0 & 1 & 0 & 0 \\ 0 & 0 & 1 & 0 \\ 0 & x & -y & 1 \end{pmatrix},$$

it is clear that both $\bar{H}w_2H$ and $\bar{H}w_2wH$ are relevant.

By a similar argument, there exist precisely two relevant double cosets $\bar{H}ww_2H$ and $\bar{H}ww_2wH$, among the double cosets of the form $\bar{H}ww_2m_2H$ where $m_2 \in M$.

Finally let us consider the double cosets of the form $\bar{H}rH$ where $r = n_+ww_2m_2$, $m_2 \in M$. We recall the lower Bruhat decomposition for $\mathrm{GL}_2(F)$,

$$\mathrm{GL}_2(F) = \bar{B}_0 \cup \bar{N}_0 \begin{pmatrix} 0 & 1 \\ -1 & 0 \end{pmatrix} \bar{B}_0$$

where $\bar{B}_0$ denotes the lower Borel subgroup of $\mathrm{GL}_2(F)$ and $\bar{N}_0$ denotes the unipotent radical of $\bar{B}_0$. We also note that

$$n_+ww_2 \begin{pmatrix} 1 & 0 & 0 & 0 \\ x & 1 & 0 & 0 \\ 0 & 0 & 1 & -x \\ 0 & 0 & 0 & 1 \end{pmatrix} = n_+ \begin{pmatrix} 1 & 0 & 0 & 0 \\ 0 & 1 & 0 & 0 \\ 0 & x & 1 & 0 \\ x & 0 & 0 & 1 \end{pmatrix} n_+^{-1} n_+ww_2 \in \bar{H}n_+ww_2.$$

Hence we may assume that m_2 is of the form

(4.17) $$m_2 = \begin{pmatrix} 1_2 & 0 \\ 0 & \lambda \cdot 1_2 \end{pmatrix}$$

or

(4.18) $$m_2 = w \begin{pmatrix} \begin{pmatrix} 1 & 0 \\ x & 1 \end{pmatrix} & 0 \\ 0 & \lambda \cdot \begin{pmatrix} 1 & -x \\ 0 & 1 \end{pmatrix} \end{pmatrix}$$

for some $x \in F$.

Suppose that m_2 is of the form (4.17). Then we have

$$r^{-1} \begin{pmatrix} 1 & 0 & 0 & 0 \\ 0 & 1 & 0 & 0 \\ y & -y & 1 & 0 \\ -y & y & 0 & 1 \end{pmatrix} r = \begin{pmatrix} 1 & 0 & 0 & 0 \\ 0 & 1 & 0 & -\lambda y \\ 0 & 0 & 1 & 0 \\ 0 & 0 & 0 & 1 \end{pmatrix}.$$

Hence $\bar{H}rH$ is not a relevant double coset.

Suppose that m_2 is of the form (4.18). Then we have

$$r^{-1} \begin{pmatrix} 1 & 0 & 0 & 0 \\ 0 & 1 & 0 & 0 \\ y & -y & 1 & 0 \\ -y & y & 0 & 1 \end{pmatrix} r = \begin{pmatrix} 1 & 0 & -\lambda y & \lambda xy \\ 0 & 1 & \lambda xy & -\lambda x^2 y \\ 0 & 0 & 1 & 0 \\ 0 & 0 & 0 & 1 \end{pmatrix}.$$

Hence when $\bar{H}rH$ is relevant, we have $x = -\lambda^{-1}$.

Conversely when

$$r = n_+ w w_2 w \begin{pmatrix} \begin{pmatrix} 1 & 0 \\ -\lambda^{-1} & 1 \end{pmatrix} & 0 \\ 0 & \lambda \cdot \begin{pmatrix} 1 & \lambda^{-1} \\ 0 & 1 \end{pmatrix} \end{pmatrix},$$

we have

$$r^{-1} \begin{pmatrix} a & 0 & 0 & 0 \\ 0 & b & 0 & 0 \\ 0 & 0 & b & 0 \\ 0 & 0 & 0 & a \end{pmatrix} \begin{pmatrix} 1 & 0 & 0 & 0 \\ 0 & 1 & 0 & 0 \\ u & v & 1 & 0 \\ v & w & 0 & 1 \end{pmatrix} r = \begin{pmatrix} b & 0 & 0 & 0 \\ 0 & b & 0 & 0 \\ \frac{2(b-a)}{\lambda^2} & \frac{a-b}{\lambda} & a & 0 \\ \frac{a-b}{\lambda} & 0 & 0 & a \end{pmatrix}$$

$$\begin{pmatrix} 1 & 0 & 0 & 0 \\ 0 & 1 & 0 & 0 \\ \frac{u+2v+w}{\lambda^3} & -\frac{u+2v+w}{\lambda^2} & 1 & 0 \\ -\frac{u+2v+w}{\lambda^2} & \frac{u+2v+w}{\lambda} & 0 & 1 \end{pmatrix} \begin{pmatrix} 1 & 0 & -\lambda u & -u \\ 0 & 1 & -u & -\lambda^{-1}u \\ 0 & 0 & 1 & 0 \\ 0 & 0 & 0 & 1 \end{pmatrix}$$

$$\begin{pmatrix} 1+\frac{u+v}{\lambda} & -u-v & 0 & 0 \\ \frac{u+v}{\lambda^2} & 1-\frac{u+v}{\lambda} & 0 & 0 \\ 0 & 0 & 1-\frac{u+v}{\lambda} & -\frac{u+v}{\lambda^2} \\ 0 & 0 & u+v & 1+\frac{u+v}{\lambda} \end{pmatrix}.$$

It is easily seen from this that we have $r^{-1}\bar{H}r \cap H = Z$ and $\bar{H}rH$ is indeed a relevant double coset.

Finally we note that

$$w \begin{pmatrix} \begin{pmatrix} 1 & 0 \\ -\lambda^{-1} & 1 \end{pmatrix} & 0 \\ 0 & \lambda \cdot \begin{pmatrix} 1 & \lambda^{-1} \\ 0 & 1 \end{pmatrix} \end{pmatrix} = \begin{pmatrix} \lambda^{-1} & 0 & 0 & 0 \\ 0 & 1 & 0 & 0 \\ 0 & 0 & \lambda & 0 \\ 0 & 0 & 0 & 1 \end{pmatrix} n_+ w \begin{pmatrix} 1 & 0 & 0 & 0 \\ 0 & \lambda & 0 & 0 \\ 0 & 0 & \lambda & 0 \\ 0 & 0 & 0 & 1 \end{pmatrix}$$

and hence we have
$$\bar{H} r H = \bar{H} n_+ w w_2 \begin{pmatrix} \lambda^{-1} & 0 & 0 & 0 \\ 0 & 1 & 0 & 0 \\ 0 & 0 & \lambda & 0 \\ 0 & 0 & 0 & 1 \end{pmatrix} n_+ w H.$$

□

4.3. Evaluation of the Novodvorsky orbital integral

DEFINITION 4.7. Let Ξ be the characteristic function of $\mathcal{K}$, the maximal compact subgroup of G defined by $\mathcal{K} = \mathrm{GSp}_4(\mathcal{O}_F)$.

Then for a relevant double coset $\bar{H} r H$ in G, we define the *Novodvorsky orbital integral* $\mathcal{N}(r)$ by

$$(4.19) \qquad \mathcal{N}(r) = \int_{\bar{H}/\bar{H} \cap r H r^{-1}} \int_H \Xi(\bar{h} r h)\, \psi(\bar{h})\, \theta(h)\, dh\, d\bar{h}.$$

In particular when
$$r = (h, \lambda) = \begin{pmatrix} h & 0 \\ 0 & \lambda \cdot {}^t h^{-1} \end{pmatrix} \quad \text{where} \quad \lambda \in F^\times,\, h \in \mathrm{GL}_2(F),$$
we write $\mathcal{N}(h, \lambda)$ for $\mathcal{N}(r)$.

For $x \in F \setminus \{0, 1\}$, let us define a matrix h_x by
$$h_x = \begin{pmatrix} 1 & x \\ 1 & 1 \end{pmatrix}.$$
Then we simply write $\mathcal{N}(x, \lambda)$ for $\mathcal{N}(h_x, \lambda)$.

Now the goal of this section is to evaluate $\mathcal{N}(x, \lambda)$ explicitly. In the proof of Proposition 4.6, it has been shown that $\bar{H} \cap r H r^{-1} = Z$ for $r = (h, \lambda)$ where h is A-regular. Hence when h is A-regular, we have

$$(4.20) \quad \mathcal{N}(h, \lambda) = \int_{\bar{U}} \int_{F^\times} \int_{F^\times} \int_{F^\times} \int_U$$
$$\Xi \left[\begin{pmatrix} 1_2 & 0 \\ Y & 1_2 \end{pmatrix} \begin{pmatrix} a & 0 & 0 & 0 \\ 0 & 1 & 0 & 0 \\ 0 & 0 & 1 & 0 \\ 0 & 0 & 0 & a \end{pmatrix} \begin{pmatrix} h & 0 \\ 0 & \lambda \cdot {}^t h^{-1} \end{pmatrix} \begin{pmatrix} b & 0 & 0 & 0 \\ 0 & c & 0 & 0 \\ 0 & 0 & c & 0 \\ 0 & 0 & 0 & b \end{pmatrix} \begin{pmatrix} 1_2 & X \\ 0 & 1_2 \end{pmatrix} \right]$$
$$\chi_E(bc)\, \psi \left[\mathrm{tr} \left\{ \begin{pmatrix} 0 & 1 \\ 1 & 0 \end{pmatrix} (X + Y) \right\} \right] d^\times a\, d^\times b\, d^\times c\, dX\, dY.$$

Here $d^\times a$ denotes the Haar measure on $F^\times$ such that $\mathcal{O}_F^\times$ has measure 1.

First we observe some elementary properties of $\mathcal{N}(x, \lambda)$.

PROPOSITION 4.8. 1. Let $w_0 = \begin{pmatrix} 0 & 1 \\ 1 & 0 \end{pmatrix}$. Then we have
$$\mathcal{N}(x, \lambda) = \mathcal{N}(w_0 h_x, \lambda) = \mathcal{N}(w_0 h_x w_0, \lambda).$$

2. When $\mathrm{ord}(x)$ is odd, we have
$$\mathcal{N}(x, \lambda) = 0.$$

3. When $\mathrm{ord}(x)$ is even, we have the functional equation
$$(4.21) \qquad \mathcal{N}(x, \lambda) = \mathcal{N}(x^{-1}, \lambda x^{-1}).$$

PROOF. 1. Since

$$\begin{pmatrix} 1_2 & 0 \\ Y & 1_2 \end{pmatrix} \begin{pmatrix} a & 0 & 0 & 0 \\ 0 & 1 & 0 & 0 \\ 0 & 0 & 1 & 0 \\ 0 & 0 & 0 & a \end{pmatrix} \begin{pmatrix} w_0 & 0 \\ 0 & w_0 \end{pmatrix}$$
$$= \begin{pmatrix} w_0 & 0 \\ 0 & w_0 \end{pmatrix} \begin{pmatrix} 1_2 & 0 \\ w_0 Y w_0 & 1_2 \end{pmatrix} \begin{pmatrix} a & 0 & 0 & 0 \\ 0 & 1 & 0 & 0 \\ 0 & 0 & 1 & 0 \\ 0 & 0 & 0 & a \end{pmatrix}^{-1} \begin{pmatrix} a & 0 & 0 & 0 \\ 0 & a & 0 & 0 \\ 0 & 0 & a & 0 \\ 0 & 0 & 0 & a \end{pmatrix}$$

and the function Ξ is bi $\mathcal{K}$-invariant, we have

$$\mathcal{N}(w_0 h_x, \lambda) = \int_{\bar{U}} \int_{F^\times} \int_{F^\times} \int_{F^\times} \int_{U}$$
$$\Xi\left[\begin{pmatrix} 1_2 & 0 \\ w_0 Y w_o & 1_2 \end{pmatrix} \begin{pmatrix} a & 0 & 0 & 0 \\ 0 & 1 & 0 & 0 \\ 0 & 0 & 1 & 0 \\ 0 & 0 & 0 & a \end{pmatrix}^{-1} \begin{pmatrix} h_x & 0 \\ 0 & \lambda \cdot {}^t h_x^{-1} \end{pmatrix} \begin{pmatrix} ab & 0 & 0 & 0 \\ 0 & ac & 0 & 0 \\ 0 & 0 & ac & 0 \\ 0 & 0 & 0 & ab \end{pmatrix} \begin{pmatrix} 1_2 & X \\ 0 & 1_2 \end{pmatrix} \right]$$
$$\chi_E(bc)\, \psi\left[\mathrm{tr}\{w_0(X+Y)\}\right] d^\times a\, d^\times b\, d^\times c\, dX\, dY.$$

It is clear from this that we have

$$\mathcal{N}(w_0 h_x, \lambda) = \mathcal{N}(h_x, \lambda).$$

The other equality is proved similarly.
2. By the previous assertion, we have

$$\mathcal{N}(x, \lambda) = \mathcal{N}(w_0 h_x w_0, \lambda) = \mathcal{N}\left(\begin{pmatrix} 1 & 1 \\ x & 1 \end{pmatrix}, \lambda \right).$$

Here we note that

$$\begin{pmatrix} 1 & 1 \\ x & 1 \end{pmatrix} = \begin{pmatrix} x^{-1} & 0 \\ 0 & 1 \end{pmatrix} \begin{pmatrix} 1 & x \\ 1 & 1 \end{pmatrix} \begin{pmatrix} x & 0 \\ 0 & 1 \end{pmatrix}.$$

Hence from (4.20), we have

$$\mathcal{N}\left(\begin{pmatrix} 1 & 1 \\ x & 1 \end{pmatrix}, \lambda \right) = \chi_E(x)^{-1} \mathcal{N}(x, \lambda).$$

Therefore

$$\mathcal{N}(x, \lambda) = \chi_E(x)^{-1} \mathcal{N}(x, \lambda).$$

Hence $\mathcal{N}(x, \lambda) = 0$ when $\mathrm{ord}(x)$ is odd.
3. By the first assertion, we have

$$\mathcal{N}(x, \lambda) = \mathcal{N}(w_0 h_x, \lambda) = \mathcal{N}\left(\begin{pmatrix} 1 & 1 \\ 1 & x \end{pmatrix}, \lambda \right).$$

Then we note that

$$\begin{pmatrix} 1 & 1 \\ 1 & x \end{pmatrix} = \begin{pmatrix} 1 & x^{-1} \\ 1 & 1 \end{pmatrix} \begin{pmatrix} 1 & 0 \\ 0 & x \end{pmatrix}.$$

Hence we have

$$\begin{pmatrix} \begin{pmatrix} 1 & 1 \\ 1 & x \end{pmatrix} & 0 \\ 0 & \lambda \cdot {}^t \begin{pmatrix} 1 & 1 \\ 1 & x \end{pmatrix}^{-1} \end{pmatrix} = \begin{pmatrix} h_{x^{-1}} & 0 \\ 0 & \lambda x^{-1} \cdot {}^t h_{x^{-1}}^{-1} \end{pmatrix} \begin{pmatrix} 1 & 0 & 0 & 0 \\ 0 & x & 0 & 0 \\ 0 & 0 & x & 0 \\ 0 & 0 & 0 & 1 \end{pmatrix},$$

and, from (4.20), we have

$$\mathcal{N}\left(\begin{pmatrix} 1 & 1 \\ 1 & x \end{pmatrix}, \lambda \right) = \chi_E(x)^{-1} \cdot \mathcal{N}(x, \lambda x^{-1}) = \mathcal{N}(x^{-1}, \lambda x^{-1})$$

4.3. EVALUATION OF THE NOVODVORSKY ORBITAL INTEGRAL

since ord (x) is even. $\square$

Let us simplify the integral (4.20) further. Since Ξ is bi $\mathcal{K}$-invariant, by moving matrices of the form

$$\begin{pmatrix} \varepsilon & 0 & 0 & 0 \\ 0 & 1 & 0 & 0 \\ 0 & 0 & 1 & 0 \\ 0 & 0 & 0 & \varepsilon \end{pmatrix}, \quad \begin{pmatrix} \varepsilon_1 & 0 & 0 & 0 \\ 0 & \varepsilon_2 & 0 & 0 \\ 0 & 0 & \varepsilon_2 & 0 \\ 0 & 0 & 0 & \varepsilon_1 \end{pmatrix}, \quad \text{where} \quad \varepsilon, \varepsilon_1, \varepsilon_2 \in \mathcal{O}_F^\times,$$

across to the left, and, to the right, respectively in (4.20), we have

$$(4.22) \qquad \mathcal{N}(x,\lambda) = \sum_{i,j,k \in \mathbb{Z}} (-1)^{j+k} \mathcal{I}\left(x, \lambda, \varpi^i, \varpi^j, \varpi^k\right)$$

where

$$(4.23) \quad \mathcal{I}(x,\lambda,a,b,c) = \int_{\bar{U}} \int_{U}$$

$$\Xi \left[\begin{pmatrix} 1_2 & 0 \\ Y & 1_2 \end{pmatrix} \begin{pmatrix} \begin{pmatrix} ab & acx \\ b & c \end{pmatrix} & 0 \\ 0 & \frac{\lambda}{1-x}\begin{pmatrix} c & -b \\ -acx & ab \end{pmatrix} \end{pmatrix} \begin{pmatrix} 1_2 & X \\ 0 & 1_2 \end{pmatrix} \right]$$

$$\psi \left[\mathrm{tr} \left\{ \begin{pmatrix} 0 & 1 \\ 1 & 0 \end{pmatrix} (X+Y) \right\} \right] dX\, dY$$

for $a, b, c \in F^\times$.

In order to evaluate the integral (4.23), let us prove the following lemma.

LEMMA 4.9. *Let* $g = (g_{ij}) \in G$.
Then we have $g \in \mathcal{K}U$ *if and only if the following conditions are satisfied:*
1. $\lambda(g)$, *the similitude of* g, *belongs to* $\mathcal{O}_F^\times$.
2. $\max_{1 \le i \le 4} \{|g_{i1}|\} \le 1$, $\max_{1 \le i \le 4} \{|g_{i2}|\} \le 1$.
3. $\max_{1 \le k < \ell \le 4} \{|A_{k\ell}|\} = 1$, *where* $A_{k\ell} = \det \begin{pmatrix} g_{k1} & g_{k2} \\ g_{\ell 1} & g_{\ell 2} \end{pmatrix}$.

PROOF. Suppose that $g = ku$ where

$$k = \begin{pmatrix} k_1 & k_2 \\ k_3 & k_4 \end{pmatrix} \in \mathcal{K}, \quad u = \begin{pmatrix} 1_2 & X \\ 0 & 1_2 \end{pmatrix} \in U.$$

Then we have

$$g = \begin{pmatrix} k_1 & k_1 X + k_2 \\ k_3 & k_3 X + k_4 \end{pmatrix}.$$

Since the 4 by 2 matrix $\begin{pmatrix} k_1 \\ k_3 \end{pmatrix}$ (mod ϖ) $\in \mathrm{Mat}_{4 \times 2}(\mathcal{O}_F/\varpi\mathcal{O}_F)$ is of rank 2, the conditions 2 and 3 hold. Since $\lambda(g) = \lambda(k) \in \mathcal{O}_F^\times$, the condition 1 also holds.

Conversely suppose that $g \in G$ satisfies the conditions 1, 2 and 3. Let

$$g = k \begin{pmatrix} h & 0 \\ 0 & \lambda \cdot {}^t h^{-1} \end{pmatrix} \begin{pmatrix} 1_2 & X \\ 0 & 1_2 \end{pmatrix}$$

be the Iwasawa decomposition of g with respect to the Siegel parabolic subgroup P. By the condition 1, we have $\lambda \in \mathcal{O}_F^\times$. Hence by replacing k by $k \begin{pmatrix} 1_2 & 0 \\ 0 & \lambda \cdot 1_2 \end{pmatrix}$, we may assume that $\lambda = 1$.

For a column vector $v = {}^t(v_1, v_2, v_3, v_4) \in F^4$, let us define its norm $\|v\|$ by
$$\|v\| = \max_{1 \le i \le 4} \{|v_i|\}.$$
For each i $(1 \le i \le 4)$, let e_i denote the column vector in F^4 whose i-th entry is 1 but the other entries are all zero. Then by the conditions 2 and 3 we have

(4.24) $\qquad 1 = \|ge_i\| = \|\begin{pmatrix} h & 0 \\ 0 & {}^t h^{-1} \end{pmatrix} e_i\| \quad (i = 1, 2),$

since
$$\|\kappa v\| = \|v\|$$
for $\kappa \in \mathcal{K}$ and $v \in F^4$. From (4.24), we have $h \in \mathrm{Mat}_{2 \times 2}(\mathcal{O}_F)$.

Let us define a norm mapping $\|\ \|$ similarly on $\wedge^2 F^4$, using the coefficients with respect to the basis $\{e_i \wedge e_j\}_{1 \le i < j \le 4}$. Then the condition 3 implies that
$$1 = \|g(e_1 \wedge e_2)\| = \|\begin{pmatrix} h & 0 \\ 0 & {}^t h^{-1} \end{pmatrix}(e_1 \wedge e_2)\| = |\det h|.$$
Thus we have $h \in \mathrm{GL}_2(\mathcal{O}_F)$ and hence $g \in \mathcal{K}U$. $\square$

Now by Lemma 4.9, in the integral (4.23), we have
$$\begin{pmatrix} 1 & 0 & 0 & 0 \\ 0 & 1 & 0 & 0 \\ r & s & 1 & 0 \\ s & t & 0 & 1 \end{pmatrix} \begin{pmatrix} \begin{pmatrix} ab & acx \\ b & c \end{pmatrix} & 0 \\ 0 & \frac{\lambda}{1-x}\begin{pmatrix} c & -b \\ -acx & ab \end{pmatrix} \end{pmatrix} \in \mathcal{K}U$$
if and only if

(4.25) $\qquad |abc\lambda| = 1,$

(4.26) $\qquad \max\{|ab|, |b|, |b(ar+s)|, |b(as+t)|\} \le 1,$

(4.27) $\qquad \max\{|acx|, |c|, |c(arx+s)|, |c(asx+t)|\} \le 1,$

(4.28) $\qquad |abc(1-x)| \cdot \max\{1, |r|, |s|, |t|, |rt - s^2|\} = 1.$

From (4.26), we have $|b| \le 1$. Also from (4.25) and (4.27), we have
$$|acx| = |b^{-1}\lambda^{-1}x| \le 1.$$
Hence we have
$$0 \le \mathrm{ord}(b) \le \mathrm{ord}(x) - \mathrm{ord}(\lambda).$$
Similarly we have $|c| \le 1$ from (4.27), and, from (4.25) and (4.26) we have
$$|ab| = |c^{-1}\lambda^{-1}| \le 1.$$
Hence we have
$$0 \le \mathrm{ord}(c) \le -\mathrm{ord}(\lambda).$$
Thus (4.23) becomes a finite sum

(4.29) $\qquad \mathcal{N}(x, \lambda) = \sum_{\substack{i+j+k = -\mathrm{ord}(\lambda) \\ 0 \le j \le \mathrm{ord}(x) - \mathrm{ord}(\lambda) \\ 0 \le k \le -\mathrm{ord}(\lambda)}} (-1)^{j+k} \mathcal{I}(x, \lambda, \varpi^i, \varpi^j, \varpi^k).$

From (4.29) we immediately have the following proposition.

PROPOSITION 4.10. $\mathcal{N}(x, \lambda)$ *vanishes unless* $\mathrm{ord}(\lambda) \le 0$ *and* $\mathrm{ord}(\lambda) \le \mathrm{ord}(x)$.

Now we observe some symmetry among the terms in (4.29).

4.3. EVALUATION OF THE NOVODVORSKY ORBITAL INTEGRAL

LEMMA 4.11. *Let $M = \operatorname{ord}(x) - \operatorname{ord}(\lambda)$ and $N = -\operatorname{ord}(\lambda)$. Then for $0 \leq j \leq M$ and $0 \leq k \leq N$, we have*

1. $\mathcal{I}\left(x, \lambda, \varpi^{N-j-k}, \varpi^j, \varpi^k\right) = \mathcal{I}\left(x, \lambda, \varpi^{k-j}, \varpi^j, \varpi^{N-k}\right)$
2. $\mathcal{I}\left(x, \lambda, \varpi^{N-j-k}, \varpi^j, \varpi^k\right) = \mathcal{I}\left(x, \lambda, \varpi^{N-M+j-k}, \varpi^{M-j}, \varpi^k\right)$.
3. $\mathcal{I}\left(x, \lambda, \varpi^{N-j-k}, \varpi^j, \varpi^k\right) = \mathcal{I}\left(x, \lambda, \varpi^{j+k-M}, \varpi^{M-j}, \varpi^{N-k}\right)$.

PROOF. First we recall that

$$(4.30) \quad \mathcal{I}\left(x, \lambda, \varpi^i, \varpi^j, \varpi^k\right) = \int_{\bar{U}} \int_{U}$$

$$\Xi\left[\begin{pmatrix} 1_2 & 0 \\ Y & 1_2 \end{pmatrix} \begin{pmatrix} \begin{pmatrix} \varpi^{N-k} & \varpi^{N-j}x \\ \varpi^j & \varpi^k \end{pmatrix} & 0 \\ 0 & \frac{\lambda}{1-x}\begin{pmatrix} \varpi^k & -\varpi^j \\ -\varpi^{N-j}x & \varpi^{N-k} \end{pmatrix} \end{pmatrix} \begin{pmatrix} 1_2 & X \\ 0 & 1_2 \end{pmatrix}\right]$$

$$\psi\left[\operatorname{tr}\{w_0(X+Y)\}\right] dX \, dY$$

where $w_0 = \begin{pmatrix} 0 & 1 \\ 1 & 0 \end{pmatrix}$.

For the first assertion, we note that

$$w_0{}^t\begin{pmatrix} \varpi^{N-k} & \varpi^{N-j}x \\ \varpi^j & \varpi^k \end{pmatrix} w_0 = \begin{pmatrix} \varpi^k & \varpi^{N-j}x \\ \varpi^j & \varpi^{N-k} \end{pmatrix}$$

and

$$\Xi\left[\begin{pmatrix} w_0 & 0 \\ 0 & w_0 \end{pmatrix} {}^tg \begin{pmatrix} w_0 & 0 \\ 0 & w_0 \end{pmatrix}\right] = \Xi(g).$$

Hence we may rewrite (4.30) as

$$\mathcal{I}\left(x, \lambda, \varpi^i, \varpi^j, \varpi^k\right) = \int_{\bar{U}} \int_{U}$$

$$\Xi\left[\begin{pmatrix} 1_2 & 0 \\ w_0 X w_0 & 1_2 \end{pmatrix} \begin{pmatrix} \begin{pmatrix} \varpi^k & \varpi^{N-j}x \\ \varpi^j & \varpi^{N-k} \end{pmatrix} & 0 \\ 0 & \frac{\lambda}{1-x}\begin{pmatrix} \varpi^k & -\varpi^j \\ -\varpi^{N-j}x & \varpi^{N-k} \end{pmatrix} \end{pmatrix} \begin{pmatrix} 1_2 & w_0 Y w_0 \\ 0 & 1_2 \end{pmatrix}\right]$$

$$\psi\left[\operatorname{tr}\{w_0(X+Y)\}\right] dX \, dY$$

and it is easily seen that we have

$$\mathcal{I}\left(x, \lambda, \varpi^{N-j-k}, \varpi^j, \varpi^k\right) = \mathcal{I}\left(x, \lambda, \varpi^{k-j}, \varpi^j, \varpi^{N-k}\right).$$

Let us prove the second assertion. Since $M - N = \operatorname{ord}(x)$, we may write $x = \varpi^{M-N}\varepsilon_x$ where $\varepsilon_x \in \mathcal{O}_F^\times$. Then we note that

$$\begin{pmatrix} \varepsilon_x & 0 \\ 0 & 1 \end{pmatrix} {}^t\begin{pmatrix} \varpi^{N-k} & \varpi^{N-j}x \\ \varpi^j & \varpi^k \end{pmatrix} \begin{pmatrix} \varepsilon_x & 0 \\ 0 & 1 \end{pmatrix}^{-1} = \begin{pmatrix} \varpi^{N-k} & \varpi^{-M+N+j}x \\ \varpi^{M-j} & \varpi^k \end{pmatrix}$$

and

$$\Xi\left[\begin{pmatrix} \varepsilon_x & 0 & 0 & 0 \\ 0 & 1 & 0 & 0 \\ 0 & 0 & 1 & 0 \\ 0 & 0 & 0 & \varepsilon_x \end{pmatrix} {}^tg \begin{pmatrix} \varepsilon_x & 0 & 0 & 0 \\ 0 & 1 & 0 & 0 \\ 0 & 0 & 1 & 0 \\ 0 & 0 & 0 & \varepsilon_x \end{pmatrix}^{-1}\right] = \Xi(g).$$

Hence by an argument similar to the one for the previous assertion, we have

$$\mathcal{I}\left(x, \lambda, \varpi^{N-j-k}, \varpi^j, \varpi^k\right) = \mathcal{I}\left(x, \lambda, \varpi^{N-M+j-k}, \varpi^{M-j}, \varpi^k\right).$$

The third assertion is just a combination of the previous two assertions. □

COROLLARY 4.12. *When* $\mathrm{ord}\,(\lambda)$ *is odd, we have*
$$\mathcal{N}(x,\lambda) = 0.$$

PROOF. By the lemma, we have
$$\mathcal{N}(x,\lambda) = \sum_{\substack{0 \le j \le M \\ 0 \le k \le N}} (-1)^{j+k} \mathcal{I}\left(x, \lambda, \varpi^{N-j-k}, \varpi^j, \varpi^k\right)$$
$$= \sum_{\substack{0 \le j \le M \\ 0 \le k \le N}} (-1)^{j+k} \mathcal{I}\left(x, \lambda, \varpi^{k-j}, \varpi^j, \varpi^{N-k}\right) = (-1)^N \mathcal{N}(x,\lambda)$$

where $N = -\mathrm{ord}\,(\lambda)$. Thus when $\mathrm{ord}\,(\lambda)$ is odd, we have $\mathcal{N}(x,\lambda) = 0$. □

Now we are ready to state our main theorem in this chapter.

THEOREM 4.13. *Let* $\lambda \in F^\times$ *and* $x \in F \setminus \{0,1\}$.
1. *Then* $\mathcal{N}(x,\lambda)$ *vanishes unless*
$$\mathrm{ord}\,(\lambda),\,\mathrm{ord}\,(x) \text{ are both even}, \quad \text{and}, \quad \mathrm{ord}\,(\lambda) \le \min\{0, \mathrm{ord}\,(x)\}.$$

2. *We have the functional equation*
$$\mathcal{N}(x,\lambda) = \mathcal{N}\left(x^{-1}, \lambda x^{-1}\right).$$

3. *Suppose that* $\lambda \in \mathcal{O}_F^\times$.
   (a) *When* $|1-x| = 1$ *and* $\mathrm{ord}\,(x)$ *is even, we have*
   $$\mathcal{N}(x,\lambda) = 1.$$
   (b) *When* $|1-x| < 1$, *we have*
   $$\mathcal{N}(x,\lambda) = |1-x|^{-1} \cdot \mathcal{K}\ell\left(2(1-x)^{-1}, -2\lambda(1-x)^{-1}\right).$$

4. *Suppose that* $\mathrm{ord}\,(\lambda) = -2n$ *where* $n > 0$, *and,* $\mathrm{ord}\,(x) = 2m$ *where* $m \ge 0$. *Let us write* $\lambda = \varpi^{-2n} \varepsilon_\lambda$ *and* $x = \varpi^{2m} \varepsilon_x$.
   (a) *When* $m \ge n$, *we have*
   $$\mathcal{N}(x,\lambda) = q^{2n}\left\{(-1)^n \mathcal{K}\ell\left(2\varpi^{-n}, -2\varpi^{-n}\varepsilon_\lambda\right) + 1 + q^{-1}\right\}.$$
   (b) *When* $0 \le m < n$, *we have*
$$\mathcal{N}(x,\lambda) = q^{2n}|1-x|^{-1}(-1)^n$$
$$\left\{\mathcal{K}\ell\left(\frac{2\varpi^{-n}}{1-x}, \frac{-2\varpi^{-n}\varepsilon_\lambda}{1-x}\right) + (-1)^m \cdot \mathcal{K}\ell\left(\frac{2\varpi^{m-n}}{1-x}, \frac{-2\varpi^{m-n}\varepsilon_\lambda\varepsilon_x}{1-x}\right)\right\}.$$

The first two assertions have been proved already. We shall prove the last two assertions in the following subsections.

4.3.1. Evaluation of $\mathcal{N}(x,\lambda)$ **when** $\lambda \in \mathcal{O}_F^\times$ **and** $x \in \mathcal{O}_F \setminus \{0,1\}$. In this case the summation in (4.29) becomes

(4.31)
$$\mathcal{N}(x,\lambda) = \sum_{0 \le j \le \mathrm{ord}(x)} (-1)^j \mathcal{I}\left(x, \lambda, \varpi^{-j}, \varpi^j, 1\right)$$

4.3. EVALUATION OF THE NOVODVORSKY ORBITAL INTEGRAL   97

where

(4.32) $\mathcal{I}\left(x, \lambda, \varpi^{-j}, \varpi^{j}, 1\right) = \int_{\bar{U}} \int_{U}$

$$\Xi \left[\begin{pmatrix} 1_2 & 0 \\ Y & 1_2 \end{pmatrix} \begin{pmatrix} \begin{pmatrix} 1 & \varpi^{-j}x \\ \varpi^{j} & 1 \end{pmatrix} & 0 \\ 0 & \frac{\lambda}{1-x}\begin{pmatrix} 1 & -\varpi^{j} \\ -\varpi^{-j}x & 1 \end{pmatrix} \end{pmatrix} \begin{pmatrix} 1_2 & X \\ 0 & 1_2 \end{pmatrix} \right]$$

$$\psi\left[\mathrm{tr}\left\{\begin{pmatrix} 0 & 1 \\ 1 & 0 \end{pmatrix}(X+Y)\right\}\right] dX\, dY.$$

Suppose that $|1 - x| = 1$. Then we have

$$\begin{pmatrix} 1 & \varpi^{-j}x \\ \varpi^{j} & 1 \end{pmatrix} \in \mathrm{GL}_2\left(\mathcal{O}_F\right)$$

and it is readily seen that for $Y \in \mathrm{Sym}^2(F)$, we have

$$\begin{pmatrix} 1_2 & 0 \\ Y & 1_2 \end{pmatrix} \begin{pmatrix} \begin{pmatrix} 1 & \varpi^{-j}x \\ \varpi^{j} & 1 \end{pmatrix} & 0 \\ 0 & \frac{\lambda}{1-x}\begin{pmatrix} 1 & -\varpi^{j} \\ -\varpi^{-j}x & 1 \end{pmatrix} \end{pmatrix} \in \mathcal{K}U$$

if and only if $Y \in \mathrm{Sym}^2(\mathcal{O}_F)$. Then for $X \in \mathrm{Sym}^2(F)$, we have

$$\begin{pmatrix} \begin{pmatrix} 1 & \varpi^{-j}x \\ \varpi^{j} & 1 \end{pmatrix} & 0 \\ 0 & \frac{\lambda}{1-x}\begin{pmatrix} 1 & -\varpi^{j} \\ -\varpi^{-j}x & 1 \end{pmatrix} \end{pmatrix} \begin{pmatrix} 1_2 & X \\ 0 & 1_2 \end{pmatrix} \in \mathcal{K}$$

if and only if $X \in \mathrm{Sym}^2(\mathcal{O}_F)$. Hence

$$\mathcal{I}\left(x, \lambda, \varpi^{-j}, \varpi^{j}, 1\right) = 1$$

and thus

$$\mathcal{N}(x, \lambda) = \sum_{0 \leq j \leq \mathrm{ord}(x)} (-1)^j = 1$$

since $\mathrm{ord}(x)$ is even.

Now suppose that $|1 - x| < 1$. Then $\mathrm{ord}(x) = 0$ and hence

(4.33) $\mathcal{N}(x, \lambda) = \mathcal{I}(x, \lambda, 1, 1, 1)$

$$= \int_{\bar{U}} \int_{U} \Xi\left[\begin{pmatrix} 1_2 & 0 \\ Y & 1_2 \end{pmatrix} \begin{pmatrix} \begin{pmatrix} 1 & x \\ 1 & 1 \end{pmatrix} & 0 \\ 0 & \frac{\lambda}{1-x}\begin{pmatrix} 1 & -1 \\ -x & 1 \end{pmatrix} \end{pmatrix} \begin{pmatrix} 1_2 & X \\ 0 & 1_2 \end{pmatrix} \right]$$

$$\psi\left[\mathrm{tr}\left\{\begin{pmatrix} 0 & 1 \\ 1 & 0 \end{pmatrix}(X+Y)\right\}\right] dX\, dY.$$

LEMMA 4.14. *Suppose that* $|1 - x| < 1$ *and* $\lambda \in \mathcal{O}_F^\times$. *Then for* $r, s, t \in F$, *we have*

(4.34) $$\begin{pmatrix} 1 & 0 & 0 & 0 \\ 0 & 1 & 0 & 0 \\ r & s & 1 & 0 \\ s & t & 0 & 1 \end{pmatrix} \begin{pmatrix} \begin{pmatrix} 1 & x \\ 1 & 1 \end{pmatrix} & 0 \\ 0 & \frac{\lambda}{1-x}\begin{pmatrix} 1 & -1 \\ -x & 1 \end{pmatrix} \end{pmatrix} \in \mathcal{K}U$$

if and only if

(4.35) $\quad\quad |(1-x)s| = 1, \quad |r+s| \leq 1, \quad |s+t| \leq 1.$

PROOF. Suppose that (4.34) holds. Then by (4.26), we have
$$\max\{|r+s|, |s+t|\} \leq 1.$$
Then since
$$(4.36) \qquad (1-x)\left(rt - s^2\right) = (1-x)\, r \cdot (s+t) - (1-x)\, s \cdot (r+s),$$
we have
$$|(1-x)\left(rt - s^2\right)| \leq \max\{|(1-x)\, r|, |(1-x)\, s|\}.$$
Hence (4.28) implies that
$$(4.37) \qquad \max\{|(1-x)\, r|, |(1-x)\, s|, |(1-x)\, t|\} = 1.$$
Since
$$(4.38) \qquad (1-x)\, r + (1-x)\, s = (1-x)(r+s) \in \varpi \mathcal{O}_F$$
and
$$(4.39) \qquad (1-x)\, s + (1-x)\, t = (1-x)(s+t) \in \varpi \mathcal{O}_F,$$
we must have $|(1-x)\, s| = 1$ in order for (4.37) to hold.

Suppose conversely that (4.35) holds. Then (4.26) clearly holds. From (4.38) and (4.39), we have
$$|(1-x)\, r| = |(1-x)\, t| = 1.$$
Hence from (4.36), we also have
$$|(1-x)\left(rt - s^2\right)| \leq 1$$
and thus (4.28) holds. Since
$$rx + s = -(1-x)\, r + (r+s), \qquad sx + t = -(1-x)\, s + (s+t),$$
(4.27) also holds. Hence we have (4.34). $\square$

Let us compute $\mathcal{N}(x, \lambda)$. Suppose that (4.35) holds. Then we note that we have
$$\begin{pmatrix} 1 & 0 & 0 & 0 \\ 0 & 1 & 0 & 0 \\ r & s & 1 & 0 \\ s & t & 0 & 1 \end{pmatrix} \left(\begin{pmatrix} \begin{pmatrix} 1 & x \\ 1 & 1 \end{pmatrix} & 0 \\ 0 & \frac{\lambda}{1-x} \begin{pmatrix} 1 & -1 \\ -x & 1 \end{pmatrix} \end{pmatrix} \right)$$
$$= \begin{pmatrix} 1 & 0 & 0 & 0 \\ 0 & 1 & 0 & 0 \\ r+s & 0 & 1 & 0 \\ 0 & sx+t & 0 & 1 \end{pmatrix} \left(\begin{pmatrix} \begin{pmatrix} 1 & x \\ 1 & 1 \end{pmatrix} & 0 \\ 0 & (1-x)\, s \end{pmatrix} \quad \begin{pmatrix} 0 & (1-x)\, s \\ (1-x)\, s & 0 \end{pmatrix} \quad \frac{\lambda}{1-x} \begin{pmatrix} 1 & -1 \\ -x & 1 \end{pmatrix} \right)$$
where $r+s \in \mathcal{O}_F$ and $sx + t \in \mathcal{O}_F$. Thus we have
$$\left(\begin{pmatrix} \begin{pmatrix} 1 & x \\ 1 & 1 \end{pmatrix} & 0 \\ \begin{pmatrix} 0 & (1-x)\, s \\ (1-x)\, s & 0 \end{pmatrix} & \frac{\lambda}{1-x} \begin{pmatrix} 1 & -1 \\ -x & 1 \end{pmatrix} \end{pmatrix} \right) = \begin{pmatrix} k_1 & k_2 \\ k_3 & k_4 \end{pmatrix} \begin{pmatrix} 1_2 & -X \\ 0 & 1_2 \end{pmatrix}$$
for some $\begin{pmatrix} k_1 & k_2 \\ k_3 & k_4 \end{pmatrix} \in \mathcal{K}$ and $X \in \mathrm{Sym}^2(F)$. Here we have
$$k_3 = \begin{pmatrix} 0 & (1-x)\, s \\ (1-x)\, s & 0 \end{pmatrix} \in \mathrm{GL}_2(\mathcal{O}_F)$$

and
$$-k_3 X + k_4 = \frac{\lambda}{1-x}\begin{pmatrix} 1 & -1 \\ -x & 1 \end{pmatrix}.$$

Thus we have
$$X = k_3^{-1} k_4 - \frac{\lambda}{(1-x)^2 s}\begin{pmatrix} -x & 1 \\ 1 & -1 \end{pmatrix}$$

where $k_3^{-1} k_4 \in \mathrm{Mat}_{2\times 2}(\mathcal{O}_F)$. Hence

$$\mathcal{N}(x, \lambda) = \int_{|(1-x)s|=1, |r+s|\leq 1, |s+t|\leq 1}$$
$$\psi\left[\mathrm{tr}\left\{\begin{pmatrix} 0 & 1 \\ 1 & 0 \end{pmatrix}\left(\begin{pmatrix} r & s \\ s & t \end{pmatrix} - \frac{\lambda}{(1-x)^2 s}\begin{pmatrix} -x & 1 \\ 1 & -1 \end{pmatrix}\right)\right\}\right] dr\, ds\, dt$$
$$= \int_{|(1-x)s|=1} \psi\left[2s - \frac{2\lambda}{(1-x)^2 s}\right] ds$$
$$= |1-x|^{-1} \int_{\mathcal{O}_F^\times} \psi\left[2(1-x)^{-1}\varepsilon - 2\lambda(1-x)^{-1}\varepsilon^{-1}\right] d\varepsilon$$
$$= |1-x|^{-1} \cdot \mathcal{K}\ell\left(2(1-x)^{-1}, -2\lambda(1-x)^{-1}\right).$$

4.3.2. Evaluation of $\mathcal{N}(x, \lambda)$ when $\lambda \in F \setminus \mathcal{O}_F$ and $x \in \mathcal{O}_F \setminus \{0, 1\}$. Let us write $\mathcal{I}_{j,k}$ for $\mathcal{I}(x, \lambda, \varpi^{2n-j-k}, \varpi^j, \varpi^k)$. Then we have

(4.40) $$\mathcal{N}(x, \lambda) = \sum_{\substack{0\leq j\leq 2m+2n \\ 0\leq k\leq 2n}} (-1)^{j+k} \mathcal{I}_{j,k}$$

where $2m = \mathrm{ord}(x)$ and $2n = -\mathrm{ord}(\lambda)$.

In (4.40), let us call $\mathcal{I}_{j,k}$ an *boundary term* when
$$j = 0, 2m+2n, \quad\text{or,}\quad k = 0, 2n$$
and let us call $\mathcal{I}_{j,k}$ an *interior term* otherwise.

Let us evaluate the boundary terms first.

PROPOSITION 4.15. *1. Suppose that $j = 0$ and $0 \leq k \leq 2n$. Then we have*
$$\mathcal{I}_{0,k} = |1-x|^{-1} q^{2n} \cdot \mathcal{K}\ell\left(2\varpi^{-k}(1-x)^{-1}, -2\varpi^{k-2n}\varepsilon_\lambda(1-x)^{-1}\right).$$

2. Suppose that $0 < j < 2m+2n$ and $k = 0$. Then we have
$$\mathcal{I}_{j,0} = |1-x|^{-1} q^{2n} \cdot \mathcal{K}\ell\left(2\varpi^{2m-j}(1-x)^{-1}, -2\varpi^{j-2n}\varepsilon_\lambda \varepsilon_x (1-x)^{-1}\right).$$

PROOF. Since the proofs are entirely similar, here we prove only the first assertion.

First we note that since the Kloosterman sum
$$\mathcal{K}\ell\left(2\varpi^{-k}(1-x)^{-1}, -2\varpi^{k-2n}\varepsilon_\lambda(1-x)^{-1}\right)$$
is invariant under $k \mapsto 2n - k$ and we also have $\mathcal{I}_{0,k} = \mathcal{I}_{0,2n-k}$ by Lemma 4.11, we may assume that $0 \leq k < 2n$.

We recall that

$$(4.41) \quad \mathcal{I}_{0,k} = \int_{\bar{U}} \int_U$$
$$\Xi \left[\begin{pmatrix} 1_2 & 0 \\ Y & 1_2 \end{pmatrix} \begin{pmatrix} \begin{pmatrix} \varpi^{2n-k} & \varpi^{2n}x \\ 1 & \varpi^k \end{pmatrix} & 0 \\ 0 & \frac{\lambda}{1-x}\begin{pmatrix} \varpi^k & -1 \\ -\varpi^{2n}x & \varpi^{2n-k} \end{pmatrix} \end{pmatrix} \begin{pmatrix} 1_2 & X \\ 0 & 1_2 \end{pmatrix} \right]$$
$$\psi \left[\operatorname{tr}\left\{ \begin{pmatrix} 0 & 1 \\ 1 & 0 \end{pmatrix} (X+Y) \right\} \right] dX\, dY.$$

Let us show that for $r, s, t \in F$, we have

$$(4.42) \quad \begin{pmatrix} 1 & 0 & 0 & 0 \\ 0 & 1 & 0 & 0 \\ r & s & 1 & 0 \\ s & t & 0 & 1 \end{pmatrix} \begin{pmatrix} \begin{pmatrix} \varpi^{2n-k} & \varpi^{2n}x \\ 1 & \varpi^k \end{pmatrix} & 0 \\ 0 & \frac{\lambda}{1-x}\begin{pmatrix} \varpi^k & -1 \\ -\varpi^{2n}x & \varpi^{2n-k} \end{pmatrix} \end{pmatrix} \in \mathcal{K}U$$

if and only if

$$(4.43) \quad |\varpi^{2n}(1-x)r| = 1, \quad |\varpi^{2n-k}r+s| \le 1, \quad |\varpi^{2n-k}s+t| \le 1.$$

In our case, by Lemma 4.9, the condition (4.42) holds if and only if

$$(4.44) \quad \max\left\{ |\varpi^{2n-k}r+s|, |\varpi^{2n-k}s+t|, |\varpi^{2n}xr+\varpi^k s|, |\varpi^{2n}xs+\varpi^k t| \right\} \le 1$$

and

$$(4.45) \quad \max\{ |\varpi^{2n}(1-x)r|, |\varpi^{2n}(1-x)s|,$$
$$|\varpi^{2n}(1-x)t|, |\varpi^{2n}(1-x)(rt-s^2)| \} = 1.$$

Suppose that (4.44) and (4.45) hold. Then from (4.44) we have

$$|\varpi^{2n-k}r+s| \le 1, \quad |\varpi^{2n-k}s+t| \le 1.$$

Next we note the Plücker relations, i.e.

$$(4.46) \quad \varpi^{2n}(1-x)(rt-s^2) \cdot 1 + \varpi^{2n}(1-x)s \cdot (\varpi^{2n-k}r+s)$$
$$- \varpi^{2n}(1-x)r \cdot (\varpi^{2n-k}s+t) = 0,$$

$$(4.47) \quad \varpi^{2n}(1-x)s \cdot \varpi^{2n-k} + \varpi^{2n}(1-x)t \cdot 1$$
$$- \varpi^{2n}(1-x) \cdot (\varpi^{2n-k}s+t) = 0,$$

$$(4.48) \quad \varpi^{2n}(1-x)r \cdot \varpi^{2n-k} + \varpi^{2n}(1-x)s \cdot 1$$
$$- \varpi^{2n}(1-x) \cdot (\varpi^{2n-k}r+s) = 0.$$

Then since $n > 0$ and $2n - k > 0$, we have

$$|\varpi^{2n}(1-x)t| < 1 \quad \text{and} \quad |\varpi^{2n}(1-x)s| < 1$$

from (4.47) and (4.48) respectively. If $|\varpi^{2n}(1-x)r| < 1$, then we also have $|\varpi^{2n}(1-x)(rt-s^2)| < 1$ from (4.46). Then (4.45) does not hold. Hence (4.43) holds.

Suppose conversely that (4.43) holds. We have $|\varpi^{2n}(1-x)s| < 1$ from (4.48) since $n > 0$ and $2n - k > 0$. Then we have

$$|\varpi^{2n}(1-x)(rt-s^2)| \le 1, \quad |\varpi^{2n}(1-x)t| < 1$$

4.3. EVALUATION OF THE NOVODVORSKY ORBITAL INTEGRAL

from (4.46) and (4.47), respectively. Hence (4.45) holds. As for (4.44), it is enough for us to note the following two identities,

$$\varpi^{2n} x r + \varpi^k s = \varpi^k \cdot \left(\varpi^{2n-k} r + s \right) - \varpi^{2n} (1-x) r,$$
$$\varpi^{2n} x s + \varpi^k t = \varpi^k \cdot \left(\varpi^{2n-k} s + t \right) - \varpi^{2n} (1-x) s.$$

Now let us evaluate (4.41). Under (4.43), we have

$$\begin{pmatrix} 1 & 0 & 0 & 0 \\ 0 & 1 & 0 & 0 \\ r & s & 1 & 0 \\ s & t & 0 & 1 \end{pmatrix} \left(\begin{pmatrix} \varpi^{2n-k} & \varpi^{2n} x \\ 1 & \varpi^k \end{pmatrix} \quad 0 \\ \frac{\lambda}{1-x} \begin{pmatrix} \varpi^k & -1 \\ -\varpi^{2n} x & \varpi^{2n-k} \end{pmatrix} \right)$$

$$= \begin{pmatrix} 1 & 0 & 0 & 0 \\ 0 & 1 & 0 & 0 \\ 0 & s' & 1 & 0 \\ s' & t' & 0 & 1 \end{pmatrix} \left(\begin{pmatrix} \varpi^{2n-k} & \varpi^{2n} x \\ 1 & \varpi^k \end{pmatrix} \quad 0 \\ \varpi^{2n}(1-x) r \begin{pmatrix} 0 & -1 \\ 0 & \varpi^{2n-k} \end{pmatrix} \quad \frac{\lambda}{1-x} \begin{pmatrix} \varpi^k & -1 \\ -\varpi^{2n} x & \varpi^{2n-k} \end{pmatrix} \right),$$

where $s' = \varpi^{2n-k} r + s \in \mathcal{O}_F$ and $t' = \left(\varpi^{2n-k} s + t \right) - \varpi^{2n-k} \left(\varpi^{2n-k} r + s \right) \in \mathcal{O}_F$. Thus we may write

$$\left(\begin{pmatrix} \varpi^{2n-k} & \varpi^{2n} x \\ 1 & \varpi^k \end{pmatrix} \quad 0 \\ \varpi^{2n}(1-x) r \begin{pmatrix} 0 & -1 \\ 0 & \varpi^{2n-k} \end{pmatrix} \quad \frac{\lambda}{1-x} \begin{pmatrix} \varpi^k & -1 \\ -\varpi^{2n} x & \varpi^{2n-k} \end{pmatrix} \right) = \begin{pmatrix} k_1 & k_2 \\ k_3 & k_4 \end{pmatrix} \begin{pmatrix} 1_2 & -X \\ 0 & 1_2 \end{pmatrix}$$

where $\begin{pmatrix} k_1 & k_2 \\ k_3 & k_4 \end{pmatrix} \in \mathcal{K}$. Then we have

$$k_3 = \varpi^{2n} (1-x) r \begin{pmatrix} 0 & -1 \\ 0 & \varpi^{2n-k} \end{pmatrix}$$

and hence

$$(4.49) \qquad -\varpi^{2n}(1-x)r \begin{pmatrix} 0 & -1 \\ 0 & \varpi^{2n-k} \end{pmatrix} X + k_4 = \frac{\lambda}{1-x} \begin{pmatrix} \varpi^k & -1 \\ -\varpi^{2n} x & \varpi^{2n-k} \end{pmatrix}.$$

Let $X = \begin{pmatrix} u & v \\ v & w \end{pmatrix}$. Then by comparing the $(1,1)$-entry of the both sides of (4.49), we have

$$\varpi^{2n}(1-x) r \cdot v \equiv \frac{\lambda \varpi^k}{1-x} \pmod{\mathcal{O}_F}.$$

Since $\varpi^{2n}(1-x) r \in \mathcal{O}_F^\times$, we have

$$v \equiv \frac{\varpi^{k-2n} \lambda}{(1-x)^2 r} \pmod{\mathcal{O}_F}.$$

Thus (4.41) becomes

$$\mathcal{I}_{0,k} = \int_{|\varpi^{2n}(1-x)r|=1, |\varpi^{2n-k}r+s|\leq 1, |\varpi^{2n-k}s+t|\leq 1} \psi \left[2s + \frac{2\varpi^{k-2n}\lambda}{(1-x)^2 r} \right] dr\, ds\, dt$$

$$= \int_{|\varpi^{2n}(1-x)r|=1} \psi \left[-2\varpi^{2n-k} r + \frac{2\varpi^{k-2n}\lambda}{(1-x)^2 r} \right] dr$$

$$= |1-x|^{-1} q^{2n} \cdot \mathcal{K}\ell \left(2\varpi^{-k}(1-x)^{-1}, -2\varpi^{k-2n}\varepsilon_\lambda (1-x)^{-1} \right).$$

$\square$

Now let us evaluate the interior terms. Let
$$A_{j,k} = \begin{pmatrix} \varpi^{2n-k} & \varpi^{2n-j}x \\ \varpi^j & \varpi^k \end{pmatrix}.$$
Then we recall that

(4.50) $\mathcal{I}_{j,k} = \displaystyle\int_{\bar{U}} \int_{U} \Xi \left[\begin{pmatrix} 1_2 & 0 \\ Y & 1_2 \end{pmatrix} \begin{pmatrix} A_{j,k} & 0 \\ 0 & \varepsilon_\lambda \cdot {}^t A_{j,k}^{-1} \end{pmatrix} \begin{pmatrix} 1_2 & X \\ 0 & 1_2 \end{pmatrix} \right]$
$\psi \left[\operatorname{tr} \left\{ \begin{pmatrix} 0 & 1 \\ 1 & 0 \end{pmatrix} (X+Y) \right\} \right] dX \, dY.$

PROPOSITION 4.16. *Suppose that $0 < j < 2m + 2n$ and $0 < k < 2n$. Then we have*
$$\mathcal{I}_{j,k} = \mathcal{K}_{\operatorname{spl}} (A_{j,k}, -\varepsilon_\lambda).$$
Here $\mathcal{K}_{\operatorname{spl}}(\cdot, \cdot)$ denotes the split Kloosterman sum as in Definition 3.3.

PROOF. By the assumption, we have $A_{j,k} \in \varpi \cdot \operatorname{Mat}_{2 \times 2}(\mathcal{O}_F)$. Hence in (4.50), for $Y \in \operatorname{Sym}^2(F)$, we have
$$\begin{pmatrix} 1_2 & 0 \\ Y & 1_2 \end{pmatrix} \begin{pmatrix} A_{j,k} & 0 \\ 0 & \varepsilon_\lambda \cdot {}^t A_{j,k}^{-1} \end{pmatrix} \in \mathcal{K} U$$
if and only if

(4.51) $\qquad Y \in \mathcal{X}_{j,k} = \left\{ S \in \operatorname{Sym}^2(F) \mid S A_{j,k} \in \operatorname{GL}_2(\mathcal{O}_F) \right\}.$

by Lemma 4.9. When (4.51) holds, we have
$$\begin{pmatrix} 1_2 & 0 \\ Y & 1_2 \end{pmatrix} \begin{pmatrix} A_{j,k} & 0 \\ 0 & \varepsilon_\lambda \cdot {}^t A_{j,k}^{-1} \end{pmatrix} = \begin{pmatrix} k_1 & k_2 \\ k_3 & k_4 \end{pmatrix} \begin{pmatrix} 1_2 & -X \\ 0 & 1_2 \end{pmatrix}$$
for some $\begin{pmatrix} k_1 & k_2 \\ k_3 & k_4 \end{pmatrix} \in \mathcal{K}$ and $X \in \operatorname{Sym}^2(F)$. Then we have
$$Y A_{j,k} = k_3, \quad -k_3 X + k_4 = \varepsilon_\lambda \cdot {}^t A_{j,k}^{-1}.$$
Hence
$$X \equiv -\varepsilon_\lambda \cdot A_{j,k}^{-1} Y^{-1} \, {}^t A_{j,k}^{-1} \pmod{\mathcal{O}_F}$$
since $k_3^{-1} k_4 \in \operatorname{Mat}_{2 \times 2}(\mathcal{O}_F)$. Thus
$$\mathcal{I}_{j,k} = \int_{\mathcal{X}_{j,k}} \psi \left[\operatorname{tr} \left\{ \begin{pmatrix} 0 & 1 \\ 1 & 0 \end{pmatrix} \left(Y - \varepsilon_\lambda \cdot A_{j,k}^{-1} Y^{-1} {}^t A_{j,k}^{-1} \right) \right\} \right] dY = \mathcal{K}_{\operatorname{spl}}(A_{j,k}, -\varepsilon_\lambda).$$
$\square$

Hence by Theorem 3.5, we have the following proposition.

PROPOSITION 4.17. *Suppose that $0 < j < 2m + 2n$ and $0 < k < 2n$. Then we have*
$$\mathcal{I}_{j,k} = |1-x|^{-1} q^{2n}$$
$$\left\{ \mathcal{K}\ell \left(\frac{2\varpi^{-k}}{1-x}, \frac{-2\varpi^{k-2n}\varepsilon_\lambda}{1-x} \right) + \mathcal{K}\ell \left(\frac{2\varpi^{2m-j}}{1-x}, \frac{-2\varpi^{j-2n}\varepsilon_\lambda \varepsilon_x}{1-x} \right) \right\}.$$

4.3. EVALUATION OF THE NOVODVORSKY ORBITAL INTEGRAL

Finally let us evaluate $\mathcal{N}(x, \lambda)$. We recall that

$$\mathcal{N}(x, \lambda) = \sum_{\substack{0 \leq j \leq 2m+2n \\ 0 \leq k \leq 2n}} (-1)^{j+k} \mathcal{I}_{j,k}.$$

By Lemma 4.20, we have

$$\mathcal{I}_{j,k} = \mathcal{I}_{2m+2n-j,k} = \mathcal{I}_{j,2n-k}.$$

Hence by Proposition 4.15 and Proposition 4.17, we have

$$(4.52) \quad \mathcal{N}(x, \lambda) = |1-x|^{-1} q^{2n} \sum_{j=0}^{2m+2n} (-1)^j \mathcal{K}\ell \left(\frac{2\varpi^{2m-j}}{1-x}, \frac{-2\varpi^{j-2n} \varepsilon_\lambda \varepsilon_x}{1-x} \right)$$

$$+ |1-x|^{-1} q^{2n} \sum_{k=0}^{2n} (-1)^k \mathcal{K}\ell \left(\frac{2\varpi^{-k}}{1-x}, \frac{-2\varpi^{k-2n} \varepsilon_\lambda}{1-x} \right).$$

Since $|1-x| \leq 1$, by Proposition 2.7, we have

$$\mathcal{K}\ell \left(\frac{2\varpi^{-k}}{1-x}, \frac{-2\varpi^{k-2n} \varepsilon_\lambda}{1-x} \right) = 0$$

unless $k = n$.

Suppose that $m \geq n > 0$. Then we have $x \equiv 0 \pmod{\varpi^n}$. Hence by Proposition 2.7, we have

$$\mathcal{K}\ell \left(2\varpi^{2m-j}, -2\varpi^{j-2n} \varepsilon_\lambda \varepsilon_x \right) = \begin{cases} 0, & \text{when } 0 \leq j < 2n-1 \\ -q^{-1}, & \text{when } j = 2n-1 \\ 1 - q^{-1}, & \text{when } 2n \leq j \leq m+n. \end{cases}$$

Therefore

$$\mathcal{N}(x, \lambda) = q^{2n} \left\{ (-1)^n \mathcal{K}\ell \left(2\varpi^{-n}, -2\varpi^{-n} \varepsilon_\lambda \right) + 1 + q^{-1} \right\}.$$

Suppose that $0 \leq m < n$. Then by Proposition 2.7, we have

$$\mathcal{K}\ell \left(\frac{2\varpi^{2m-j}}{1-x}, \frac{-2\varpi^{j-2n} \varepsilon_\lambda \varepsilon_x}{1-x} \right) = 0$$

unless $j = m + n$. Hence we have

$$\mathcal{N}(x, \lambda) = |1-x|^{-1} q^{2n}$$

$$\left\{ (-1)^n \mathcal{K}\ell \left(\frac{2\varpi^{-n}}{1-x}, \frac{-2\varpi^{-n} \varepsilon_\lambda}{1-x} \right) + (-1)^{m-n} \mathcal{K}\ell \left(\frac{2\varpi^{m-n}}{1-x}, \frac{-2\varpi^{m-n} \varepsilon_\lambda \varepsilon_x}{1-x} \right) \right\}.$$

Thus we have finished the proof of Theorem 4.13.

CHAPTER 5

Evaluation of the Bessel Orbital Integral

In this chapter we shall evaluate the Bessel orbital integral explicitly for the main relevant double cosets.

5.1. Relevant double cosets

In this section we shall go back and forth between the usual realization of $G = \mathrm{GSp}_4(F)$ and its another realization as G_1, the quaternion similitude unitary group of degree two over D_1. We recall that

$$G_1 = \left\{ g \in \mathrm{GL}_2(D_1) \mid g^* \begin{pmatrix} 0 & 1 \\ 1 & 0 \end{pmatrix} g = \mu(g) \begin{pmatrix} 0 & 1 \\ 1 & 0 \end{pmatrix}, \mu(g) \in \mathrm{GL}_1(F) \right\}$$

where

$$g^* = \begin{pmatrix} \bar{a} & \bar{c} \\ \bar{b} & \bar{d} \end{pmatrix} \text{ for } g = \begin{pmatrix} a & b \\ c & d \end{pmatrix},$$

$$D_1 = \left\{ \begin{pmatrix} u & v \\ v^\sigma & u^\sigma \end{pmatrix} \mid u, v \in E \right\}$$

and $D_1 \ni x \mapsto \bar{x} \in D_1$ denotes the canonical involution of D_1, namely

$$\bar{x} = \begin{pmatrix} u^\sigma & -v \\ -v^\sigma & u \end{pmatrix} \text{ for } x = \begin{pmatrix} u & v \\ v^\sigma & u^\sigma \end{pmatrix}.$$

We also recall that R_1 denotes the upper Bessel subgroup of G_1 defined by

$$R_1 = \left\{ \begin{pmatrix} a & 0 & 0 & 0 \\ 0 & a^\sigma & 0 & 0 \\ 0 & 0 & a & 0 \\ 0 & 0 & 0 & a^\sigma \end{pmatrix} \begin{pmatrix} 1_2 & X \\ 0 & 1_2 \end{pmatrix} \mid a, b \in E^\times, X \in D_1^- \right\}$$

where $D_1^- = \{x \in D_1 \mid \mathrm{tr}(x) = 0\}$. Then the lower Bessel subgroup $\bar{R}_1$ is defined as

$$\bar{R}_1 = \{{}^t r \mid r \in R_1\}.$$

Let P_1 denote the upper Siegel parabolic subgroup of G_1. Then P_1 has the Levi decomposition $P_1 = M_1 U_1$ where

$$M_1 = \left\{ \begin{pmatrix} A & 0 \\ 0 & \lambda \cdot \bar{A}^{-1} \end{pmatrix} \mid A \in D_1^\times, \lambda \in \mathrm{GL}_1(F) \right\}$$

$$U_1 = \left\{ \begin{pmatrix} 1_2 & X \\ 0 & 1_2 \end{pmatrix} \mid X \in D_1^- \right\}.$$

5.1. RELEVANT DOUBLE COSETS

Then the lower Siegel parabolic subgroup $\bar{P}_1$ of G_1 is given by the Levi decomposition $\bar{P}_1 = M_1 \bar{U}_1$ where

$$\bar{U}_1 = \left\{ \begin{pmatrix} 1_2 & 0 \\ Y & 1_2 \end{pmatrix} \mid Y \in D_1^- \right\}.$$

Let us denote the Levi part of R_1 by T_1, i.e.

$$T_1 = \left\{ \begin{pmatrix} a & 0 & 0 & 0 \\ 0 & a^\sigma & 0 & 0 \\ 0 & 0 & a & 0 \\ 0 & 0 & 0 & a^\sigma \end{pmatrix} \mid a \in E^\times \right\}.$$

Then we have $R_1 = T_1 U_1$ and $\bar{R}_1 = T_1 \bar{U}_1$. We often identify T_1 with the subgroup of $D_1^\times$ given by

$$\left\{ \begin{pmatrix} a & 0 \\ 0 & a^\sigma \end{pmatrix} \mid a \in E^\times \right\}.$$

We note that the two realizations, G and G_1, are conjugate in $\mathrm{GL}_4(E)$, namely $G = \alpha G_1 \alpha^{-1}$ where

$$\alpha = \begin{pmatrix} 1 & 0 & 0 & 0 \\ 0 & 1 & 0 & 0 \\ 0 & 0 & 0 & 1 \\ 0 & 0 & -1 & 0 \end{pmatrix} \begin{pmatrix} 1 & 1 & 0 & 0 \\ \eta & -\eta & 0 & 0 \\ 0 & 0 & 1 & 1 \\ 0 & 0 & \eta & -\eta \end{pmatrix}.$$

Then the upper Bessel subgroup R and the lower Bessel subgroup $\bar{R}$ in the ordinary realization G are given by $R = TU$ and $\bar{R} = T\bar{U}$ where

$$T = \left\{ \begin{pmatrix} \begin{pmatrix} a & b \\ bd & a \end{pmatrix} & 0 \\ 0 & \begin{pmatrix} a & -bd \\ -b & a \end{pmatrix} \end{pmatrix} \mid a, b \in F \text{ such that } a^2 - b^2 d \in \mathrm{GL}_1(F) \right\}.$$

We often identify T with the subgroup of $\mathrm{GL}_2(F)$ given by

$$\left\{ \begin{pmatrix} a & b \\ bd & a \end{pmatrix} \mid a, b \in F \text{ such that } a^2 - b^2 d \in \mathrm{GL}_1(F) \right\}.$$

Then we note that

(5.1) $$T = \left\{ g \in \mathrm{GL}_2(F) \mid {}^t g \begin{pmatrix} -d & 0 \\ 0 & 1 \end{pmatrix} g = \det g \cdot \begin{pmatrix} -d & 0 \\ 0 & 1 \end{pmatrix} \right\}.$$

We recall that a character τ of R_1 and a character ξ of $\bar{R}_1$ are defined by

$$\tau \left[\begin{pmatrix} a & 0 & 0 & 0 \\ 0 & a^\sigma & 0 & 0 \\ 0 & 0 & a & 0 \\ 0 & 0 & 0 & a^\sigma \end{pmatrix} \begin{pmatrix} 1_2 & X \\ 0 & 1_2 \end{pmatrix} \right] = \Omega(a) \psi \left[\mathrm{tr} \left(\begin{pmatrix} -\eta & 0 \\ 0 & \eta \end{pmatrix} X \right) \right],$$

$$\xi \left[\begin{pmatrix} a & 0 & 0 & 0 \\ 0 & a^\sigma & 0 & 0 \\ 0 & 0 & a & 0 \\ 0 & 0 & 0 & a^\sigma \end{pmatrix} \begin{pmatrix} 1_2 & 0 \\ Y & 1_2 \end{pmatrix} \right] = \Omega'(a) \psi \left[\mathrm{tr} \left(\begin{pmatrix} -\eta^{-1} & 0 \\ 0 & \eta^{-1} \end{pmatrix} Y \right) \right],$$

respectively, where Ω denotes a character of $E^\times$ and $\Omega'(a) = \Omega(a^\sigma)$ for $a \in E^\times$. But since Ω is unramified and $E^\times = F^\times \mathcal{O}_E^\times$, we have $\Omega = \Omega'$.

We note that in the other realization, τ and ξ are given, respectively, by

$$\tau\left[\begin{pmatrix}\begin{pmatrix}a & b \\ bd & a\end{pmatrix} & 0 \\ 0 & \begin{pmatrix}a & -bd \\ -b & a\end{pmatrix}\end{pmatrix}\begin{pmatrix}1_2 & S \\ 0 & 1_2\end{pmatrix}\right] = \Omega(a+\eta b)\psi\left[\text{tr}\left(\begin{pmatrix}-d & 0 \\ 0 & 1\end{pmatrix}S\right)\right],$$

$$\xi\left[\begin{pmatrix}\begin{pmatrix}a & b \\ bd & a\end{pmatrix} & 0 \\ 0 & \begin{pmatrix}a & -bd \\ -b & a\end{pmatrix}\end{pmatrix}\begin{pmatrix}1_2 & 0 \\ T & 1_2\end{pmatrix}\right] = \Omega(a+\eta b)\psi\left[\text{tr}\left(\begin{pmatrix}-d^{-1} & 0 \\ 0 & 1\end{pmatrix}T\right)\right].$$

DEFINITION 5.1. We call a double coset $\bar{R}sR$ in G *relevant* if the mapping $\bar{R}sR \ni \bar{r}sr \mapsto \xi(\bar{r})\tau(r) \in \mathbb{C}^\times$ is well defined, i.e.

(5.2) $$\xi(\bar{r}) = \tau(s^{-1}\bar{r}s) \quad \text{for any} \quad \bar{r} \in \bar{R} \cap sRs^{-1}.$$

Let us describe the relevant double cosets explicitly. First we recall the Bruhat decomposition

$$G = \bar{P}P \cup \bar{P}w_1 P \cup \bar{P}w_2 P$$

where $w_1 = \begin{pmatrix} 0 & 0 & 1 & 0 \\ 0 & 0 & 0 & 1 \\ -1 & 0 & 0 & 0 \\ 0 & -1 & 0 & 0 \end{pmatrix}$ and $w_2 = \begin{pmatrix} 1 & 0 & 0 & 0 \\ 0 & 0 & 0 & 1 \\ 0 & 0 & 1 & 0 \\ 0 & -1 & 0 & 0 \end{pmatrix}.$

PROPOSITION 5.2. 1. *All the* $(\bar{R}, R)$-*double cosets in* $\bar{P}P$ *are relevant.*
2. *There exist two relevant double cosets in* $\bar{P}w_1 P$, *namely*

$$\bar{R}\begin{pmatrix} 0 & 0 & 0 & 1 \\ 0 & 0 & -1 & 0 \\ 0 & -d & 0 & 0 \\ d & 0 & 0 & 0 \end{pmatrix} R \quad \text{and} \quad \bar{R}\begin{pmatrix} 0 & 0 & 0 & 1 \\ 0 & 0 & 1 & 0 \\ 0 & -d & 0 & 0 \\ -d & 0 & 0 & 0 \end{pmatrix} R.$$

3. *The relevant double cosets in* $\bar{P}w_2 P$ *are the double cosets represented as*

$$\bar{R}w_2 \begin{pmatrix} a & 0 & 0 & 0 \\ 0 & 1 & 0 & 0 \\ 0 & 0 & -a^{-1} & 0 \\ 0 & 0 & 0 & -1 \end{pmatrix} R, \quad a \in F^\times.$$

PROOF. 1. Suppose that $\bar{R}sR \subset \bar{P}P$. Then it is clear that we may take s to be of the form

$$s = \begin{pmatrix} h & 0 \\ 0 & \lambda \cdot {}^t h^{-1} \end{pmatrix}.$$

Since $s^{-1}\bar{U}s = \bar{U}$, we have $\bar{R} \cap sRs^{-1} \subset T$. Let $T' = \bar{R} \cap sRs^{-1}$. Here we note that when we identify T with $E^\times$, we have $T' = F^\times K$ where $K = T' \cap \mathcal{O}_E^\times$. Suppose that $t' \in T'$ and let us write $t' = zk$ where $z \in F^\times$ and $k \in K$. Then since Ω is unramified, we have $\xi(t') = \Omega(t') = \Omega(z)$. On the other hand, we have $\tau(s^{-1}t's) = \Omega(z)\Omega(s^{-1}ks)$. Here $s^{-1}ks$ is an element of the compact subgroup $s^{-1}Ks$ of T. Since any compact subgroup of $E^\times$ is contained in $\mathcal{O}_E^\times$, we have $\Omega(s^{-1}ks) = 1$. Thus we have $\xi(t') = \tau(s^{-1}t's)$ for $t' \in T' = \bar{R} \cap sRs^{-1}$. Hence the double coset $\bar{R}sR$ is relevant.
2. Suppose that $\bar{R}sR$ is a relevant double coset in $\bar{P}w_1 P$. Since $\bar{P}w_1 P = \bar{U}w_1 P$, we may take s to be of the form

$$s = w_1 \begin{pmatrix} h & 0 \\ 0 & \lambda \cdot {}^t h^{-1} \end{pmatrix}.$$

Then since
$$s^{-1}\begin{pmatrix} 1_2 & 0 \\ Y & 1_2 \end{pmatrix} s = \begin{pmatrix} 1_2 & -\lambda \cdot h^{-1}Y^t h^{-1} \\ 0 & 1_2 \end{pmatrix},$$
we have
$$\psi\left[\operatorname{tr}\left(\begin{pmatrix} -d^{-1} & 0 \\ 0 & 1 \end{pmatrix}Y\right)\right] = \psi\left[-\lambda \cdot \operatorname{tr}\left(\begin{pmatrix} -d & 0 \\ 0 & 1 \end{pmatrix} h^{-1}Y^t h^{-1}\right)\right]$$
for any $Y \in \operatorname{Sym}^2(F)$. Thus we have
$$^t(w_0 h)\begin{pmatrix} -d & 0 \\ 0 & 1 \end{pmatrix}(w_0 h) = d\lambda \cdot \begin{pmatrix} -d & 0 \\ 0 & 1 \end{pmatrix} \quad \text{where} \quad w_0 = \begin{pmatrix} 0 & 1 \\ -1 & 0 \end{pmatrix}.$$
By taking the determinant of the both sides, we have $\lambda = \pm d^{-1}\det h$. Hence by (5.1), we have $w_0 h \in T$ or $\begin{pmatrix} 1 & 0 \\ 0 & -1 \end{pmatrix} w_0 h \in T$. When $w_0 h \in T$, let us write $w_0 h = g$. Then we have
$$s = w_1 \begin{pmatrix} w_0^{-1} & 0 \\ 0 & d^{-1 t}w_0 \end{pmatrix}\begin{pmatrix} g & 0 \\ 0 & \det g \cdot {}^t g^{-1} \end{pmatrix} \in \bar{R}\begin{pmatrix} 0 & 0 & 0 & 1 \\ 0 & 0 & -1 & 0 \\ 0 & -d & 0 & 0 \\ d & 0 & 0 & 0 \end{pmatrix} R.$$

It is easily seen that this double coset is indeed relevant. The other case is similar.

3. Suppose that $\bar{R}sR$ is a relevant double coset in $\bar{P}w_2 P$. Then it is clear that we may take s to be of the form $m_1 w_2 m_2$ where $m_1, m_2 \in M$. Then we note that
$$\operatorname{GL}_2(F) = TB_0 = \bar{B}_0 T$$
where B_0 and $\bar{B}_0$ denote the upper and the lower Borel subgroup of $\operatorname{GL}_2(F)$ respectively. Hence by an argument similar to the one given in the proof of Proposition 4.1, we may assume that $m_1 = 1_4$ and that m_2 is of the form
$$m_2 = \begin{pmatrix} a & 0 & 0 & 0 \\ 0 & 1 & 0 & 0 \\ 0 & 0 & \lambda & 0 \\ 0 & 0 & 0 & \lambda a \end{pmatrix}$$
since
$$w_2 \begin{pmatrix} 1 & 0 & 0 & 0 \\ y & 1 & 0 & 0 \\ 0 & 0 & 1 & -y \\ 0 & 0 & 0 & 1 \end{pmatrix} w_2^{-1} = \begin{pmatrix} 1 & 0 & 0 & 0 \\ 0 & 1 & 0 & 0 \\ 0 & -y & 1 & 0 \\ -y & 0 & 0 & 1 \end{pmatrix} \in \bar{U}.$$
Then for $s = w_2 m_2$ we have
$$s^{-1} \begin{pmatrix} 1 & 0 & 0 & 0 \\ 0 & 1 & 0 & 0 \\ x & y & 1 & 0 \\ y & z & 0 & 1 \end{pmatrix} s = \begin{pmatrix} 1 & 0 & 0 & 0 \\ -ay & 1 & 0 & -\lambda az \\ \lambda^{-1}ax & 0 & 1 & ay \\ 0 & 0 & 0 & 1 \end{pmatrix}.$$
Hence when $\bar{R}sR$ is relevant, we have
$$\psi\left[\operatorname{tr}\left(\begin{pmatrix} -d^{-1} & 0 \\ 0 & 1 \end{pmatrix}\begin{pmatrix} 0 & 0 \\ 0 & z \end{pmatrix}\right)\right] = \psi\left[\operatorname{tr}\left(\begin{pmatrix} -d & 0 \\ 0 & 1 \end{pmatrix}\begin{pmatrix} 0 & 0 \\ 0 & -\lambda az \end{pmatrix}\right)\right]$$
for any $z \in F$. Thus we have $\lambda = -a^{-1}$. It is easily seen that such a double coset is indeed relevant. $\square$

Let us parameterize the relevant double cosets in $\bar{P}P$. In order for that and also in the evaluation of the orbital integrals, we employ the G_1 realization of the symplectic similitude group.

LEMMA 5.3. *We have*

$$(5.3) \qquad \bar{P}_1 P_1 = \bigcup_{\substack{A \in T_1 \backslash D_1^\times / T_1 \\ \mu \in F^\times}} \bar{R}_1 \begin{pmatrix} A & 0 \\ 0 & \mu \cdot \bar{A}^{-1} \end{pmatrix} R_1.$$

PROOF. Only the disjointness of the decomposition is non-trivial. Suppose that

$$\bar{R}_1 \begin{pmatrix} A & 0 \\ 0 & \mu \cdot \bar{A}^{-1} \end{pmatrix} R = \bar{R}_1 \begin{pmatrix} A' & 0 \\ 0 & \mu' \cdot \bar{A}'^{-1} \end{pmatrix} R.$$

Then we have

$$\begin{pmatrix} A & 0 \\ 0 & \mu \cdot \bar{A}^{-1} \end{pmatrix} = \begin{pmatrix} h_1 & 0 \\ 0 & \det h_1 \cdot {}^t h_1^{-1} \end{pmatrix} \begin{pmatrix} A' & 0 \\ 0 & \mu' \cdot \bar{A}'^{-1} \end{pmatrix} \begin{pmatrix} h_2 & 0 \\ 0 & \det h_1 \cdot {}^t h_2^{-1} \end{pmatrix}$$

for some $h_1, h_2 \in T_1$. Thus $A = A'$. By taking the similitude of the both sides, we have $\mu = \det h_1 \cdot \det h_2 \cdot \mu_1$. On the other hand, $A = h_1 A h_2$ implies that $\det h_1 \cdot \det h_2 = 1$. Hence we have $\mu = \mu'$. □

In order to parameterize the relevant double cosets in $\bar{P}_1 P_1$, let us recall some facts concerning the $T_1 \backslash D_1^\times / T_1$ double coset decomposition following Jacquet [**J1**], mutatis mutandis. We define a mapping $X_1 : D_1^\times \to F \cup \{\infty\}$ by

$$X_1 \begin{pmatrix} a & b \\ b^\sigma & a^\sigma \end{pmatrix} = \frac{bb^\sigma}{aa^\sigma}.$$

Then since

$$X_1 \left[\begin{pmatrix} a_1 & 0 \\ 0 & a_1^\sigma \end{pmatrix} \begin{pmatrix} a & b \\ b^\sigma & a^\sigma \end{pmatrix} \begin{pmatrix} a_2 & 0 \\ 0 & a_2^\sigma \end{pmatrix} \right] = X_1 \begin{pmatrix} a_1 a_2 a & a_1 a_2^\sigma b \\ a_1^\sigma a_2 b^\sigma & a_1^\sigma a_2^\sigma a^\sigma \end{pmatrix} = X_1 \begin{pmatrix} a & b \\ b^\sigma & a^\sigma \end{pmatrix},$$

the mapping X_1 is indeed from $T_1 \backslash D_1^\times / T_1$ to $F \cup \{\infty\}$.

LEMMA 5.4. *The mapping* $X_1 : T_1 \backslash D_1^\times / T_1 \to F \cup \{\infty\}$ *is injective.*

PROOF. When $X_1 \begin{pmatrix} a & b \\ b^\sigma & a^\sigma \end{pmatrix} = 0, \infty$, we have $b = 0, a = 0$, respectively, and then

$$\begin{pmatrix} a & 0 \\ 0 & a^\sigma \end{pmatrix} \in T_1, \quad \begin{pmatrix} 0 & b \\ b^\sigma & 0 \end{pmatrix} = \begin{pmatrix} b & 0 \\ 0 & b^\sigma \end{pmatrix} \begin{pmatrix} 0 & 1 \\ 1 & 0 \end{pmatrix} \in T_1 \begin{pmatrix} 0 & 1 \\ 1 & 0 \end{pmatrix}.$$

Suppose that $X_1 \begin{pmatrix} a & b \\ b^\sigma & a^\sigma \end{pmatrix} \neq 0, \infty$ and $X_1 \begin{pmatrix} a & b \\ b^\sigma & a^\sigma \end{pmatrix} = X_1 \begin{pmatrix} a_1 & b_1 \\ b_1^\sigma & a_1^\sigma \end{pmatrix}$. Then we have $bb^\sigma (aa^\sigma)^{-1} = b_1 b_1^\sigma (a_1 a_1^\sigma)^{-1}$. By Hilbert's Theorem 90, there exists $c \in E^\times$ such that $ba^{-1} = b_1 a_1^{-1} c^\sigma c^{-1}$. Then we have

$$\begin{pmatrix} a & b \\ b^\sigma & a^\sigma \end{pmatrix} = \begin{pmatrix} aa_1^{-1} c^{-1} & 0 \\ 0 & (aa_1^{-1} c^{-1})^\sigma \end{pmatrix} \begin{pmatrix} a_1 & b_1 \\ b_1^\sigma & a_1^\sigma \end{pmatrix} \begin{pmatrix} c & 0 \\ 0 & c^\sigma \end{pmatrix}.$$

□

DEFINITION 5.5. We say that a double coset $T_1 h T_1$ in $D_1^\times$ is T_1-*regular* when $X_1(h) \neq 0, \infty$.

We say that a double coset $T_1 h T_1$ in $D_1^\times$ is T_1-*singular* when $X_1(h) = 0, \infty$.

5.1. RELEVANT DOUBLE COSETS

LEMMA 5.6. *The mapping X_1 induces a bijection between the T_1-regular double coset in $D_1^\times$ and $N_{E/F}(E^\times) \setminus \{1\}$.*

PROOF. Suppose that $h = \begin{pmatrix} a & b \\ b^\sigma & a^\sigma \end{pmatrix}$ is T_1-regular. Then $X_1(h) = N_{E/F}(b/a)$. Since h is invertible, we have $aa^\sigma - bb^\sigma \neq 0$ and hence $X_1(h) \neq 1$.

Conversely, for a given $x \in N_{E/F}(E^\times) \setminus \{1\}$, take $b \in E^\times$ such that $bb^\sigma = x$. Then we have

$$\begin{pmatrix} 1 & b \\ b^\sigma & 1 \end{pmatrix} \in D_1^\times \quad \text{and} \quad X_1 \begin{pmatrix} 1 & b \\ b^\sigma & 1 \end{pmatrix} = x.$$

□

As for the stabilizer of a T_1-regular double coset, we have the following lemma.

LEMMA 5.7. *Suppose that $h = \begin{pmatrix} a & b \\ b^\sigma & a^\sigma \end{pmatrix} \in D_1^\times$ is T_1-regular. Then for $a_1, a_2 \in E^\times$,*

$$\begin{pmatrix} a_1 & 0 \\ 0 & a_1^\sigma \end{pmatrix} \begin{pmatrix} a & b \\ b^\sigma & a^\sigma \end{pmatrix} \begin{pmatrix} a_2 & 0 \\ 0 & a_2^\sigma \end{pmatrix} = \begin{pmatrix} a & b \\ b^\sigma & a^\sigma \end{pmatrix} \begin{pmatrix} \lambda & 0 \\ 0 & \lambda \end{pmatrix}$$

for some $\lambda \in F^\times$ implies that $a_1, a_2 \in F^\times$.

PROOF. We have

$$\begin{pmatrix} a_1 a_2 a & a_1 a_2^\sigma b \\ a_1^\sigma a_2 b^\sigma & a_1^\sigma a_2^\sigma a^\sigma \end{pmatrix} = \begin{pmatrix} a & b \\ b^\sigma & a^\sigma \end{pmatrix} \begin{pmatrix} \lambda & 0 \\ 0 & \lambda \end{pmatrix}.$$

Since h is T_1-regular, we have $a \neq 0$ and $b \neq 0$. Hence $a_1 a_2 = a_1 a_2^\sigma = \lambda \in F^\times$. Thus $a_2 = a_2^\sigma$, i.e. $a_2 \in F^\times$ and $a_1 = \lambda a_2^{-1} \in F^\times$. □

For $A \in D_1^\times$ and $\mu \in F^\times$, let us denote by (A, μ), the element of G_1 defined by

$$(A, \mu) = \begin{pmatrix} A & 0 \\ 0 & \mu \cdot \bar{A}^{-1} \end{pmatrix}.$$

Then we may summarize our $(\bar{R}_1, R_1)$-double coset decomposition in $\bar{P}_1 P_1$ as the following proposition.

PROPOSITION 5.8. *1. The $(\bar{R}_1, R_1)$-double cosets in $\bar{P}_1 P_1$ are represented by (A, μ) where $\mu \in F^\times$ and*

$$A = \begin{pmatrix} 1 & 0 \\ 0 & 1 \end{pmatrix}, \quad \begin{pmatrix} 0 & 1 \\ 1 & 0 \end{pmatrix}$$

or

(5.4) $\qquad A = \begin{pmatrix} 1 & u \\ u^\sigma & 1 \end{pmatrix} \quad \text{where} \quad u \in E^\times \quad \text{such that} \quad uu^\sigma \neq 1.$

2. When $s = (A, \mu)$ where A is of the form (5.4), we have

$$\bar{R}_1 \cap s R_1 s^{-1} = Z_1$$

where Z_1 denotes the center of G_1.

5.2. Evaluation of the Bessel orbital integral

DEFINITION 5.9. Let Ξ_1 be the characteristic function of $\mathcal{K}_1$, the maximal compact subgroup of G_1 defined by $\mathcal{K}_1 = G_1 \cap \mathrm{GL}_4(\mathcal{O}_E)$.

For $u \in E^\times$ such that $uu^\sigma \neq 1$, let us define a matrix A_u by
$$A_u = \begin{pmatrix} 1 & u \\ u^\sigma & 1 \end{pmatrix}.$$

Then for $\mu \in F^\times$, we define the Bessel orbital integral $\mathcal{B}(u, \mu)$ by

(5.5) $\quad \mathcal{B}(u,\mu) = \Omega(u) \int_{\bar{R}_1/Z_1} \int_{R_1} \Xi_1 \left[\bar{r} \begin{pmatrix} A_u & 0 \\ 0 & \mu \cdot \bar{A}_u^{-1} \end{pmatrix} r \right] \xi(\bar{r}) \tau(r) \, d\bar{r} \, dr.$

The goal of this section is evaluate $\mathcal{B}(u, \mu)$ explicitly.

First we note that since $T_1 = Z_1(T_1 \cap \mathcal{K}_1)$, the integral (5.5) simplifies to

(5.6) $\quad \mathcal{B}(u,\mu) = \Omega(u) \int_{Z_1} \int_{\bar{U}_1} \int_{U_1} \Xi_1 \left[z \begin{pmatrix} 1_2 & 0 \\ Y & 1_2 \end{pmatrix} \begin{pmatrix} A_u & 0 \\ 0 & \mu \cdot \bar{A}_u^{-1} \end{pmatrix} \begin{pmatrix} 1_2 & X \\ 0 & 1_2 \end{pmatrix} \right]$
$$\Omega(z) \psi \left[\mathrm{tr} \left(\begin{pmatrix} -\eta^{-1} & 0 \\ 0 & \eta^{-1} \end{pmatrix} (dX + Y) \right) \right] dX \, dY \, d^\times z.$$

Now we observe some properties of $\mathcal{B}(u, \mu)$.

PROPOSITION 5.10. *1. The integral $\mathcal{B}(u,\mu)$ vanishes unless $\mathrm{ord}(\mu)$ is even and $\mathrm{ord}(\mu) \leq \min\{0, \mathrm{ord}(uu^\sigma)\}$.*

Suppose that $\mathrm{ord}(\mu) = -2n$ where $n \geq 0$. Let us write $\mu = \varpi^{-2n} \varepsilon_\mu$. Then we have

(5.7) $\quad \mathcal{B}(u,\mu) = \Omega(u) \Omega(\varpi)^n \int_{\bar{U}_1} \int_{U_1} \psi \left[\mathrm{tr} \left(\begin{pmatrix} -\eta^{-1} & 0 \\ 0 & \eta^{-1} \end{pmatrix} (d\varepsilon_\mu X + Y) \right) \right]$
$$\Xi_1 \left[\begin{pmatrix} 1_2 & 0 \\ Y & 1_2 \end{pmatrix} \begin{pmatrix} \varpi^n A_u & 0 \\ 0 & \varpi^{-n} \bar{A}_u^{-1} \end{pmatrix} \begin{pmatrix} 1_2 & X \\ 0 & 1_2 \end{pmatrix} \right] dX \, dY.$$

2. We have the functional equation

(5.8) $\quad \mathcal{B}(u,\mu) = \Omega(u) \mathcal{B}\left(u^{-1}, \mu(uu^\sigma)^{-1}\right).$

PROOF. 1. Suppose that in (5.6) we have
$$z \begin{pmatrix} 1_2 & 0 \\ Y & 1_2 \end{pmatrix} \begin{pmatrix} A_u & 0 \\ 0 & \mu \cdot \bar{A}_u^{-1} \end{pmatrix} \begin{pmatrix} 1_2 & X \\ 0 & 1_2 \end{pmatrix} \in \mathcal{K}_1.$$

Then by taking the similitude, we have $z^2\mu \in \mathcal{O}_E^\times$. Hence $\mathrm{ord}(\mu)$ is even. Also we have $zA_u \in \mathrm{Mat}_{2\times 2}(\mathcal{O}_E)$ and hence $|z| \leq 1$ and $|zu| \leq 1$. Thus $|\mu| = |z|^{-2} \geq 1$ and $|\mu^{-1} uu^\sigma| = |z^2 uu^\sigma| \leq 1$.

Suppose that $\mathrm{ord}(\mu) = -2n$ where $n \geq 0$. Then as we have shown, the integral (5.6) is supported on $z \in \varpi^n \mathcal{O}_F^\times$. Hence we have

$$\mathcal{B}(u,\mu) = \Omega(u) \Omega(\varpi)^n \int_{\bar{U}_1} \int_{U_1} \psi \left[\mathrm{tr} \left(\begin{pmatrix} -\eta^{-1} & 0 \\ 0 & \eta^{-1} \end{pmatrix} (dX + Y) \right) \right]$$
$$\Xi_1 \left[\begin{pmatrix} 1_2 & 0 \\ Y & 1_2 \end{pmatrix} \begin{pmatrix} \varpi^n A_u & 0 \\ 0 & \varpi^{-n} \varepsilon_\mu \bar{A}_u^{-1} \end{pmatrix} \begin{pmatrix} 1_2 & X \\ 0 & 1_2 \end{pmatrix} \right] dX \, dY.$$

Then a change of variable $X \mapsto \varepsilon_\mu X$ yields (5.7).

2. We rewrite (5.6) as

$$\mathcal{B}(u,\mu) = \Omega(u) \int_{Z_1} \int_{\bar{U}_1} \int_{U_1} \Omega(z) \psi \left[\text{tr} \left(\begin{pmatrix} -\eta^{-1} & 0 \\ 0 & \eta^{-1} \end{pmatrix} (dX + Y) \right) \right]$$

$$\Xi_1 \left[z \begin{pmatrix} 1_2 & 0 \\ Y & 1_2 \end{pmatrix} \begin{pmatrix} 1_2 & 0 \\ 0 & \mu \det A_u^{-1} \cdot 1_2 \end{pmatrix} \begin{pmatrix} 1_2 & A_u X A_u^{-1} \\ 0 & 1_2 \end{pmatrix} \begin{pmatrix} A_u & 0 \\ 0 & A_u \end{pmatrix} \right] dX\, dY\, d^\times z.$$

As we have shown (3.34), we have

$$A_u \begin{pmatrix} -\eta^{-1} & 0 \\ 0 & \eta^{-1} \end{pmatrix} A_u^{-1} = -A_{u^{-\sigma}} \begin{pmatrix} -\eta^{-1} & 0 \\ 0 & \eta^{-1} \end{pmatrix} A_{u^{-\sigma}}^{-1}.$$

We also note that $\det A_u = 1 - uu^\sigma = -uu^\sigma \cdot \det A_{u^{-\sigma}}$. Thus by a change of variable $X \mapsto A_u^{-1} X A_u$, we have

$$\mathcal{B}(u,\mu) = \int_{Z_1} \int_{\bar{U}_1} \int_{U_1} \Omega(uz) \psi \left[\text{tr} \left(\begin{pmatrix} -\eta^{-1} & 0 \\ 0 & \eta^{-1} \end{pmatrix} (-dA_{u^{-\sigma}}^{-1} X A_{u^{-\sigma}} + Y) \right) \right]$$

$$\Xi_1 \left[z \begin{pmatrix} 1_2 & 0 \\ Y & 1_2 \end{pmatrix} \begin{pmatrix} 1_2 & 0 \\ 0 & -\frac{\mu}{uu^\sigma} \det A_{u^{-\sigma}}^{-1} \cdot 1_2 \end{pmatrix} \begin{pmatrix} 1_2 & X \\ 0 & 1_2 \end{pmatrix} \begin{pmatrix} A_u & 0 \\ 0 & A_u \end{pmatrix} \right] dX\, dY\, d^\times z.$$

Further by a change of variable $X \mapsto -A_{u^{-\sigma}} X A_{u^{-\sigma}}^{-1}$, we have

$$\mathcal{B}(u,\mu) = \int_{Z_1} \int_{\bar{U}_1} \int_{U_1} dX\, dY\, d^\times z \cdot \Omega(uz) \psi \left[\text{tr} \left(\begin{pmatrix} -\eta^{-1} & 0 \\ 0 & \eta^{-1} \end{pmatrix} (dX + Y) \right) \right]$$

$$\Xi_1 \left[z \begin{pmatrix} 1_2 & 0 \\ Y & 1_2 \end{pmatrix} \begin{pmatrix} A_{u^{-\sigma}} & 0 \\ 0 & \frac{\mu}{uu^\sigma} \bar{A}_{u^{-\sigma}}^{-1} \end{pmatrix} \begin{pmatrix} 1_2 & X \\ 0 & 1_2 \end{pmatrix} \begin{pmatrix} A_{u^{-\sigma}}^{-1} A_u & 0 \\ 0 & -A_{u^{-\sigma}}^{-1} A_u \end{pmatrix} \right].$$

Here we note that

$$A_{u^{-\sigma}}^{-1} A_u = \begin{pmatrix} 0 & u \\ u^\sigma & 0 \end{pmatrix} = \varpi^m \begin{pmatrix} 0 & \varepsilon_u \\ \varepsilon_u^\sigma & 0 \end{pmatrix}$$

where $m = \text{ord}(u)$. Hence

$$\begin{pmatrix} A_{u^{-\sigma}}^{-1} A_u & 0 \\ 0 & -A_{u^{-\sigma}}^{-1} A_u \end{pmatrix} \in \varpi^m \cdot \mathcal{K}_1.$$

Since Ξ_1 is right $\mathcal{K}_1$-invariant and Ω is unramified, we have

$$\mathcal{B}(u,\mu) = \int_{Z_1} \int_{\bar{U}_1} \int_{U_1} dX\, dY\, d^\times z \cdot \Omega(\varpi^m z) \psi \left[\text{tr} \left(\begin{pmatrix} -\eta^{-1} & 0 \\ 0 & \eta^{-1} \end{pmatrix} (dX + Y) \right) \right]$$

$$\Xi_1 \left[\varpi^m z \begin{pmatrix} 1_2 & 0 \\ Y & 1_2 \end{pmatrix} \begin{pmatrix} A_{u^{-\sigma}} & 0 \\ 0 & \frac{\mu}{uu^\sigma} \bar{A}_{u^{-\sigma}}^{-1} \end{pmatrix} \begin{pmatrix} 1_2 & X \\ 0 & 1_2 \end{pmatrix} \right].$$

Hence

$$\mathcal{B}(u,\lambda) = \Omega(u^\sigma) \mathcal{B}\left(u^{-\sigma}, \mu(uu^\sigma)^{-1}\right) = \Omega(u) \mathcal{B}\left(u^{-1}, \mu(uu^\sigma)^{-1}\right).$$

$\square$

Now we state our main theorem in this chapter.

THEOREM 5.11. *Let $\mu \in F^\times$ and $u \in E^\times$ such that $uu^\sigma \neq 1$.*
1. *The Bessel orbital integral $\mathcal{B}(u,\mu)$ vanishes unless $\text{ord}(\mu)$ is even and*

$$\text{ord}(\mu) \leq \min\{0, \text{ord}(uu^\sigma)\}.$$

2. We have the functional equation
$$\mathcal{B}(u,\mu) = \Omega(u)\,\mathcal{B}\left(u^{-1}, \mu(uu^\sigma)^{-1}\right).$$

3. Suppose that $\mu \in \mathcal{O}_F^\times$.
   (a) When $|1 - uu^\sigma| = 1$, we have
   $$\mathcal{B}(u,\mu) = \Omega(u).$$
   (b) When $|1 - uu^\sigma| < 1$, we have
   $$\mathcal{B}(u,\lambda) = |1 - uu^\sigma|^{-1} \cdot \mathcal{K}\ell\left(2(1-uu^\sigma)^{-1}, -2\mu(1-uu^\sigma)^{-1}\right).$$

4. Suppose that $\mathrm{ord}(\mu) = -2n$ where $n > 0$, and, $\mathrm{ord}(u) = m$ where $m \geq 0$. Let us write $\mu = \varpi^{-2n}\varepsilon_\mu$ and $u = \varpi^m \varepsilon_u$.
   (a) When $m \geq n$, we have
   $$\mathcal{B}(u,\mu) = \Omega(\varpi)^{m+n} q^{2n} \left\{(-1)^n \cdot \mathcal{K}\ell(2\varpi^{-n}, -2\varpi^{-n}\varepsilon_\mu) + 1 + q^{-1}\right\}.$$
   (b) When $0 \leq m < n$, we have
   $$\mathcal{B}(u,\mu) = \Omega(\varpi)^{m+n} |1-uu^\sigma|^{-1}(-1)^n$$
   $$\left\{\mathcal{K}\ell\left(\frac{2\varpi^{-n}}{1-uu^\sigma}, \frac{-2\varpi^{-n}\varepsilon_\mu}{1-uu^\sigma}\right) + (-1)^m \cdot \mathcal{K}\ell\left(\frac{2\varpi^{m-n}}{1-uu^\sigma}, \frac{-2\varpi^{m-n}\varepsilon_\mu\varepsilon_u\varepsilon_u^\sigma}{1-uu^\sigma}\right)\right\}.$$

The first two assertions have been proved already. We shall prove the last two assertions in the following subsections.

5.2.1. Evaluation of $\mathcal{B}(u,\mu)$ when $\mu \in \mathcal{O}_F^\times$ and $u \in \mathcal{O}_E$. In this case (5.7) becomes

(5.9) $$\mathcal{B}(u,\mu) = \Omega(u) \int_{\bar{U}_1}\int_{U_1} \psi\left[\mathrm{tr}\left(\begin{pmatrix}-\eta^{-1} & 0 \\ 0 & \eta^{-1}\end{pmatrix}(d\varepsilon_\mu X + Y)\right)\right]$$
$$\Xi_1\left[\begin{pmatrix}1_2 & 0 \\ Y & 1_2\end{pmatrix}\begin{pmatrix}A_u & 0 \\ 0 & \bar{A}_u^{-1}\end{pmatrix}\begin{pmatrix}1_2 & X \\ 0 & 1_2\end{pmatrix}\right] dX\,dY.$$

LEMMA 5.12. *In* (5.9), *let us consider the condition*

(5.10) $$\begin{pmatrix}1_2 & 0 \\ Y & 1_2\end{pmatrix}\begin{pmatrix}A_u & 0 \\ 0 & \bar{A}_u^{-1}\end{pmatrix} \in \mathcal{K}_1 U_1$$

for $Y = \begin{pmatrix} v & w \\ w^\sigma & -v \end{pmatrix} \in D_1^-$, *i.e.* $v, w \in E$ *and* $v^\sigma = -v$.

   (a) When $|1-uu^\sigma| = 1$, the condition (5.10) holds if and only if
(5.11) $$|v| \leq 1 \text{ and } |w| \leq 1.$$
   (b) When $|1-uu^\sigma| < 1$, the condition (5.10) holds if and only if
(5.12) $$|(1-uu^\sigma)v| = 1 \text{ and } |uv+w| \leq 1.$$

PROOF. Suppose that $|1-uu^\sigma| = 1$. Then we have $A_u \in \mathrm{GL}_2(\mathcal{O}_E)$. Hence the conditions (5.10) and (5.11) are clearly equivalent.

Suppose that $|1-uu^\sigma| < 1$. First we note that then $u \in \mathcal{O}_E^\times$. Hence by Lemma 4.9, the condition (5.10) holds if and only if

(5.13) $$\max\{|v + u^\sigma w|, |uv + w|\} \leq 1$$

5.2. EVALUATION OF THE BESSEL ORBITAL INTEGRAL

and
(5.14) $$\max\left\{|(1-uu^\sigma)v|, |(1-uu^\sigma)w|, |(1-uu^\sigma)(v^2+ww^\sigma)|\right\} = 1.$$

Here we note the Plücker relation
$$(1-uu^\sigma)(v^2+ww^\sigma) = (1-uu^\sigma)v\cdot(v+u^\sigma w) + (1-uu^\sigma)w\cdot(uv+w)^\sigma.$$
Hence when we assume (5.13), we have
$$|(1-uu^\sigma)(v^2+ww^\sigma)| \leq \max\{|(1-uu^\sigma)v|, |(1-uu^\sigma)w|\}.$$
Therefore the condition (5.10) is equivalent to (5.13) and
(5.15) $$\max\{|(1-uu^\sigma)v|, |(1-uu^\sigma)w|\} = 1.$$

Here we recall two more Plücker relations
(5.16) $$(1-uu^\sigma)v + (1-uu^\sigma)w\cdot u^\sigma - (1-uu^\sigma)(v+u^\sigma w) = 0,$$
(5.17) $$(1-uu^\sigma)w + (1-uu^\sigma)v\cdot u - (1-uu^\sigma)\cdot(uv+w) = 0.$$

Suppose that (5.13) and (5.15) hold. Then by (5.16), we must have
$$|(1-uu^\sigma)v| = 1$$
since $u \in \mathcal{O}_E^\times$ and $1 - uu^\sigma \in \varpi\mathcal{O}_F$.

Suppose conversely that (5.12) holds. Then we have $|(1-uu^\sigma)w| \leq 1$ from (5.17). Since $v + u^\sigma w = (1-uu^\sigma)v + u^\sigma(uv+w)$, we also have $|v+u^\sigma w| \leq 1$. Thus both (5.13) and (5.15) hold. □

Let us evaluate $\mathcal{B}(u, \mu)$. When $|1-uu^\sigma| = 1$, it is clear from (5.11) that we have $\mathcal{B}(u,\mu) = \Omega(u)$.

Suppose that $|1-uu^\sigma| < 1$. For $v, w \in E$ such that $v^\sigma = -v$, let us suppose that (5.12) holds. Then since

$$\begin{pmatrix} 1 & 0 & 0 & 0 \\ 0 & 1 & 0 & 0 \\ v & w & 1 & 0 \\ w^\sigma & -v & 0 & 1 \end{pmatrix} \left(\begin{pmatrix} 1 & u \\ u^\sigma & 1 \end{pmatrix} \quad 0 \atop 0 \quad \frac{1}{1-uu^\sigma}\begin{pmatrix} 1 & u \\ u^\sigma & 1 \end{pmatrix} \right)$$
$$= \begin{pmatrix} 1 & 0 & 0 & 0 \\ 0 & 1 & 0 & 0 \\ 0 & uv+w & 1 & 0 \\ -u^\sigma v + w^\sigma & 0 & 0 & 1 \end{pmatrix} \left(\begin{pmatrix} 1 & u \\ u^\sigma & 1 \end{pmatrix} \quad 0 \atop (1-uu^\sigma)v\begin{pmatrix} 1 & 0 \\ 0 & -1 \end{pmatrix} \quad \frac{1}{1-uu^\sigma}\begin{pmatrix} 1 & u \\ u^\sigma & 1 \end{pmatrix} \right),$$

we may write
$$\left(\begin{pmatrix} 1 & u \\ u^\sigma & 1 \end{pmatrix} \quad 0 \atop (1-uu^\sigma)v\begin{pmatrix} 1 & 0 \\ 0 & -1 \end{pmatrix} \quad \frac{1}{1-uu^\sigma}\begin{pmatrix} 1 & u \\ u^\sigma & 1 \end{pmatrix} \right) = \begin{pmatrix} k_1 & k_2 \\ k_3 & k_4 \end{pmatrix}\begin{pmatrix} 1_2 & -X \\ 0 & 1_2 \end{pmatrix}$$

for some $\begin{pmatrix} k_1 & k_2 \\ k_3 & k_4 \end{pmatrix} \in \mathcal{K}_1$ and $X \in D_1^-$. Here
$$k_3 = (1-uu^\sigma)v\begin{pmatrix} 1 & 0 \\ 0 & -1 \end{pmatrix} \in \mathrm{GL}_2(\mathcal{O}_E), \quad -k_3 X + k_4 = \frac{1}{1-uu^\sigma}\begin{pmatrix} 1 & u \\ u^\sigma & 1 \end{pmatrix}.$$

Thus we have
$$X \equiv -(1-uu^\sigma)^{-2}v^{-1}\begin{pmatrix} 1 & u \\ -u^\sigma & -1 \end{pmatrix} \pmod{\mathcal{O}_E}.$$

Hence (5.9) becomes

$$\mathcal{B}(u,\mu) = \int_{|(1-uu^\sigma)v|=1, |uv+w|\leq 1}$$
$$\psi\left[\mathrm{tr}\left(\begin{pmatrix} -\eta^{-1} & 0 \\ 0 & \eta^{-1} \end{pmatrix}\left(\frac{-d\mu}{(1-uu^\sigma)^2 v}\begin{pmatrix} 1 & u \\ -u^\sigma & -1 \end{pmatrix} + \begin{pmatrix} v & w \\ w^\sigma & -v \end{pmatrix}\right)\right)\right] dv\, dw$$
$$= \int_{|(1-uu^\sigma)v|=1, |uv+w|\leq 1} \psi\left[-2\eta^{-1}v + 2\mu(1-uu^\sigma)^{-2}\left(\eta^{-1}v\right)^{-1}\right] dv\, dw.$$

Then by a change of variable $v = (1-uu^\sigma)^{-1}\eta\varepsilon$, we have

$$\mathcal{B}(u,\mu) = |1-uu^\sigma|^{-1}\int_{\mathcal{O}_F^\times} \psi\left(-2(1-uu^\sigma)^{-1}\varepsilon + 2\mu(1-uu^\sigma)^{-1}\varepsilon^{-1}\right) d\varepsilon$$
$$= |1-uu^\sigma|^{-1} \cdot \mathcal{K}\ell\left(2(1-uu^\sigma)^{-1}, -2\mu(1-uu^\sigma)^{-1}\right).$$

5.2.2. Evaluation of $\mathcal{B}(u,\mu)$ when $\mu \in F \setminus \mathcal{O}_F$ and $u \in \mathcal{O}_E$. By Theorem 3.10, it is enough for us to prove the following proposition.

PROPOSITION 5.13. *Suppose that $\mu = \varpi^{-2n}\varepsilon_\mu$ where $n > 0$ and $\varepsilon_\mu \in \mathcal{O}_F^\times$. Then for $u \in \mathcal{O}_E$ such that $u \neq 0$ and $uu^\sigma \neq 1$, we have*

$$\mathcal{B}(u,\mu) = \Omega(u)\Omega(\varpi)^n \mathcal{K}_{\mathrm{an}}\left(\varpi^n A_u, d^{-1}\varepsilon_\mu\right).$$

PROOF. Since $\varpi^n A_u \in \varpi \cdot \mathrm{Mat}_{2\times 2}(\mathcal{O}_E)$, we have

$$\begin{pmatrix} 1_2 & 0 \\ Y & 1_2 \end{pmatrix}\begin{pmatrix} \varpi^n A_u & 0 \\ 0 & \varpi^{-n}\bar{A}_u^{-1} \end{pmatrix} \in \mathcal{K}_1 U_1$$

in (5.7) if and only if

(5.18) $$Y \in \mathcal{Y}_B = \{Z \in D_1^- \mid YB \in \mathrm{GL}_2(\mathcal{O}_E)\}$$

where $B = \varpi^n A_u$, by Lemma 4.9. When (5.18) holds, we have

$$\begin{pmatrix} 1_2 & 0 \\ Y & 1_2 \end{pmatrix}\begin{pmatrix} B & 0 \\ 0 & \bar{B}^{-1} \end{pmatrix} = \begin{pmatrix} k_1 & k_2 \\ k_3 & k_4 \end{pmatrix}\begin{pmatrix} 1_2 & -X \\ 0 & 1_2 \end{pmatrix}$$

for some $\begin{pmatrix} k_1 & k_2 \\ k_3 & k_4 \end{pmatrix} \in \mathcal{K}_1$ and $X \in D_1^-$. Then we have

$$k_3 = YB \in \mathrm{GL}_2(\mathcal{O}_E), \quad -k_3 X + k_4 = \bar{B}^{-1} = \det B^{-1} \cdot B.$$

Hence

$$X \equiv -\det B^{-1} \cdot B^{-1} Y^{-1} B \pmod{\mathcal{O}_E}.$$

Thus (5.7) becomes

$$\mathcal{B}(u,\mu)$$
$$= \Omega(u)\Omega(\varpi)^n \int_{\mathcal{Y}_B} \psi\left[\mathrm{tr}\left\{\begin{pmatrix} -\eta^{-1} & 0 \\ 0 & \eta^{-1} \end{pmatrix}(Y - d\varepsilon_\mu \det B^{-1} \cdot B^{-1}Y^{-1}B)\right\}\right] dY.$$

Then by a change of variable $Y \mapsto -dY$, we have

$$\mathcal{B}(u,\mu) = \Omega(u)\Omega(\varpi)^n$$
$$\int_{\mathcal{Y}_B} \psi\left[\mathrm{tr}\left\{\begin{pmatrix} \eta & 0 \\ 0 & -\eta \end{pmatrix}(Y - d^{-1}\varepsilon_\mu \det B^{-1} \cdot B^{-1}Y^{-1}B)\right\}\right] dY$$
$$= \Omega(u)\Omega(\varpi)^n \mathcal{K}_{\mathrm{an}}(B, d^{-1}\varepsilon_\mu)$$
$$= \Omega(u)\Omega(\varpi)^n \mathcal{K}_{\mathrm{an}}(\varpi^n A_u, d^{-1}\varepsilon_\mu).$$

□

CHAPTER 6

Evaluation of the Quadratic Orbital Integral

In this chapter we evaluate the quadratic orbital integral for the main relevant double cosets explicitly.

6.1. Double coset decomposition

We recall that we regard $G = \mathrm{GSp}(4)$ and its subgroups as algebraic groups over F. The goal of this section is to describe explicitly the double coset decomposition $G_F \backslash G_E / H_E$, where H denotes the upper Novodvorsky subgroup of G.

We start with some preliminary consideration.

6.1.1. $G_F \backslash G_E / P_E$ **double coset decomposition.** Here P denotes the upper Siegel parabolic subgroup of G.

PROPOSITION 6.1. *For $\epsilon \in F^\times$, let*

(6.1) $$\mathcal{C}_\epsilon = \begin{pmatrix} 1 & 0 & -\epsilon & 0 \\ 0 & 1 & 0 & 1 \\ -\epsilon^{-1}\eta & 0 & -\eta & 0 \\ 0 & \eta & 0 & -\eta \end{pmatrix} \in G_F.$$

Then we have

(6.2) $$G_E = G_F P_E \cup G_F \begin{pmatrix} 1 & 0 & 0 & 0 \\ 0 & 1 & 0 & 1 \\ 0 & 0 & -2\eta & 0 \\ 0 & \eta & 0 & -\eta \end{pmatrix} P_E \cup \left(\bigcup_{\epsilon \in F^\times / N_{E/F}(E^\times)} G_F \mathcal{C}_\epsilon P_E \right).$$

PROOF. For this decomposition, we utilize the isomorphism between $\mathrm{PGSp}(4)$ and $\mathrm{SO}(3,2)$. Let us recall the isomorphism. Let F^4 denote the four dimensional F vector space of column vectors. For each i ($1 \leq i \leq 4$), let e_i denote the column vector whose i-th entry is 1 but the other entries are all zero. Let $V = \wedge^2 F^4$. Then V is endowed with a non-degenerate symmetric bilinear form $B : V \times V \to F$ defined by

$$v \wedge w = B(v, w)(e_1 \wedge e_2 \wedge e_3 \wedge e_4).$$

Then we have a homomorphism $\Phi : G \to \mathrm{SO}(V)$ defined by

$$\Phi(g)v = \frac{1}{\lambda(g)} \wedge^2 (g) v \quad \text{for} \quad v \in V = \wedge^2 F^4.$$

Here we recall that $\lambda(g)$ denotes the similitude of g. Let $v_0 = e_1 \wedge e_3 + e_2 \wedge e_4 \in V$. Then it is readily seen that

$$\Phi(g)v_0 = v_0 \quad \text{for} \quad g \in G.$$

Thus let

$$W = v_0^\perp = \{ v \in V \mid B(v, v_0) = 0 \}$$

and let us denote the homomorphism $G \ni g \mapsto \Phi(g)|_W \in \mathrm{SO}(W)$ by ϕ. Then the homomorphism ϕ induces the isomorphism

$$\mathrm{PGSp}(4) = G/Z \simeq \mathrm{SO}(W) = \mathrm{SO}(3,2)$$

where Z denotes the center of G.

Let $G^* = \mathrm{SO}(W)$ and let $P^* = \phi(P)$. Then we note that

$$P^* = \{h \in G^* \mid hf_1 = \mu f_1, \ \mu \in \mathrm{GL}(1)\}$$

where $f_1 = e_1 \wedge e_2 \in W$. Since $\ker \phi = Z$, it is clear that if we have the double coset decomposition

(6.3) $$G_E^* = \bigcup_{i \in I} G_F^* \alpha_i P_E^*,$$

then we obtain the double coset decomposition

$$G_E = \bigcup_{i \in I} G_F \beta_i P_E$$

by taking $\beta_i \in G_E$ such that $\phi(\beta_i) = \alpha_i$ for each $i \in I$.

Let us consider (6.3). Let $\mathcal{X}$ denote the set of one dimensional isotropic E-subspaces in W_E. Then G_E^* acts on $\mathcal{X}$ from the left and the action is transitive by Witt's theorem. Since $Ef_1 \in \mathcal{X}$, we may identify $\mathcal{X}$ with G_E^*/P_E^*. Hence our task is to determine the G_F^*-orbits in $\mathcal{X}$.

Let $w \in W_E$ be a non-zero isotropic vector. Let us write $w = w_1 + \eta w_2$, where $w_1, w_2 \in W_F$. Then we have

$$0 = B(w,w) = \{B(w_1, w_1) + dB(w_2, w_2)\} + 2\eta B(w_1, w_2),$$

i.e.

(6.4) $$\begin{cases} B(w_1, w_1) = -dB(w_2, w_2), \\ B(w_1, w_2) = 0. \end{cases}$$

Suppose that $w_1 = 0$. Then we have $w_2 \neq 0$ and $B(w_2, w_2) = 0$. Hence by Witt's theorem there exists $h \in G_F^*$ such that $w_2 = hf_1$. Hence $w = \eta w_2 \in h(Ef_1)$.

Suppose that $w_1 \neq 0$ and $B(w_1, w_1) = 0$. When $w_2 \in Fw_1$, we have $w = w_1 + \eta w_2 \in Ew_1$. By Witt's theorem, there exists $h \in G_F^*$ such that $w_1 = hf_1$ and hence $w \in h(Ef_1)$. When $w_2 \notin Fw_1$, the set $\{w_1, w_2\}$ spans an isotropic two dimensional F-subspace in W_F. Hence by Witt's theorem there exists $h \in G_F^*$ such that $w_1 = hf_1$ and $w_2 = hf_2$, where $f_2 = e_1 \wedge e_4$. Then $w = h(f_1 + \eta f_2)$. Here we note that

$$\phi \begin{pmatrix} 1 & 0 & 0 & 0 \\ 0 & 1 & 0 & 1 \\ 0 & 0 & -2\eta & 0 \\ 0 & \eta & 0 & -\eta \end{pmatrix} (e_1 \wedge e_2) = (e_1 \wedge e_2) + \eta(e_1 \wedge e_4).$$

Suppose that $w_1 \neq 0$ and $B(w_1, w_1) \neq 0$. Then by (6.4), we have

$$\begin{pmatrix} B(w_1, w_1) & B(w_1, w_2) \\ B(w_1, w_2) & B(w_2, w_2) \end{pmatrix} = B(w_1, w_1) \cdot \begin{pmatrix} 1 & 0 \\ 0 & -d^{-1} \end{pmatrix} \in \mathrm{GL}_2(F).$$

We note that when we replace w by αw where $\alpha = a + \eta b \in E^\times$, $a, b \in F$, we have

$$\alpha w = (aw_1 + bdw_2) + \eta(bw_1 + aw_2)$$

and
$$B(aw_1 + bdw_2, aw_1 + bdw_2) = (a^2 - b^2 d) \cdot B(w_1, w_1) = N_{E/F}(\alpha) \cdot B(w_1, w_1).$$

Therefore such G_F^*-orbits are parametrized by $F^\times / N_{E/F}(E^\times)$, by taking, for each $\xi \in F^\times$, a pair of vectors $w_1, w_2 \in W_F$ such that

$$\begin{pmatrix} B(w_1, w_1) & B(w_1, w_2) \\ B(w_1, w_2) & B(w_2, w_2) \end{pmatrix} = \begin{pmatrix} \xi & 0 \\ 0 & -d^{-1}\xi \end{pmatrix}.$$

Then we note that

$$\phi(\mathcal{C}_\epsilon)(e_1 \wedge e_2) = -\frac{1}{2\eta}(\mathcal{C}_\epsilon e_1 \wedge \mathcal{C}_\epsilon e_2) = w_1 + \eta w_2$$

where $w_1 = -\frac{1}{2}(e_1 \wedge e_4 + \epsilon^{-1} e_2 \wedge e_3)$ and $w_2 = -\frac{1}{2}(d^{-1} e_1 \wedge e_2 - \epsilon^{-1} e_3 \wedge e_4)$. Here we have

$$\begin{pmatrix} B(w_1, w_1) & B(w_1, w_2) \\ B(w_1, w_2) & B(w_2, w_2) \end{pmatrix} = \frac{1}{2\epsilon} \begin{pmatrix} 1 & 0 \\ 0 & -d^{-1} \end{pmatrix}.$$

Thus we have finished the proof of the proposition. □

6.1.2. $D_\epsilon^\times \backslash \mathrm{GL}_2(E) / A_E$ double coset decomposition. As a prerequisite to the $G_F \backslash G_E / H_E$ double coset decomposition, here we review the $D_\epsilon^\times \backslash \mathrm{GL}_2(E) / A_E$ double coset decomposition following Jacquet [**J2**], mutatis mutandis.

We recall that for $\epsilon \in F^\times$,

$$D_\epsilon = \left\{ \begin{pmatrix} a & \epsilon b \\ b^\sigma & a^\sigma \end{pmatrix} \mid a, b \in E \right\}.$$

In order to investigate the double coset decomposition, we would like to construct a mapping $D_\epsilon^\times \backslash \mathrm{GL}_2(E) / A_E \to E \cup \{\infty\}$. For that purpose, let us first define a mapping $H_\epsilon : \mathrm{GL}_2(E) \to \mathrm{GL}_2(E)$ by $H_\epsilon(g) = g^{-i} g$ where i denotes the involution of $\mathrm{GL}_2(E)$ defined by

(6.5) $\qquad g^i = \begin{pmatrix} 0 & \epsilon \\ 1 & 0 \end{pmatrix} g^\sigma \begin{pmatrix} 0 & \epsilon \\ 1 & 0 \end{pmatrix}^{-1}$, i.e. $\begin{pmatrix} a & b \\ c & d \end{pmatrix}^i = \begin{pmatrix} d^\sigma & \epsilon c^\sigma \\ \epsilon^{-1} b^\sigma & a^\sigma \end{pmatrix}.$

It is clear from (6.5) that

(6.6) $\qquad\qquad\qquad g^i = g$, i.e. $H(g) = 1_2 \iff g \in D_\epsilon^\times.$

LEMMA 6.2. 1. For $g_1, g_2 \in \mathrm{GL}_2(E)$, we have

$$D_\epsilon^\times g_1 = D_\epsilon^\times g_2 \iff H_\epsilon(g_1) = H_\epsilon(g_2).$$

2. For $g, g_2 \in \mathrm{GL}_2(E)$, we have

$$D_\epsilon^\times g_1 A_E = D_\epsilon^\times g_2 A_E \iff H_\epsilon(g_1) = H_\epsilon(g_2 a) \quad \text{for some} \ a \in A_E.$$

PROOF. Suppose that $D_\epsilon^\times g_1 = D_\epsilon^\times g_2$. Then there exists $h \in D_\epsilon^\times$ such that $hg_2 = g_1$. Then we have

$$H_\epsilon(g_1) = (g_2^{-i} h^{-i})(hg_2) = g_2^{-i} g_2 = H_\epsilon(g_2).$$

Conversely when $H_\epsilon(g_1) = H_\epsilon(g_2)$, we have

$$g_1 g_2^{-1} = g_1^i H_\epsilon(g_1) g_2^{-1} = g_1^i H_\epsilon(g_2) g_2^{-1} = g_1^i g_2^{-i} = (g_1 g_2^{-1})^i.$$

Hence $g_1 g_2^{-1} \in D_\epsilon^\times$ by (6.6).

The second assertion follows immediately from the first one. □

6.1. DOUBLE COSET DECOMPOSITION

DEFINITION 6.3. We define a mapping $Y_\epsilon : D_\epsilon^\times \backslash \mathrm{GL}_2(E) / A_E \to E \cup \{\infty\}$ by

$$Y_\epsilon(D_\epsilon^\times g A_E) = \frac{qr}{ps} \quad \text{where} \quad H_\epsilon(g) = g^{-i}g = \begin{pmatrix} p & q \\ r & s \end{pmatrix}.$$

Here we remark that Y_ϵ is well defined since the mapping

$$\mathrm{GL}_2(E) \ni \begin{pmatrix} s & t \\ u & v \end{pmatrix} \mapsto \frac{tu}{sv} \in E \cup \{\infty\}$$

is constant on each (A_E, A_E)-double coset in $\mathrm{GL}_2(E)$ and $H_\epsilon(ga) = a^{-i} H_\epsilon(g) a$ for $a \in A_E$.

DEFINITION 6.4. We call a double coset $D_\epsilon^\times g A_E$ *regular* when $Y_\epsilon(g) \neq 0, \infty$. We call a double coset $D_\epsilon^\times g A_E$ *singular* when $Y_\epsilon(g) = 0, \infty$.

PROPOSITION 6.5. 1. *Suppose that* $\epsilon \notin N_{E/F}(E^\times)$. *Then* $D_\epsilon^\times A_E$ *is the only singular double coset. The regular double cosets are represented as* $D_\epsilon^\times \begin{pmatrix} 1 & y \\ 0 & 1 \end{pmatrix} A_E$ *where* $y \in E^\times$.

2. *Suppose that* $\epsilon = uu^\sigma$ *for* $u \in E^\times$. *Then there exist four singular double cosets, which are given by*

$$D_\epsilon^\times A_E, \quad D_\epsilon^\times \begin{pmatrix} 1 & u \\ 0 & 1 \end{pmatrix} A_E, \quad D_\epsilon^\times \begin{pmatrix} u & -u \\ 1 & 1 \end{pmatrix} A_E, \quad D_\epsilon^\times \begin{pmatrix} u & -u \\ 1 & 1 \end{pmatrix} \begin{pmatrix} 1 & 1 \\ 0 & 1 \end{pmatrix} A_E.$$

The regular double cosets are represented as $D_\epsilon^\times \begin{pmatrix} 1 & y \\ 0 & 1 \end{pmatrix} A_E$ *where* $y \in E^\times$ *such that* $yy^\sigma \neq \epsilon$.

PROOF. 1. Let B be the upper Borel subgroup of $\mathrm{GL}(2)$. Then since $D_\epsilon^\times$ acts transitively on $\mathbf{P}^1(E)$, we have $\mathrm{GL}_2(E) = D_\epsilon^\times B_E$. Hence any double coset is represented by $\begin{pmatrix} 1 & y \\ 0 & 1 \end{pmatrix}$ where $y \in E$. Then the rest is clear since

$$Y_\epsilon \begin{pmatrix} 1 & y \\ 0 & 1 \end{pmatrix} = \frac{-\epsilon^{-1} yy^\sigma}{1 - \epsilon^{-1} yy^\sigma}. \tag{6.7}$$

2. Suppose that $\begin{pmatrix} c \\ d \end{pmatrix} \in E^2 \setminus \{0\}$. If $cc^\sigma - \epsilon dd^\sigma \neq 0$, then

$$\begin{pmatrix} c \\ d \end{pmatrix} = \begin{pmatrix} c & \epsilon d^\sigma \\ d & c^\sigma \end{pmatrix} \begin{pmatrix} 1 \\ 0 \end{pmatrix}.$$

If $cc^\sigma - \epsilon dd^\sigma = 0$, then we have $d \neq 0$ and $\epsilon = N_{E/F}(cd^{-1})$. Hence there exists $\alpha \in E^\times$ such that $cd^{-1} = u\alpha\alpha^{-\sigma}$. Thus

$$\begin{pmatrix} c \\ d \end{pmatrix} = d\alpha^{-\sigma} \begin{pmatrix} \alpha & 0 \\ 0 & \alpha^\sigma \end{pmatrix} \begin{pmatrix} u \\ 1 \end{pmatrix}.$$

Hence we have

$$\mathrm{GL}_2(E) = D_\epsilon^\times B_E \cup D_\epsilon^\times \begin{pmatrix} u & -u \\ 1 & 1 \end{pmatrix} B_E.$$

For the $(D_\epsilon^\times, A_E)$-double cosets in $D_\epsilon^\times B_E$, the singular ones are given by $D_\epsilon^\times A_E$ and $D_\epsilon^\times \begin{pmatrix} 1 & x \\ 0 & 1 \end{pmatrix} A_E$ where $x \in E$ such that $xx^\sigma = \epsilon$, by (6.7). As for the latter,

there exists $a \in E^\times$ such that $x = uaa^{-\sigma}$ and then
$$\begin{pmatrix} 1 & x \\ 0 & 1 \end{pmatrix} = \begin{pmatrix} a & 0 \\ 0 & a^\sigma \end{pmatrix} \begin{pmatrix} 1 & u \\ 0 & 1 \end{pmatrix} \begin{pmatrix} a & 0 \\ 0 & a^\sigma \end{pmatrix}^{-1} \in D_\epsilon^\times \begin{pmatrix} 1 & u \\ 0 & 1 \end{pmatrix} A_E.$$

Let us take a double coset of the form $D_\epsilon^\times \begin{pmatrix} u & -u \\ 1 & 1 \end{pmatrix} \begin{pmatrix} 1 & y \\ 0 & 1 \end{pmatrix} A_E$. Then we have
$$H_\epsilon \left[\begin{pmatrix} u & -u \\ 1 & 1 \end{pmatrix} \begin{pmatrix} 1 & y \\ 0 & 1 \end{pmatrix} \right] = \begin{pmatrix} 0 & -u \\ u^{-\sigma} & u^{-\sigma}(y + y^\sigma) \end{pmatrix}.$$

Hence this is a singular double coset. Then we note that by Lemma 6.2 we have
$$D_\epsilon^\times \begin{pmatrix} u & -u \\ 1 & 1 \end{pmatrix} \begin{pmatrix} 1 & y \\ 0 & 1 \end{pmatrix} A_E = D_\epsilon^\times \begin{pmatrix} u & -u \\ 1 & 1 \end{pmatrix} \begin{pmatrix} 1 & z \\ 0 & 1 \end{pmatrix} A_E$$
if and only if
$$(6.8) \quad H_\epsilon \left[\begin{pmatrix} u & -u \\ 1 & 1 \end{pmatrix} \begin{pmatrix} 1 & y \\ 0 & 1 \end{pmatrix} \right] = H_\epsilon \left[\begin{pmatrix} u & -u \\ 1 & 1 \end{pmatrix} \begin{pmatrix} 1 & y \\ 0 & 1 \end{pmatrix} \begin{pmatrix} a_1 & 0 \\ 0 & a_2 \end{pmatrix} \right]$$
for some $a_1, a_2 \in E^\times$. Since
$$H_\epsilon \left[\begin{pmatrix} u & -u \\ 1 & 1 \end{pmatrix} \begin{pmatrix} 1 & y \\ 0 & 1 \end{pmatrix} \begin{pmatrix} a_1 & 0 \\ 0 & a_2 \end{pmatrix} \right] = \begin{pmatrix} 0 & -a_2 a_2^{-\sigma} u \\ a_1 a_1^{-\sigma} u^{-\sigma} & a_1^{-\sigma} a_2 u^{-\sigma}(z + z^\sigma) \end{pmatrix},$$
we have (6.8) if and only if $z + z^\sigma = a(y + y^\sigma)$ for some $a \in F^\times$. Since $\operatorname{tr}_{E/F}$ is a surjective F-linear mapping, we have
$$D_\epsilon^\times \begin{pmatrix} u & -u \\ 1 & 1 \end{pmatrix} B_E = D_\epsilon^\times \begin{pmatrix} u & -u \\ 1 & 1 \end{pmatrix} A_E \cup D_\epsilon^\times \begin{pmatrix} u & -u \\ 1 & 1 \end{pmatrix} \begin{pmatrix} 1 & 1 \\ 0 & 1 \end{pmatrix} A_E.$$
□

PROPOSITION 6.6. *Let $\mathcal{E}$ be a set of representatives of $F^\times / N_{E/F}(E^\times)$. Then the mapping*
$$D_\epsilon^\times \begin{pmatrix} 1 & y \\ 0 & 1 \end{pmatrix} A_E \mapsto Y_\epsilon \begin{pmatrix} 1 & y \\ 0 & 1 \end{pmatrix} = \frac{-\epsilon^{-1} y y^\sigma}{1 - \epsilon^{-1} y y^\sigma}$$
induces a bijection between the set of regular $(D_\epsilon^\times, A_E)$ double cosets in $\mathrm{GL}_2(E)$ and $F \setminus \{0, 1\}$, when ϵ runs over $\mathcal{E}$.

PROOF. Since
$$Y_\epsilon \begin{pmatrix} 1 & y \\ 0 & 1 \end{pmatrix}^{-1} = \frac{1 - \epsilon^{-1} y y^\sigma}{-\epsilon^{-1} y y^\sigma} = 1 - \epsilon (y y^\sigma)^{-1},$$
the mapping is surjective.

Suppose that $Y_\epsilon \begin{pmatrix} 1 & y \\ 0 & 1 \end{pmatrix} = Y_{\epsilon'} \begin{pmatrix} 1 & z \\ 0 & 1 \end{pmatrix}$, i.e. $\frac{-\epsilon^{-1} y y^\sigma}{1 - \epsilon^{-1} y y^\sigma} = \frac{-\epsilon'^{-1} z z^\sigma}{1 - \epsilon'^{-1} z z^\sigma}$. Then we have $\epsilon^{-1} y y^\sigma = \epsilon'^{-1} z z^\sigma$ and hence $\epsilon = \epsilon'$. Then we have $y y^\sigma = z z^\sigma$ and there exists $\alpha \in E^\times$ such that $y = \alpha \alpha^{-\sigma} z$. Hence
$$\begin{pmatrix} 1 & y \\ 0 & 1 \end{pmatrix} = \begin{pmatrix} \alpha & 0 \\ 0 & \alpha^\sigma \end{pmatrix} \begin{pmatrix} 1 & z \\ 0 & 1 \end{pmatrix} \begin{pmatrix} \alpha & 0 \\ 0 & \alpha^\sigma \end{pmatrix}^{-1} \in D_\epsilon^\times \begin{pmatrix} 1 & z \\ 0 & 1 \end{pmatrix} A_E.$$
Therefore the mapping is also injective. □

Finally let us consider the stabilizer of the double coset $D_\epsilon^\times g A_E$, i.e.
$$Z(g) = \left\{ (h, a) \in D_\epsilon^\times \times A_E \mid hga^{-1} = g \right\}.$$

LEMMA 6.7. *Let $g \in \mathrm{GL}_2(E)$.*
1. *When g is regular, we have*
$$Z(g) = \{(z, z) \mid z \in F^\times\}.$$

2. *When $g = 1_2$, we have*
$$Z(g) = \{(t, t) \mid t \in T\} \quad \text{where} \quad T = \left\{ \begin{pmatrix} a & 0 \\ 0 & a^\sigma \end{pmatrix} \mid a \in E^\times \right\}.$$

3. *Suppose that $\epsilon = uu^\sigma$ for $u \in E^\times$.*
    (a) *When $g = \begin{pmatrix} 1 & u \\ 0 & 1 \end{pmatrix}$ or $\begin{pmatrix} u & -u \\ 1 & 1 \end{pmatrix} \begin{pmatrix} 1 & 1 \\ 0 & 1 \end{pmatrix}$, we have*
    $$Z(g) = \{(z, z) \mid z \in F^\times\}.$$

    (b) *When $g = \begin{pmatrix} u & -u \\ 1 & 1 \end{pmatrix}$, we have*
    $$Z(g) = \{(gag^{-1}, a) \mid a \in A_F\}.$$

PROOF. By Lemma 6.2, for $a = \begin{pmatrix} a_1 & 0 \\ 0 & a_2 \end{pmatrix} \in A_E$, there exists an $h \in D_\epsilon^\times$ such that $(h, a) \in Z(g)$ if and only if $H_\epsilon(g) = H_\epsilon(ga)$, i.e.

(6.9) $\quad \begin{pmatrix} p & q \\ r & s \end{pmatrix} = \begin{pmatrix} a_1 a_2^{-\sigma} p & a_2 a_2^{-\sigma} q \\ a_1 a_1^{-\sigma} r & a_1^{-\sigma} a_2 s \end{pmatrix} \quad \text{where} \quad H_\epsilon(g) = \begin{pmatrix} p & q \\ r & s \end{pmatrix}.$

When g is regular, then p, q, r, s are all different from zero. Hence (6.9) holds if and only if $a_1 = a_2 \in F^\times$.

When $g = 1_2$, we have $H_\epsilon(g) = 1_2$. Hence (6.9) holds if and only if $a_1^\sigma = a_2$.

Suppose that $\epsilon = uu^\sigma$. Then we have
$$H_\epsilon \begin{pmatrix} 1 & u \\ 0 & 1 \end{pmatrix} = \begin{pmatrix} 1 & u \\ -u^{-1} & 0 \end{pmatrix}, \quad H_\epsilon \left[\begin{pmatrix} u & -u \\ 1 & 1 \end{pmatrix} \begin{pmatrix} 1 & 1 \\ 0 & 1 \end{pmatrix} \right] = \begin{pmatrix} 0 & -u \\ u^{-\sigma} & 2u^{-\sigma} \end{pmatrix},$$
$$H_\epsilon \begin{pmatrix} u & -u \\ 1 & 1 \end{pmatrix} = \begin{pmatrix} 0 & -u \\ u^{-\sigma} & 0 \end{pmatrix}.$$

Thus the assertion clearly follows. □

6.1.3. $G_F \backslash G_E / H_E$ double coset decomposition. We shall parameterize the (G_F, H_E) double cosets in $G_F C_\epsilon H_E$ in this subsection.

First we remark the following lemma.

LEMMA 6.8. *Let us define a mapping $\theta : G_E \to C_E$ by*
$$\theta(g) = g^{-\sigma} g.$$

1. *For $g_1, g_2 \in G_E$, we have*
$$G_F g_1 = G_F g_2 \iff \theta(g_1) = \theta(g_2).$$

2. *For $g_1, g_2 \in G_E$, we have*
$$G_F g_1 H_E = G_F g_2 H_E \iff \theta(g_1) = \theta(g_2 h) \quad \text{for some} \quad h \in H_E.$$

PROOF. The second assertion follows from the first one immediately. As for the first assertion, we observe that

$$\theta\left(g_1 g_2^{-1}\right) = g_2^\sigma \left(g_1^{-\sigma} g_1\right) g_2^{-1} = g_2^\sigma \theta\left(g_1\right) g_2^{-1}.$$

Thus it is clear that

$$\theta\left(g_1\right) = \theta\left(g_2\right) \Longleftrightarrow \theta\left(g_1 g_2^{-1}\right) = 1_4.$$

Thus we obtain the assertion since $G_F = \{g \in G_E \mid \theta(g) = 1_4\}$. □

PROPOSITION 6.9. *For $\epsilon \in F^\times$, we have*

(6.10) $$G_F \mathcal{C}_\epsilon P_E = \bigcup_{m,\lambda} G_F \mathcal{C}_\epsilon \begin{pmatrix} m & 0 \\ 0 & \lambda \cdot {}^t m^{-1} \end{pmatrix} H_E$$

where m runs over the set of representatives of $D_\epsilon^\times \backslash \mathrm{GL}_2(E)/A_E$ and λ runs over the set of representatives of $E^\times/\mathcal{O}_E^1$ where $\mathcal{O}_E^1 = \{x \in \mathcal{O}_E^\times \mid xx^\sigma = 1\}$.

PROOF. First we observe that we have

(6.11) $$\theta\left[\mathcal{C}_\epsilon \begin{pmatrix} m & 0 \\ 0 & \lambda \cdot {}^t m^{-1} \end{pmatrix} \begin{pmatrix} 1_2 & X \\ 0 & 1_2 \end{pmatrix}\right]$$
$$= \begin{pmatrix} 1_2 & -X^\sigma \\ 0 & 1_2 \end{pmatrix} \begin{pmatrix} {}^t M^{-1} & 0 \\ 0 & \Lambda \cdot M \end{pmatrix} \begin{pmatrix} 1_2 & 0 \\ \left(\begin{smallmatrix} -\epsilon & 0 \\ 0 & 1 \end{smallmatrix}\right)^{-1} X \left(\begin{smallmatrix} -\epsilon & 0 \\ 0 & 1 \end{smallmatrix}\right)^{-1} & 1_2 \end{pmatrix} \mathcal{C}_\epsilon^{-\sigma} \mathcal{C}_\epsilon,$$

where

(6.12) $$\begin{cases} M = \lambda^{-1} {}^t m^\sigma \begin{pmatrix} -\epsilon & 0 \\ 0 & 1 \end{pmatrix}^{-1} m \begin{pmatrix} -\epsilon & 0 \\ 0 & 1 \end{pmatrix} \\ \Lambda = \lambda \lambda^{-\sigma}. \end{cases}$$

Here we note that for $m = \begin{pmatrix} a & b \\ c & d \end{pmatrix}$, by (6.5), we have

$$\begin{pmatrix} -\epsilon & 0 \\ 0 & 1 \end{pmatrix} {}^t m^\sigma \begin{pmatrix} -\epsilon & 0 \\ 0 & 1 \end{pmatrix}^{-1} = \begin{pmatrix} a^\sigma & -\epsilon c^\sigma \\ -\epsilon^{-1} b^\sigma & d^\sigma \end{pmatrix} = \det m^\sigma \cdot m^{-i}.$$

Hence in (6.12)

$$M = \lambda^{-1} \det m^\sigma \cdot \begin{pmatrix} -\epsilon & 0 \\ 0 & 1 \end{pmatrix}^{-1} H_\epsilon(m) \begin{pmatrix} -\epsilon & 0 \\ 0 & 1 \end{pmatrix}.$$

Thus for $m_1, m_2 \in \mathrm{GL}_2(E)$ and $\lambda_1, \lambda_2 \in E^\times$, we have

$$G_F \mathcal{C}_\epsilon \begin{pmatrix} m_1 & 0 \\ 0 & \lambda_1 \cdot {}^t m_1^{-1} \end{pmatrix} H_E = G_F \mathcal{C}_\epsilon \begin{pmatrix} m_2 & 0 \\ 0 & \lambda_2 \cdot {}^t m_2^{-1} \end{pmatrix} H_E$$

if and only if there exist $a, b \in E^\times$ such that

(6.13) $$\lambda_1^{-1} \det m_1^\sigma \cdot H_\epsilon(m_1) = \lambda_2^{-1} (ab)^{-1} (ab)^\sigma \det m_2^\sigma \cdot H_\epsilon\left[m_2 \begin{pmatrix} a & 0 \\ 0 & b \end{pmatrix}\right],$$

(6.14) $$\lambda_1 \lambda_1^{-\sigma} = \lambda_2 \lambda_2^{-\sigma} (ab)(ab)^{-\sigma}.$$

It is clear that (6.14) is equivalent to

(6.15) $$\lambda_1^{-1} \lambda_2 ab \in F^\times.$$

Suppose that (6.13) and (6.15) hold. Let $\alpha = \lambda_1^{-1}\lambda_2 ab \in F^\times$. Then (6.13) becomes

$$(6.16) \qquad \alpha \cdot \det m_1^\sigma \cdot H_\epsilon(m_1) = (ab)^\sigma \det m_2^\sigma \cdot H_\epsilon\left[m_2\begin{pmatrix} a & 0 \\ 0 & b \end{pmatrix}\right].$$

By taking the determinant of the both sides of (6.16), we have

$$N_{E/F}\left(\alpha \cdot \det m_1^\sigma\right) = N_{E/F}\left(ab \cdot \det m_2^\sigma\right).$$

Hence there exists $\beta \in E^\times$ such that $\alpha \cdot \det m_1^\sigma = ab \cdot \det m_2^\sigma \cdot \beta\beta^{-\sigma}$, i.e.

$$(6.17) \qquad \lambda_1^{-1} \det m_1^\sigma = \lambda_2^{-1} \det m_2^\sigma \cdot \beta\beta^{-\sigma}.$$

Then (6.13) reads

$$H_\epsilon(m_1) = (\beta ab)^{-1}(\beta ab)^\sigma H_\epsilon\left[m_2\begin{pmatrix} a & 0 \\ 0 & b \end{pmatrix}\right] = H_\epsilon\left[m_2\begin{pmatrix} \beta^{-1}b^{-1} & 0 \\ 0 & \beta^{-1}a^{-1} \end{pmatrix}\right].$$

Hence we have

$$(6.18) \qquad D_\epsilon^\times m_1 A_E = D_\epsilon^\times m_2 A_E.$$

It is clear that when $m_1 = m_2$, (6.17) implies that

$$\lambda_1 \lambda_1^\sigma = \lambda_2 \lambda_2^\sigma.$$

Thus there is no redundancy among the double cosets when m and λ run over as above.

Now suppose that we are given $m_1 \in \mathrm{GL}_2(E)$ and $\lambda_1 \in E^\times$. Let $m_2 \in \mathrm{GL}_2(E)$ be the representative of $D_\epsilon^\times \backslash \mathrm{GL}_2(E) / A_E$ so that (6.18) holds. Then there exists $a_0, b_0 \in E^\times$ such that

$$(6.19) \qquad H_\epsilon(m_1) = H_\epsilon\left[m_2\begin{pmatrix} a_0 & 0 \\ 0 & b_0 \end{pmatrix}\right].$$

Let $\lambda_0 = \lambda_1(a_0b_0)^{-1}(a_0b_0)^\sigma \det m_1^{-\sigma} \det m_2^\sigma$. Then we have

$$\lambda_1^{-1} \det m_1^\sigma \cdot H_\epsilon(m_1) = \lambda_0^{-1}(a_0b_0)^{-1}(a_0b_0)^\sigma \det m_2^\sigma \cdot H_\epsilon\left[m_2\begin{pmatrix} a_0 & 0 \\ 0 & b_0 \end{pmatrix}\right].$$

By (6.19), we have $a_0 b_0 \cdot \det\left(m_1^{-1} m_2\right) \in F^\times$. Hence

$$\lambda_1^{-1}\lambda_0 a_0 b_0 = \left\{a_0 b_0 \cdot \det\left(m_1^{-1} m_2\right)\right\}^\sigma \in F^\times.$$

Thus we have

$$G_F \mathcal{C}_\epsilon \begin{pmatrix} m_1 & 0 \\ 0 & \lambda_1 \cdot {}^t m_1^{-1} \end{pmatrix} H_E = G_F \mathcal{C}_\epsilon \begin{pmatrix} m_2 & 0 \\ 0 & \lambda_0 \cdot {}^t m_2^{-1} \end{pmatrix} H_E.$$

Now let λ_2 be the representative of $E^\times / \mathcal{O}_E^1$ so that there exists $\delta \in E^\times$ such that $\lambda_0 = \lambda_2 \delta \delta^{-\sigma}$. Then we have $\lambda_0^{-1}\lambda_2 \delta^2 = \delta\delta^\sigma \in F^\times$ and

$$\lambda_0^{-1} \det m_2^\sigma \cdot H_\epsilon(m_2) = \lambda_2^{-1}\delta^{-2}\delta^{2\sigma} \det m_2^\sigma \cdot H_\epsilon\left[m_2\begin{pmatrix} \delta & 0 \\ 0 & \delta \end{pmatrix}\right].$$

Hence we have

$$G_F \mathcal{C}_\epsilon \begin{pmatrix} m_2 & 0 \\ 0 & \lambda_0 \cdot {}^t m_2^{-1} \end{pmatrix} H_E = G_F \mathcal{C}_\epsilon \begin{pmatrix} m_2 & 0 \\ 0 & \lambda_2 \cdot {}^t m_2^{-1} \end{pmatrix} H_E$$

and we have finished the proof. $\square$

6.2. Relevant double cosets

We recall that for Ω, a character of $E^\times$, by abuse of notation, we also denote by Ω a character of H_E defined by

$$(6.20) \quad \Omega\left[\begin{pmatrix} a & 0 & 0 & 0 \\ 0 & b & 0 & 0 \\ 0 & 0 & b & 0 \\ 0 & 0 & 0 & a \end{pmatrix} \begin{pmatrix} 1_2 & X \\ 0 & 1_2 \end{pmatrix}\right] = \Omega(ab)\, \psi_E\left[\operatorname{tr}\left(\begin{pmatrix} 0 & 1 \\ 1 & 0 \end{pmatrix} X\right)\right].$$

For $r \in G_E$, let us denote by $\Delta(r)$ the stabilizer of the double coset $G_F r H_E$, i.e.

$$(6.21) \quad \Delta(r) = \left\{ (g,h) \in G_F \times H_E \mid grh^{-1} = r \right\}.$$

DEFINITION 6.10. We say that a double coset $G_F r H_E$ in G_E is *relevant* if the mapping

$$(6.22) \quad G_F r H_E \ni grh \mapsto (\omega\chi_E)(\lambda(g))\,\Omega(h) \in \mathbb{C}^\times$$

is well defined. Here $\omega = \Omega|_{F^\times}$ and $\lambda(g)$ denotes the similitude of g.

Note that the condition (6.22) is equivalent to

$$(6.23) \quad (\omega\chi_E)(\lambda(h)) = \Omega(h) \quad \text{for} \quad (g,h) \in \Delta(r)$$

since $\lambda(g) = \lambda(rhr^{-1}) = \lambda(h)$ for $(g,h) \in \Delta(r)$.

Let us take $\{1, \varpi\}$ as the set of representatives of $E^\times/\mathcal{O}_E^1$. Then the relevant double cosets are described as follows.

PROPOSITION 6.11. *1. The relevant double cosets in $G_F C_1 P_E$ are represented as*

$$G_F C_1 \begin{pmatrix} m & 0 \\ 0 & \lambda \cdot {}^t m^{-1} \end{pmatrix} H_E$$

where λ runs over the set of representatives of $E^\times/\mathcal{O}_E^1$ and $m = \begin{pmatrix} 1 & y \\ 0 & 1 \end{pmatrix}$ where y runs over the set of representatives of $E^\times/\mathcal{O}_E^1$, $y \notin \mathcal{O}_E^1$, or,

$$m = 1_2, \quad \begin{pmatrix} 1 & 1 \\ 0 & 1 \end{pmatrix}, \quad \begin{pmatrix} 1 & -1 \\ 1 & 1 \end{pmatrix} \begin{pmatrix} 1 & 1 \\ 0 & 1 \end{pmatrix}.$$

2. All the double cosets in $G_F C_\varpi P_E$ are relevant and they are represented as

$$G_F C_\varpi \begin{pmatrix} m & 0 \\ 0 & \lambda \cdot {}^t m^{-1} \end{pmatrix} H_E$$

where λ runs over the set of representatives of $E^\times/\mathcal{O}_E^1$ and $m = 1_2$, or, $\begin{pmatrix} 1 & y \\ 0 & 1 \end{pmatrix}$ where y runs over the set of representatives of $E^\times/\mathcal{O}_E^1$.

3. The only relevant double coset in $G_F P_E$ is

$$G_F \begin{pmatrix} C & 0 \\ 0 & \eta \cdot {}^t C^{-1} \end{pmatrix} H_E \quad \text{where} \quad C = \begin{pmatrix} 1 & 1 \\ \eta & -\eta \end{pmatrix}.$$

4. Let $C_0 = \begin{pmatrix} 1 & 0 & 0 & 0 \\ 0 & 1 & 0 & 1 \\ 0 & 0 & -2\eta & 0 \\ 0 & \eta & 0 & -\eta \end{pmatrix}$. Then the relevant double cosets in $G_F C_0 P_E$ are represented as

$$G_F C_0 H_E, \quad G_F C_0 \begin{pmatrix} 0 & 1 & 0 & 0 \\ 1 & 0 & 0 & 0 \\ 0 & 0 & 0 & 1 \\ 0 & 0 & 1 & 0 \end{pmatrix} H_E$$

and

$$G_F C_0 \begin{pmatrix} \begin{pmatrix} 0 & 1 \\ 1 & 0 \end{pmatrix}\begin{pmatrix} 1 & 1 \\ 0 & 1 \end{pmatrix} & 0 \\ 0 & \lambda \begin{pmatrix} 0 & 1 \\ 1 & 0 \end{pmatrix}\begin{pmatrix} 1 & 0 \\ -1 & 1 \end{pmatrix} \end{pmatrix} H_E \quad \text{where} \quad \lambda \in F^\times.$$

PROOF. 1, 2. As we have seen in the proof of Proposition 6.9,

$$\theta\left[C_\epsilon \begin{pmatrix} g & 0 \\ 0 & \mu \cdot {}^t g^{-1} \end{pmatrix} \begin{pmatrix} 1_2 & X \\ 0 & 1_2 \end{pmatrix} \right]$$
$$= \begin{pmatrix} 1_2 & -X^\sigma \\ 0 & 1_2 \end{pmatrix} \begin{pmatrix} {}^t M^{-1} & 0 \\ 0 & \Lambda \cdot M \end{pmatrix} \begin{pmatrix} 1_2 & 0 \\ \left(\begin{smallmatrix} -\epsilon & 0 \\ 0 & 1 \end{smallmatrix}\right)^{-1} X \left(\begin{smallmatrix} -\epsilon & 0 \\ 0 & 1 \end{smallmatrix}\right)^{-1} & 1_2 \end{pmatrix} C_\epsilon^{-\sigma} C_\epsilon,$$

where

$$\begin{cases} M = \mu^{-1} \det g^\sigma \cdot \begin{pmatrix} -\epsilon & 0 \\ 0 & 1 \end{pmatrix}^{-1} H_\epsilon(g) \begin{pmatrix} -\epsilon & 0 \\ 0 & 1 \end{pmatrix} \\ \Lambda = \mu \mu^{-\sigma}. \end{cases}$$

Thus we have

$$\theta\left[C_\epsilon \begin{pmatrix} m & 0 \\ 0 & \lambda \cdot {}^t m^{-1} \end{pmatrix} \begin{pmatrix} a & 0 & 0 & 0 \\ 0 & b & 0 & 0 \\ 0 & 0 & b & 0 \\ 0 & 0 & 0 & a \end{pmatrix} \begin{pmatrix} 1_2 & X \\ 0 & 1_2 \end{pmatrix} \right] = \theta\left[C_\epsilon \begin{pmatrix} m & 0 \\ 0 & \lambda \cdot {}^t m^{-1} \end{pmatrix} \right]$$

if and only if $X = 0$,

$$(ab)(ab)^{-\sigma} = 1, \quad \text{i.e.} \quad ab \in F^\times,$$

and

$$H_\epsilon\left[m \begin{pmatrix} b^{-1} & 0 \\ 0 & a^{-1} \end{pmatrix} \right] = H_\epsilon(m).$$

Then the assertion follows immediately from Lemma 6.7.

Here we note that when, in particular, the double coset $D_\epsilon^\times m A_E$ is regular, we have

(6.24) $$\Delta\left[C_\epsilon \begin{pmatrix} m & 0 \\ 0 & \lambda \cdot {}^t m^{-1} \end{pmatrix} \right] = \{(z, z) \mid z \in Z_F\}.$$

3. First we note that since

$$\begin{pmatrix} C\delta C^{-1} & 0 \\ 0 & {}^t C^{-1} {}^t \delta^{-1} {}^t C \end{pmatrix} \in G_F$$

for $\delta \in D_1^\times$, we may represent any (G_F, H_E) double coset in $G_F P_E$ as

$$G_F \begin{pmatrix} C & 0 \\ 0 & {}^tC^{-1} \end{pmatrix} \begin{pmatrix} m & 0 \\ 0 & \lambda \cdot {}^tm^{-1} \end{pmatrix} H_E$$

when m runs over the set of representatives of $D_1^\times \backslash \mathrm{GL}_2(E) / A_E$ and λ runs over $E^\times$.

Now we have

$$\theta \left[\begin{pmatrix} C & 0 \\ 0 & {}^tC^{-1} \end{pmatrix} \begin{pmatrix} g & 0 \\ 0 & \mu \cdot {}^tg^{-1} \end{pmatrix} \begin{pmatrix} 1_2 & X \\ 0 & 1_2 \end{pmatrix} \right]$$

$$= \begin{pmatrix} 1_2 & -X^\sigma \\ 0 & 1_2 \end{pmatrix} \begin{pmatrix} g^{-\sigma} & 0 \\ 0 & \mu^{-\sigma} \cdot {}^tg^\sigma \end{pmatrix} \begin{pmatrix} \begin{pmatrix} 0 & 1 \\ 1 & 0 \end{pmatrix} & 0 \\ 0 & \begin{pmatrix} 0 & 1 \\ 1 & 0 \end{pmatrix} \end{pmatrix} \begin{pmatrix} g & 0 \\ 0 & \mu \cdot {}^tg^{-1} \end{pmatrix} \begin{pmatrix} 1_2 & X \\ 0 & 1_2 \end{pmatrix}.$$

Here we note that

$$g^{-\sigma} \begin{pmatrix} 0 & 1 \\ 1 & 0 \end{pmatrix} g = \begin{pmatrix} 0 & 1 \\ 1 & 0 \end{pmatrix} H_1(g).$$

Thus

(6.25) $\theta \left[\begin{pmatrix} C & 0 \\ 0 & {}^tC^{-1} \end{pmatrix} \begin{pmatrix} g & 0 \\ 0 & \mu \cdot {}^tg^{-1} \end{pmatrix} \begin{pmatrix} 1_2 & X \\ 0 & 1_2 \end{pmatrix} \right]$

$$= \begin{pmatrix} \begin{pmatrix} 0 & 1 \\ 1 & 0 \end{pmatrix} H_1(g) & 0 \\ 0 & \mu \mu^{-\sigma} \cdot \begin{pmatrix} 0 & 1 \\ 1 & 0 \end{pmatrix} {}^t H_1(g)^{-1} \end{pmatrix} \begin{pmatrix} 1_2 & Y \\ 0 & 1_2 \end{pmatrix}$$

where

$$Y = X - \mu \mu^{-\sigma} \cdot g^{-1} C^{-\sigma} C g^\sigma X^\sigma {}^t g^\sigma {}^t C {}^t C^{-\sigma} {}^t g^{-1}.$$

Hence we have

$$\theta \left[\begin{pmatrix} Cm & 0 \\ 0 & \lambda \cdot {}^t C^{-1} {}^t m^{-1} \end{pmatrix} \right]$$

$$= \theta \left[\begin{pmatrix} Cm & 0 \\ 0 & \lambda \cdot {}^t C^{-1} {}^t m^{-1} \end{pmatrix} \begin{pmatrix} a & 0 & 0 & 0 \\ 0 & b & 0 & 0 \\ 0 & 0 & b & 0 \\ 0 & 0 & 0 & a \end{pmatrix} \begin{pmatrix} 1_2 & X \\ 0 & 1_2 \end{pmatrix} \right]$$

if and only if

(6.26) $\qquad H_1(m) = H_1 \left[m \begin{pmatrix} a & 0 \\ 0 & b \end{pmatrix} \right],$

(6.27) $\qquad (ab)(ab)^{-\sigma} = 1, \quad \text{i.e. } ab \in F^\times,$

and
(6.28)
$$\lambda^\sigma \cdot C^\sigma m X {}^t m {}^t C^\sigma = \lambda \cdot Cm^\sigma X^\sigma {}^t m^\sigma {}^t C, \quad \text{i.e.} \quad \lambda^\sigma \cdot C^\sigma m X {}^t m {}^t C^\sigma \in \mathrm{Sym}^2(F).$$

Here we note that (6.27) follows from (6.26) by taking the determinant.

By Lemma 6.7, the torus part of the stabilizer satisfies the condition (6.23) except for the case when $m = \begin{pmatrix} 1 & -1 \\ 1 & 1 \end{pmatrix}$.

As for the unipotent part, put $S = \lambda^\sigma \cdot C^\sigma m X {}^t m {}^t C^\sigma$. Then we have

$$\psi_E \left[\mathrm{tr} \left(\begin{pmatrix} 0 & 1 \\ 1 & 0 \end{pmatrix} X \right) \right] = \psi_E \left[\lambda^{-\sigma} \mathrm{tr} \left({}^t C^{-\sigma} {}^t m^{-1} \begin{pmatrix} 0 & 1 \\ 1 & 0 \end{pmatrix} m^{-1} C^{-\sigma} S \right) \right].$$

Here

(6.29)
$${}^tC^{-\sigma}\, {}^tm^{-1}\begin{pmatrix} 0 & 1 \\ 1 & 0 \end{pmatrix} m^{-1}C^{-\sigma} = \frac{1}{2d}\begin{pmatrix} d(1-y) & -\eta y \\ -\eta y & -1-y \end{pmatrix} \quad \text{when} \quad m = \begin{pmatrix} 1 & y \\ 0 & 1 \end{pmatrix},$$

and

(6.30)
$${}^tC^{-\sigma}\, {}^tm^{-1}\begin{pmatrix} 0 & 1 \\ 1 & 0 \end{pmatrix} m^{-1}C^{-\sigma} = \frac{1}{4d}\begin{pmatrix} 0 & \eta \\ \eta & -2 \end{pmatrix} \quad \text{when} \quad m = \begin{pmatrix} 1 & -1 \\ 1 & 1 \end{pmatrix}\begin{pmatrix} 1 & 1 \\ 0 & 1 \end{pmatrix}.$$

Thus we have

$$\psi_E\left[\lambda^{-\sigma}\operatorname{tr}\left({}^tC^{-\sigma}\, {}^tm^{-1}\begin{pmatrix} 0 & 1 \\ 1 & 0 \end{pmatrix} m^{-1}C^{-\sigma}S\right)\right] = 1 \quad \text{for any} \quad S \in \operatorname{Sym}^2(F)$$

if and only if

(6.31) $\operatorname{tr}_{E/F}\left(\lambda^{-\sigma}(1-y)\right) = \operatorname{tr}_{E/F}\left(\lambda^{-\sigma}\eta y\right) = \operatorname{tr}_{E/F}\left(\lambda^{-\sigma}(1+y)\right) = 0,$

$$\text{when} \quad m = \begin{pmatrix} 1 & y \\ 0 & 1 \end{pmatrix},$$

and,

(6.32) $\operatorname{tr}_{E/F}\left(\lambda^{-\sigma}\eta\right) = \operatorname{tr}_{E/F}\left(\lambda^{-\sigma}\right) = 0, \quad \text{when} \quad m = \begin{pmatrix} 1 & -1 \\ 1 & 1 \end{pmatrix}\begin{pmatrix} 1 & 1 \\ 0 & 1 \end{pmatrix}.$

Since $\lambda \neq 0$, the only case when (6.31) or (6.32) holds is when $m = 1_2$ and $\lambda = \eta\ell$ for some $\ell \in F^\times$. Then for any $\ell \in F^\times$, we have

$$\begin{pmatrix} C & 0 \\ 0 & \eta\ell \cdot {}^tC^{-1} \end{pmatrix} = \begin{pmatrix} 1_2 & 0 \\ 0 & \ell \cdot 1_2 \end{pmatrix}\begin{pmatrix} C & 0 \\ 0 & \eta \cdot {}^tC^{-1} \end{pmatrix} \in G_F\begin{pmatrix} C & 0 \\ 0 & \eta \cdot {}^tC^{-1} \end{pmatrix} H_E.$$

4. First let us determine $G_F \cap \mathcal{C}_0 P_E \mathcal{C}_0^{-1}$. Let e_i $(1 \leq i \leq 4)$ denote the column vector whose i-th entry is 1 but the other entries are all zero. Then we have

$$g \in G_F \cap \mathcal{C}_0 P_E \mathcal{C}_0^{-1} \iff g \in G_F \quad \text{and} \quad g(E\mathcal{C}_0 e_1 + E\mathcal{C}_0 e_2) \subset E\mathcal{C}_0 e_1 + E\mathcal{C}_0 e_2$$

where $\mathcal{C}_0 e_1 = e_1$ and $\mathcal{C}_0 e_2 = e_2 + \eta e_4$.

Suppose that $g \in G_F \cap \mathcal{C}_0 P_E \mathcal{C}_0^{-1}$. Then there exist $\alpha, \beta \in E$ such that

$$ge_1 = \alpha e_1 + \beta(e_2 + \eta e_4) = \alpha e_1 + \beta e_2 + \beta\eta e_4.$$

Since $ge_1 \in F^4$, we have $\alpha, \beta, \beta\eta \in F$. Thus $ge_1 = ae_1$, $a \in F^\times$ and g belongs to the maximal parabolic subgroup of G which preserves the one dimensional space spanned by e_1. Similarly there exist $\gamma, \delta \in E$ such that

(6.33) $$g(e_2 + \eta e_4) = \gamma e_1 + \delta(e_2 + \eta e_4).$$

Let us write $\gamma = s + t\eta$, $\delta = u + v\eta$ where $s, t, u, v \in F$. Then (6.23) reads

$$ge_2 + \eta g e_4 = (se_1 + ue_2 + dve_4) + \eta(te_1 + ve_2 + ue_4)$$

and hence

$$ge_2 = se_1 + ue_2 + dve_4, \quad ge_4 = te_1 + ve_2 + ue_4.$$

Thus g is of the form

(6.34)
$$g = \begin{pmatrix} a & 0 & 0 & 0 \\ 0 & u & 0 & v \\ 0 & 0 & b & 0 \\ 0 & dv & 0 & u \end{pmatrix} \begin{pmatrix} 1 & x & 0 & 0 \\ 0 & 1 & 0 & 0 \\ 0 & 0 & 1 & 0 \\ 0 & 0 & -x & 1 \end{pmatrix} \begin{pmatrix} 1 & 0 & y & z \\ 0 & 1 & z & 0 \\ 0 & 0 & 1 & 0 \\ 0 & 0 & 0 & 1 \end{pmatrix} \in G_F.$$

Suppose conversely that $g \in G_F$ is of the form (6.34). Then we have

(6.35)
$$\mathcal{C}_0^{-1} g \mathcal{C}_0 = \begin{pmatrix} a & 0 & 0 & 0 \\ 0 & u+\eta v & 0 & 0 \\ 0 & 0 & b & 0 \\ 0 & 0 & 0 & u-\eta v \end{pmatrix} \begin{pmatrix} 1 & x & 0 & 0 \\ 0 & 1 & 0 & 0 \\ 0 & 0 & 1 & 0 \\ 0 & 0 & -x & 1 \end{pmatrix} \begin{pmatrix} 1 & \eta z & -2\eta y & -\eta z \\ 0 & 1 & -\eta z & 0 \\ 0 & 0 & 1 & 0 \\ 0 & 0 & -\eta z & 1 \end{pmatrix}$$

and surely $g \in G_F \cap \mathcal{C}_0 P_E \mathcal{C}_0^{-1}$.

Now let us consider the (G_F, H_E) double cosets in $G_F \mathcal{C}_0 P_E$. By (6.35), such double cosets are represented by

$$G_F \mathcal{C}_0 \begin{pmatrix} m & 0 \\ 0 & \lambda \cdot {}^t m^{-1} \end{pmatrix} H_E \quad \text{where} \quad m \in B_0(E) \backslash \mathrm{GL}_2(E) / A_E$$

where B_0 denotes the upper Borel subgroup of $\mathrm{GL}(2)$. Thus as we have seen in the proof of Proposition 4.1, we may assume that

$$m = 1_2, w, \quad \text{or} \quad w \begin{pmatrix} 1 & 1 \\ 0 & 1 \end{pmatrix} \quad \text{where} \quad w = \begin{pmatrix} 0 & 1 \\ 1 & 0 \end{pmatrix}.$$

Suppose that $m = 1_2$. Then by (6.35), for $\lambda \in E^\times$, we have

$$\mathcal{C}_0 \begin{pmatrix} 1_2 & 0 \\ 0 & \lambda \cdot 1_2 \end{pmatrix} = \mathcal{C}_0 \begin{pmatrix} 1 & 0 & 0 & 0 \\ 0 & \lambda^\sigma & 0 & 0 \\ 0 & 0 & \lambda \lambda^\sigma & 0 \\ 0 & 0 & 0 & \lambda \end{pmatrix} \mathcal{C}_0^{-1} \mathcal{C}_0 \begin{pmatrix} 1 & 0 & 0 & 0 \\ 0 & \lambda^\sigma & 0 & 0 \\ 0 & 0 & \lambda^\sigma & 0 \\ 0 & 0 & 0 & 1 \end{pmatrix}^{-1} \in G_F \mathcal{C}_0 H_E.$$

Then again by (6.35), we have

$$\Delta(\mathcal{C}_0) = \left\{ (\mathcal{C}_0 h \mathcal{C}_0^{-1}, h) \mid h = a \begin{pmatrix} 1 & 0 & -2\eta y & 0 \\ 0 & 1 & 0 & 0 \\ 0 & 0 & 1 & 0 \\ 0 & 0 & 0 & 1 \end{pmatrix}, a \in F^\times, y \in F \right\}.$$

Thus the condition (6.23) is satisfied and the double coset $G_F \mathcal{C}_0 H_E$ is relevant.

Suppose that $m = w$. Then by (6.35), for $\lambda \in E^\times$, we have

$$\mathcal{C}_0 \begin{pmatrix} w & 0 \\ 0 & \lambda \cdot w \end{pmatrix} = \mathcal{C}_0 \begin{pmatrix} 1 & 0 & 0 & 0 \\ 0 & \lambda^\sigma & 0 & 0 \\ 0 & 0 & \lambda \lambda^\sigma & 0 \\ 0 & 0 & 0 & \lambda \end{pmatrix} \mathcal{C}_0^{-1} \mathcal{C}_0 \begin{pmatrix} w & 0 \\ 0 & w \end{pmatrix} \begin{pmatrix} \lambda^\sigma & 0 & 0 & 0 \\ 0 & 1 & 0 & 0 \\ 0 & 0 & 1 & 0 \\ 0 & 0 & 0 & \lambda^\sigma \end{pmatrix}^{-1}$$

$$\in G_F \mathcal{C}_0 \begin{pmatrix} w & 0 \\ 0 & w \end{pmatrix} H_E.$$

By (6.35), we have

$$H_E \cap \begin{pmatrix} w & 0 \\ 0 & w \end{pmatrix}^{-1} \mathcal{C}_0^{-1} G_F \mathcal{C}_0 \begin{pmatrix} w & 0 \\ 0 & w \end{pmatrix} = \left\{ a \begin{pmatrix} 1 & 0 & 0 & 0 \\ 0 & 1 & 0 & -2\eta y \\ 0 & 0 & 1 & 0 \\ 0 & 0 & 0 & 1 \end{pmatrix} \mid a \in F^\times, y \in F \right\}.$$

Hence the condition (6.23) is satisfied and the double coset $G_F \mathcal{C}_0 \begin{pmatrix} w & 0 \\ 0 & w \end{pmatrix} H_E$ is relevant.

Finally suppose that $m = w \begin{pmatrix} 1 & 1 \\ 0 & 1 \end{pmatrix}$. Put

$$r = \mathcal{C}_0 \begin{pmatrix} w \begin{pmatrix} 1 & 1 \\ 0 & 1 \end{pmatrix} & 0 \\ 0 & \lambda \cdot w \begin{pmatrix} 1 & 0 \\ -1 & 1 \end{pmatrix} \end{pmatrix}.$$

Let us determine $\Delta(r)$. Since $G_F \cap r H_E r^{-1} \subset G_F \cap \mathcal{C}_0 P_E \mathcal{C}_0^{-1}$, we may write $g \in G_F \cap r H_E r^{-1}$ as

$$g = \begin{pmatrix} a & 0 & 0 & 0 \\ 0 & u & 0 & v \\ 0 & 0 & b & 0 \\ 0 & dv & 0 & u \end{pmatrix} \begin{pmatrix} 1 & x & 0 & 0 \\ 0 & 1 & 0 & 0 \\ 0 & 0 & 1 & 0 \\ 0 & 0 & -x & 1 \end{pmatrix} \begin{pmatrix} 1 & 0 & y & z \\ 0 & 1 & z & 0 \\ 0 & 0 & 1 & 0 \\ 0 & 0 & 0 & 1 \end{pmatrix}.$$

Then by a direct computation, we have

$$r^{-1} g r = \begin{pmatrix} \begin{pmatrix} \delta - a\gamma & \delta - a\gamma - a \\ a\gamma & a\gamma + \delta \end{pmatrix} & * \\ 0 & \begin{pmatrix} \delta^\sigma + \gamma\delta^\sigma & -\gamma\delta^\sigma \\ \delta^\sigma - b + \gamma\delta^\sigma & b - \gamma\delta^\sigma \end{pmatrix} \end{pmatrix}$$

where $\gamma = x + \eta z$ and $\delta = u + \eta v$. Hence we have $\gamma = 0$ and $\delta = a = b$. Therefore

$$\Delta(r) = \left\{ (rhr^{-1}, h) \mid h = a \begin{pmatrix} 1_2 & 2\eta\lambda y \begin{pmatrix} -1 & 1 \\ 1 & -1 \end{pmatrix} \\ 0 & 1_2 \end{pmatrix}, a \in F^\times, y \in F \right\}.$$

Here we note that

$$\psi_E \left[2\eta\lambda y \cdot \mathrm{tr}\left(\begin{pmatrix} 0 & 1 \\ 1 & 0 \end{pmatrix} \begin{pmatrix} -1 & 1 \\ 1 & -1 \end{pmatrix} \right) \right] = \psi \left[4y \cdot \mathrm{tr}_{E/F}(\eta\lambda) \right].$$

Thus the double coset $G_F \mathcal{C}_0 \begin{pmatrix} w\begin{pmatrix} 1 & 1 \\ 0 & 1 \end{pmatrix} & 0 \\ 0 & \lambda \cdot w\begin{pmatrix} 1 & 0 \\ -1 & 0 \end{pmatrix} \end{pmatrix} H_E$ is relevant if and only if $\lambda \in F^\times$. □

6.3. Evaluation of the quadratic orbital integral

DEFINITION 6.12. Let Ξ_E be the characteristic function of $\mathcal{K}_E$, the maximal compact subgroup of G_E defined by $\mathcal{K}_E = \mathrm{GSp}_4(\mathcal{O}_E)$.

Then for $\epsilon \in \{1, \varpi\}$, $\lambda \in E^\times$ and

(6.36) $$\begin{cases} y \in E^\times, \ yy^\sigma \neq 1, & \text{when } \epsilon = 1 \\ y \in E^\times, & \text{when } \epsilon = \varpi, \end{cases}$$

let us define the *quadratic orbital integral* $\mathcal{Q}(\epsilon, y, \lambda)$ by

$$(6.37) \quad \mathcal{Q}(\epsilon, y, \lambda) = \Omega\left(\epsilon^{-1} y\right) \int_{G_F} \int_{Z_F \backslash H_E} \Xi_E \left(g \mathcal{C}_\epsilon \begin{pmatrix} \begin{pmatrix} 1 & y \\ 0 & 1 \end{pmatrix} & 0 \\ 0 & \lambda \cdot \begin{pmatrix} 1 & 0 \\ -y & 1 \end{pmatrix} \end{pmatrix} h \right) (\omega \chi_E)(\lambda(g)) \Omega(h) \, dg \, dh.$$

Our goal in this section is to evaluate the quadratic orbital integrals $\mathcal{Q}(\epsilon, y, \lambda)$ explicitly.

For $\lambda \in E^\times$ and $y \in E^\times$ satisfying (6.36), let

$$(6.38) \quad s_\epsilon(y) = \frac{-\epsilon^{-1} y y^\sigma}{1 - \epsilon^{-1} y y^\sigma},$$

$$(6.39) \quad t_\epsilon(y, \lambda) = \frac{\epsilon \lambda \lambda^\sigma}{1 - \epsilon^{-1} y y^\sigma} = -N_{E/F}\left(\epsilon \lambda y^{-1}\right) s_\epsilon(y).$$

Here we note that by Proposition 6.6 and Proposition 6.9, we have the following lemma.

LEMMA 6.13.

$$(6.40)$$

$$G_F \mathcal{C}_\epsilon \begin{pmatrix} \begin{pmatrix} 1 & y \\ 0 & 1 \end{pmatrix} & 0 \\ 0 & \lambda \begin{pmatrix} 1 & 0 \\ -y & 1 \end{pmatrix} \end{pmatrix} H_E = G_F \mathcal{C}_{\epsilon'} \begin{pmatrix} \begin{pmatrix} 1 & y' \\ 0 & 1 \end{pmatrix} & 0 \\ 0 & \lambda' \begin{pmatrix} 1 & 0 \\ -y' & 1 \end{pmatrix} \end{pmatrix} H_E$$

$$\iff \epsilon = \epsilon' \text{ and } s_\epsilon(y) = s_\epsilon(y'), \, t_\epsilon(y, \lambda) = t_\epsilon(y', \lambda').$$

In other words, the two invariants $s_\epsilon(y)$ and $t_\epsilon(y, \lambda)$ determine such double cosets uniquely.

In order to simplify the integral (6.37), first we note the following lemma.

LEMMA 6.14. *For $g \in G_E$, we have*

$$g \in G_F \mathcal{K}_E \iff \theta(g) = g^{-\sigma} g \in \mathrm{Mat}_{4 \times 4}(\mathcal{O}_E).$$

PROOF. It is clear that $g \in G_F \mathcal{K}_E$ implies that $\theta(g) \in \mathcal{K}_E \subset \mathrm{Mat}_{4 \times 4}(\mathcal{O}_E)$. Suppose conversely that $\theta(g) \in \mathrm{Mat}_{4 \times 4}(\mathcal{O}_E)$. Then since

$$\det(\theta(g)) = (\det g)^{-\sigma}(\det g) \in \mathcal{O}_E^\times,$$

we have $\theta(g) \in \mathcal{K}_E$.

By the Iwasawa decomposition, we may write g as

$$g = \begin{pmatrix} 1_2 & X \\ 0 & 1_2 \end{pmatrix} \begin{pmatrix} A & 0 \\ 0 & \lambda \cdot {}^t A^{-1} \end{pmatrix} k$$

where $k \in \mathcal{K}_E$, $X \in \mathrm{Sym}^2(E)$, $A = \begin{pmatrix} a & u \\ 0 & b \end{pmatrix} \in \mathrm{GL}_2(E)$, and, $\lambda \in E^\times$. Now we write $a = \varpi^\ell \varepsilon_a$, $b = \varpi^m \varepsilon_b$, and, $\lambda = \varpi^n \varepsilon_\lambda$, where $\varepsilon_a, \varepsilon_b, \varepsilon_\lambda \in \mathcal{O}_E^\times$. Then by moving

across
$$\begin{pmatrix} \varepsilon_a & 0 & 0 & 0 \\ 0 & \varepsilon_b & 0 & 0 \\ 0 & 0 & \varepsilon_\lambda \varepsilon_b^{-1} & 0 \\ 0 & 0 & 0 & \varepsilon_\lambda \varepsilon_a^{-1} \end{pmatrix} \in \mathcal{K}_E$$

to the right and

$$\begin{pmatrix} \varpi^\ell & 0 & 0 & 0 \\ 0 & \varpi^m & 0 & 0 \\ 0 & 0 & \varpi^{n-\ell} & 0 \\ 0 & 0 & 0 & \varpi^{n-m} \end{pmatrix} \in G_F$$

to the left, respectively, we may assume that $a = b = \lambda = 1$. Then since

$$k^\sigma \theta(g) k^{-1} = \begin{pmatrix} A^{-\sigma} & 0 \\ 0 & {}^t A^\sigma \end{pmatrix} \begin{pmatrix} 1_2 & -X^\sigma + X \\ 0 & 1_2 \end{pmatrix} \begin{pmatrix} A & 0 \\ 0 & {}^t A^{-1} \end{pmatrix} \in \mathcal{K}_E,$$

we have

$$A^{-\sigma} A = \begin{pmatrix} 1 & -u^\sigma + u \\ 0 & 1 \end{pmatrix} \in \mathrm{GL}_2(\mathcal{O}_E).$$

Hence we have $u = x + \eta y$ where $x \in F$ and $y \in \mathcal{O}_F$. Since

$$A = \begin{pmatrix} 1 & x \\ 0 & 1 \end{pmatrix} \begin{pmatrix} 1 & \eta y \\ 0 & 1 \end{pmatrix} \in \mathrm{GL}_2(F) \, \mathrm{GL}_2(\mathcal{O}_E),$$

we may assume that $A = 1_2$. Then similarly

$$\begin{pmatrix} 1_2 & -X^\sigma + X \\ 0 & 1_2 \end{pmatrix} \in \mathcal{K}_E$$

implies that $X = S + \eta T$ where $S \in \mathrm{Sym}^2(F)$ and $T \in \mathrm{Sym}^2(\mathcal{O}_F)$. Then we have

$$\begin{pmatrix} 1_2 & X \\ 0 & 1_2 \end{pmatrix} = \begin{pmatrix} 1_2 & S \\ 0 & 1_2 \end{pmatrix} \begin{pmatrix} 1_2 & \eta T \\ 0 & 1_2 \end{pmatrix} \in G_F \mathcal{K}_E.$$

Thus we have proved the assertion. □

Now we obtain an expression of the quadratic orbital integral as a finite sum of the quadratic Kloosterman sums studied in Chapter 3 as follows.

PROPOSITION 6.15. *Let us write $\lambda = \varpi^{\mathrm{ord}(\lambda)} \varepsilon_\lambda$ and for an integer i, let*

(6.41) $$T_i = T_i(\epsilon, y, \lambda) = \epsilon^{-1} \varpi^{-\mathrm{ord}(\lambda)} \begin{pmatrix} \varpi^i & y \\ y^\sigma & -\epsilon \varpi^{-i} \left(1 - \epsilon^{-1} y y^\sigma \right) \end{pmatrix}.$$

Then we have

(6.42)
$$\mathcal{Q}(\epsilon, y, \lambda) = \Omega\left(\epsilon^{-1} y\right) (\omega \chi_E)(\varpi)^{-\mathrm{ord}(\lambda)} \sum_{\{i \in \mathbb{Z} \mid T_i \in \mathrm{Mat}_{2 \times 2}(\mathcal{O}_E)\}} (-1)^i \mathcal{H}(T_i, \varepsilon_\lambda).$$

PROOF. Since $E^\times = F^\times \mathcal{O}_E^\times$, by moving across matrices of the form

$$\begin{pmatrix} \varepsilon & 0 & 0 & 0 \\ 0 & 1 & 0 & 0 \\ 0 & 0 & 1 & 0 \\ 0 & 0 & 0 & \varepsilon \end{pmatrix} \quad \text{where} \quad \varepsilon \in \mathcal{O}_E^\times,$$

to the right, we may rewrite (6.37) as

(6.43)
$$\mathcal{Q}(\epsilon, y, \lambda) = \Omega\left(\epsilon^{-1} y\right) \sum_{i \in \mathbb{Z}} \int_{\mathcal{X}_i} (\omega \chi_E)(\lambda(g)) \Omega(\varpi)^i \psi_E \left[\mathrm{tr}\left(\begin{pmatrix} 0 & 1 \\ 1 & 0 \end{pmatrix} X\right)\right] dX$$

where $\mathcal{X}_i$ denotes the set of matrices in $\mathrm{Sym}^2(E)$ such that

(6.44)
$$g\, \mathcal{C}_\epsilon \begin{pmatrix} \begin{pmatrix} 1 & y \\ 0 & 1 \end{pmatrix} & 0 \\ 0 & \lambda \begin{pmatrix} 1 & 0 \\ -y & 1 \end{pmatrix} \end{pmatrix} \begin{pmatrix} \varpi^i & 0 & 0 & 0 \\ 0 & 1 & 0 & 0 \\ 0 & 0 & 1 & 0 \\ 0 & 0 & 0 & \varpi^i \end{pmatrix} \begin{pmatrix} 1_2 & X \\ 0 & 1_2 \end{pmatrix} \in \mathcal{K}_E$$

for some $g \in G_F$. When (6.44) holds, we have

$$\lambda(g) \in \varpi^{-i} \lambda^{-1} \mathcal{O}_E^\times \cap F^\times.$$

Hence (6.43) reads

(6.45)
$$\mathcal{Q}(\epsilon, y, \lambda) = \Omega(y)(\omega \chi_E)(\varpi)^{-\mathrm{ord}(\lambda)} \sum_{i \in \mathbb{Z}} \int_{\mathcal{X}_i} (-1)^i \psi_E \left[\mathrm{tr}\left(\begin{pmatrix} 0 & 1 \\ 1 & 0 \end{pmatrix} X\right)\right] dX.$$

By Lemma 6.14, $\mathcal{X}_i$ is the set of matrices in $\mathrm{Sym}^2(E)$ such that

(6.46)
$$\theta \left[\mathcal{C}_\epsilon \begin{pmatrix} \begin{pmatrix} 1 & y \\ 0 & 1 \end{pmatrix} & 0 \\ 0 & \lambda \begin{pmatrix} 1 & 0 \\ -y & 1 \end{pmatrix} \end{pmatrix} \begin{pmatrix} \varpi^i & 0 & 0 & 0 \\ 0 & 1 & 0 & 0 \\ 0 & 0 & 1 & 0 \\ 0 & 0 & 0 & \varpi^i \end{pmatrix} \begin{pmatrix} 1_2 & X \\ 0 & 1_2 \end{pmatrix} \right]$$
$$= \begin{pmatrix} -X^\sigma S_i & S_i^{-\sigma} - X^\sigma S_i X \\ S_i & S_i X \end{pmatrix} \in \mathrm{Mat}_{4 \times 4}(\mathcal{O}_E),$$

where $S_i = -\varepsilon_\lambda^{-\sigma} T_i$. Since $X^\sigma S_i = -\varepsilon_\lambda^{-\sigma} X^\sigma T_i$ and $X^\sigma T_i = {}^t(T_i X)^\sigma$, we have

$$X^\sigma S_i \in \mathrm{Mat}_{2 \times 2}(\mathcal{O}_E) \iff S_i X \in \mathrm{Mat}_{2 \times 2}(\mathcal{O}_E).$$

Also we may write

$$S_i^{-\sigma} - X^\sigma S_i X = -\varepsilon_\lambda \left\{ T_i^{-\sigma} - \left(\varepsilon_\lambda^{-\sigma} X^\sigma\right) T_i \left(\varepsilon_\lambda^{-1} X\right) \right\}.$$

Thus we have $\mathcal{X}_i = \varepsilon_\lambda \mathcal{Z}_{T_i}$ where we recall that

$$\mathcal{Z}_{T_i} = \left\{ Z \in \mathrm{Sym}^2(E) \mid T_i Z \in \mathrm{Mat}_{2 \times 2}(\mathcal{O}_E),\ T_i^{-\sigma} - Z^\sigma T_i Z \in \mathrm{Mat}_{2 \times 2}(\mathcal{O}_E) \right\}.$$

Hence the proposition follows. $\square$

PROPOSITION 6.16. *The orbital integral $\mathcal{Q}(\epsilon, y, \lambda)$ vanishes unless $\mathrm{ord}(s_\epsilon(y))$ and $\mathrm{ord}(t_\epsilon(y, \lambda))$ are both even, and, $\mathrm{ord}(t_\epsilon(y, \lambda)) \leq \min\{0, \mathrm{ord}(s_\epsilon(y))\}$.*

PROOF. Suppose that $\mathcal{Q}(\epsilon, y, \lambda) \neq 0$. Then by (6.43), there exists $i \in \mathbb{Z}$ such that $T_i \in \mathrm{Mat}_{2 \times 2}(\mathcal{O}_E)$. Hence, in particular, we have $|\epsilon^{-1} \varpi^{-\mathrm{ord}(\lambda)} y| \leq 1$, i.e.

(6.47)
$$\mathrm{ord}(y) \geq \mathrm{ord}(\epsilon) + \mathrm{ord}(\lambda).$$

Thus we have $\mathrm{ord}(s_\epsilon(y)) \geq \mathrm{ord}(t_\epsilon(y, \lambda))$.

Let us show that $\mathrm{ord}(t_\epsilon(y, \lambda)) \leq 0$. Since $|\epsilon^{-1} y y^\sigma| \leq |\epsilon \lambda \lambda^\sigma|$ by (6.47), we have

(6.48)
$$|1 - \epsilon^{-1} y y^\sigma| \leq \max\{1, |\epsilon^{-1} y y^\sigma|\} \leq \max\{1, |\epsilon \lambda \lambda^\sigma|\}.$$

Suppose that $|\epsilon\lambda\lambda^\sigma| < |1 - \epsilon^{-1}yy^\sigma|$. Then (6.48) implies that $|\epsilon\lambda\lambda^\sigma| < 1$ and hence

(6.49) $$2\operatorname{ord}(\lambda) + \operatorname{ord}(\epsilon) > 0.$$

Since $|\epsilon^{-1}yy^\sigma| \leq |\epsilon\lambda\lambda^\sigma| < 1$, we have $|1 - \epsilon^{-1}yy^\sigma| = 1$. Now $T_i \in \operatorname{Mat}_{2\times 2}(\mathcal{O}_E)$ implies that

$$\epsilon^{-1}\varpi^{i-\operatorname{ord}(\lambda)} \in \mathcal{O}_E \quad \text{and} \quad \varpi^{-i-\operatorname{ord}(\lambda)}\left(1 - \epsilon^{-1}yy^\sigma\right) \in \mathcal{O}_E,$$

i.e. $\operatorname{ord}(\lambda) + \operatorname{ord}(\epsilon) \leq i \leq -\operatorname{ord}(\lambda)$. This is impossible by (6.49). Hence we have $|\epsilon\lambda\lambda^\sigma| \geq |1 - \epsilon^{-1}yy^\sigma|$, i.e. $|t_\epsilon(y,\lambda)| \geq 1$.

As for the parity of the ordinal, it is clear from (6.39) that their ordinals have the same parity. Now we rewrite T_i as

(6.50) $$T_i = \epsilon^{-1}\varpi^{-\operatorname{ord}(\lambda)} \begin{pmatrix} \varpi^i & y \\ y^\sigma & \varpi^{-i}yy^\sigma s_\epsilon(y)^{-1} \end{pmatrix}.$$

Let $\operatorname{ord}(s_\epsilon(y)) = \ell$. Since the norm map $N_{E/F} : \mathcal{O}_E^\times \to \mathcal{O}_F^\times$ is surjective, there exists $\varepsilon_1 \in \mathcal{O}_E^\times$ such that

$$yy^\sigma s_\epsilon(y)^{-1} = \varpi^{2\operatorname{ord}(y)-\ell}\varepsilon_1\varepsilon_1^\sigma.$$

By Proposition 6.15, we have

(6.51) $$\mathcal{Q}(\epsilon, y, \lambda) = \Omega\left(\epsilon^{-1}y\right)(\omega\chi_E)(\varpi)^{-\operatorname{ord}(\lambda)} \sum_{i=k}^{2\operatorname{ord}(y)-k-\ell} (-1)^i \mathcal{H}(T_i, \varepsilon_\lambda)$$

where $k = \operatorname{ord}(\lambda) + \operatorname{ord}(\epsilon)$. Here we observe that

$$\begin{pmatrix} \varepsilon_1^{-\sigma} & 0 \\ 0 & \varepsilon_1 \end{pmatrix} \begin{pmatrix} 0 & 1 \\ 1 & 0 \end{pmatrix} T_i \begin{pmatrix} 0 & 1 \\ 1 & 0 \end{pmatrix} \begin{pmatrix} \varepsilon_1^{-1} & 0 \\ 0 & \varepsilon_1^\sigma \end{pmatrix}$$

$$= \epsilon^{-1}\varpi^{-\operatorname{ord}(\lambda)} \begin{pmatrix} \varpi^{-i-\ell+2\operatorname{ord}(y)} & y^\sigma \\ y & \varpi^{i+\ell-2\operatorname{ord}(y)}yy^\sigma s_\epsilon(y)^{-1} \end{pmatrix} = T_{-i-\ell+2\operatorname{ord}(y)}^\sigma.$$

Hence by Proposition 3.14, we have

(6.52) $$\mathcal{H}(T_i, \varepsilon_\lambda) = \mathcal{H}(T_{-i-\ell+2\operatorname{ord}(y)}, \varepsilon_\lambda) \quad \text{for} \quad k \leq i \leq 2\operatorname{ord}(y) - k - \ell.$$

Thus

$$\mathcal{Q}(\epsilon, y, \lambda) = \Omega\left(\epsilon^{-1}y\right)(\omega\chi_E)(\varpi)^{-\operatorname{ord}(\lambda)} \sum_{i=k}^{2\operatorname{ord}(y)-k-\ell} (-1)^i \mathcal{H}(T_{-i-\ell+2\operatorname{ord}(y)}, \varepsilon_\lambda)$$

$$= (-1)^\ell \mathcal{Q}(\epsilon, y, \lambda).$$

Hence $\mathcal{Q}(\epsilon, y, \lambda)$ vanishes when ℓ is odd. $\square$

Here we remark that when $\epsilon = \varpi$,

(6.53) $$\operatorname{ord}(1 - \varpi^{-1}yy^\sigma) = \begin{cases} 2\operatorname{ord}(y) - 1, & \text{when } \operatorname{ord}(y) \leq 0 \\ 0, & \text{when } \operatorname{ord}(y) > 0. \end{cases}$$

Thus we have

(6.54) $$\operatorname{ord}(s_\varpi(y)) = \begin{cases} 0, & \text{when } \operatorname{ord}(y) \leq 0 \\ 2\operatorname{ord}(y) - 1, & \text{when } \operatorname{ord}(y) > 0. \end{cases}$$

In particular, $\operatorname{ord}(s_\epsilon(y)) < 0$ implies $\epsilon = 1$ by (6.54). Also, by Proposition 6.16, we have

(6.55) $$\mathcal{Q}(\varpi, y, \lambda) \text{ vanishes unless } \operatorname{ord}(y) \leq 0.$$

We also note that for T_i as (6.41), we have

(6.56) $$\det T_i = -\epsilon^{-1} \varpi^{-2\operatorname{ord}(\lambda)} = -\frac{\varepsilon_\lambda \varepsilon_\lambda^\sigma (1 - s_\epsilon(y))}{t_\epsilon(y, \lambda)}.$$

Now we state our main theorem of this chapter.

THEOREM 6.17. *1. The orbital integral $\mathcal{Q}(\epsilon, y, \lambda)$ vanishes unless $\operatorname{ord}(s_\epsilon(y))$ and $\operatorname{ord}(t_\epsilon(y, \lambda))$ are both even, and, $\operatorname{ord}(t_\epsilon(y, \lambda)) \leq \min\{0, \operatorname{ord}(s_\epsilon(y))\}$.*
2. Suppose that $\operatorname{ord}(t_\epsilon(y, \lambda)) = 0$.
  (a) Suppose that $|1 - s_\epsilon(y)| = 1$ and $\operatorname{ord}(s_\epsilon(y))$ is even. Then $\epsilon = 1$ and we have
  $$\mathcal{Q}(1, y, \lambda) = \Omega(\varpi)^m,$$
  where $\operatorname{ord}(s_1(y)) = 2m$.
  (b) Suppose that $|1 - s_\epsilon(y)| < 1$. Then $\operatorname{ord}(s_\epsilon(y)) = 0$ and we have
  $$\mathcal{Q}(\epsilon, y, \lambda) = |1 - s_\epsilon(y)|^{-1} \cdot \mathcal{K}\ell\left(\frac{2}{1 - s_\epsilon(y)}, -\frac{2\,t_\epsilon(y, \lambda)}{1 - s_\epsilon(y)}\right).$$
3. Suppose that $\operatorname{ord}(s_\epsilon(y)) = 2m$, $\operatorname{ord}(t_\epsilon(y, \lambda)) = -2n$ where $m \geq 0$ and $n > 0$. Let us write $s_\epsilon(y) = \varpi^{2m}\varepsilon_s$ and $t_\epsilon(y, \lambda) = \varpi^{-2n}\varepsilon_t$.
  (a) When $m \geq n$, then $\epsilon = 1$ and we have
  $$\mathcal{Q}(1, y, \lambda) = \Omega(\varpi)^{m+n} q^{2n} \left\{(-1)^n \cdot \mathcal{K}\ell\left(2\varpi^{-n}, -2\varpi^{-n}\varepsilon_t\right) + 1 + q^{-1}\right\}.$$
  (b) When $n > m \geq 0$, we have
  $$\mathcal{Q}(\epsilon, y, \lambda) = \Omega(\varpi)^{m+n} q^{2n} |1 - s_\epsilon(y)|^{-1} (-1)^n$$
  $$\cdot \left\{\mathcal{K}\ell\left(\frac{2\varpi^{-n}}{1 - s_\epsilon(y)}, -\frac{2\varpi^{-n}\varepsilon_t}{1 - s_\epsilon(y)}\right) + (-1)^m \cdot \mathcal{K}\ell\left(\frac{2\varpi^{m-n}}{1 - s_\epsilon(y)}, -\frac{2\varpi^{m-n}\varepsilon_s\varepsilon_t}{1 - s_\epsilon(y)}\right)\right\}.$$
4. Suppose that $\operatorname{ord}(s_1(y)) = -2m$, $\operatorname{ord}(t_1(y, \lambda)) = -2(m+n)$ where $m > 0$ and $n \geq 0$. Let us write $s_1(y) = \varpi^{-2m}\varepsilon_s$, $t_1(y, \lambda) = \varpi^{-2(m+n)}\varepsilon_t$ and $\lambda = \varpi^{\operatorname{ord}(\lambda)}\varepsilon_\lambda$.
  (a) When $n = 0$, we have
  $$\mathcal{Q}(1, y, \lambda) = 1.$$
  (b) When $m \geq n > 0$, we have
  $$\mathcal{Q}(1, y, \lambda) = \Omega(\varpi)^n q^{2n} \left\{(-1)^n \cdot \mathcal{K}\ell\left(2\varpi^{-n}, 2\varpi^{-n}\varepsilon_\lambda\varepsilon_\lambda^\sigma yy^\sigma\right) + 1 + q^{-1}\right\}.$$
  (c) When $n > m$, we have
  $$\mathcal{Q}(1, y, \lambda) = \Omega(\varpi)^n q^{2n} (-1)^n$$
  $$\cdot \left\{\mathcal{K}\ell\left(2\varpi^{-n}, 2\varpi^{-n}\varepsilon_\lambda\varepsilon_\lambda^\sigma yy^\sigma\right) + \mathcal{K}\ell\left(2\varpi^{-n+m}, 2\varpi^{-n+m}\varepsilon_\lambda\varepsilon_\lambda^\sigma yy^\sigma \varepsilon_s^{-1}\right)\right\}.$$

The first assertion has been proved already. We shall prove the other assertions in the following subsections. Before getting into the proof of the theorem we shall prove the following fuctional equation for $\mathcal{Q}(1, y, \lambda)$ as a corollary of the theorem.

COROLLARY 6.18. *Suppose that*
$$\mathrm{ord}\,(s_1(y)) = -2m, \quad \mathrm{ord}\,(t_1(y,\lambda)) = -2(m+n)$$
where $m > 0$ and $n \geq 0$. Then let us take $y' \in E^\times$ and $\lambda' \in E^\times$ such that $s_1(y') = s_1(y)^{-1}$ and $t_1(y',\lambda') = t_1(y,\lambda)s_1(y)^{-1}$.

Then we have
$$\tag{6.57} \mathcal{Q}(1,y,\lambda) = \Omega(\varpi)^{-m}\mathcal{Q}(1,y',\lambda').$$

PROOF. Let $s = s_1(y)$ and $t = t_1(y,\lambda)$. Then we note that
$$\frac{1}{1-s^{-1}} = yy^\sigma, \quad \frac{ts^{-1}}{1-s^{-1}} = -\lambda\lambda^\sigma.$$
Since $y \in \mathcal{O}_E^\times$ and $1 - s^{-1} \in \mathcal{O}_F^\times$ by the condition $s = s_1(y) \notin \mathcal{O}_F$, we have
$$\mathcal{K}\ell\left(2\varpi^{-n}, 2\varpi^{-n}\varepsilon_\lambda\varepsilon_\lambda^\sigma yy^\sigma\right) = \mathcal{K}\ell\left(2\varpi^{-n}yy^\sigma, 2\varpi^{-n}\varepsilon_\lambda\varepsilon_\lambda^\sigma\right)$$
$$= \mathcal{K}\ell\left(\frac{2\varpi^{-n}}{1-s^{-1}}, -\frac{2\varpi^{-n}\varepsilon_t\varepsilon_s^{-1}}{1-s^{-1}}\right)$$
and also
$$\mathcal{K}\ell\left(2\varpi^{-n+m}, 2\varpi^{-n+m}\varepsilon_\lambda\varepsilon_\lambda^\sigma yy^\sigma\varepsilon_s^{-1}\right) = \mathcal{K}\ell\left(2\varpi^{-n+m}yy^\sigma, 2\varpi^{-n+m}\varepsilon_\lambda\varepsilon_\lambda^\sigma\varepsilon_s^{-1}\right)$$
$$= \mathcal{K}\ell\left(\frac{2\varpi^{-n+m}}{1-s^{-1}}, -\frac{2\varpi^{-n+m}\varepsilon_s^{-1}\left(\varepsilon_t\varepsilon_s^{-1}\right)}{1-s^{-1}}\right).$$

Thus the functional equation (6.57) above follows from Theorem 6.17. □

6.3.1. Evaluation of $\mathcal{Q}(\epsilon, y, \lambda)$ when $t_\epsilon(y,\lambda) \in \mathcal{O}_F^\times$ and $s_\epsilon(y) \in \mathcal{O}_F$. First suppose that $|1 - s_\epsilon(y)| = 1$. Then since $1 - s_\epsilon(y) = \left(1-\epsilon^{-1}yy^\sigma\right)^{-1} \in \mathcal{O}_F^\times$, we have $2m = \mathrm{ord}\,(s_\epsilon(y)) = -\mathrm{ord}\,(\epsilon) + 2\,\mathrm{ord}\,(y)$. Hence $\epsilon = 1$ and $\mathrm{ord}\,(y) = m$. Then it is clear in (6.41) that $T_i \in \mathrm{Mat}_{2\times 2}(\mathcal{O}_E)$ if and only if $i = 0$. Since $T_0 \in \mathrm{GL}_2(\mathcal{O}_E)$, we have $\mathcal{Q}(1, y, \lambda) = \Omega(\varpi)^m$ by Theorem 3.16.

Now let us consider the case when $|1 - s_\epsilon(y)| < 1$. Then we have $s_\epsilon(y) \in \mathcal{O}_F^\times$. Hence $|t_\epsilon(y,\lambda)| = |N_{E/F}\left(\epsilon\lambda y^{-1}\right)s_\epsilon(y)| = 1$ implies that $\epsilon\lambda y^{-1} \in \mathcal{O}_E^\times$.

Thus from (6.50), $T_i \in \mathrm{Mat}_{2\times 2}(\mathcal{O}_E)$ if and only if $i = \mathrm{ord}\,(\lambda) + \mathrm{ord}\,(\epsilon)$, and then, the entries of $T_{\mathrm{ord}(\lambda)+\mathrm{ord}(\epsilon)}$ are all in $\mathcal{O}_E^\times$ and its determinant is in $\varpi\mathcal{O}_F$ by (6.56).

Hence by Theorem 3.16, we have
$$\mathcal{Q}(\epsilon, y, \lambda) = |1 - s_\epsilon(y)|^{-1} \cdot \mathcal{K}\ell\left(\frac{-2\,t_\epsilon(y,\lambda)}{\varepsilon_\lambda\varepsilon_\lambda^\sigma(1-s_\epsilon(y))}, \frac{-2\varepsilon_y\varepsilon_y^\sigma\,t_\epsilon(y,\lambda)}{1-s_\epsilon(y)}\right)$$
$$= |1 - s_\epsilon(y)|^{-1} \cdot \mathcal{K}\ell\left(\frac{2\varepsilon_y\varepsilon_y^\sigma\,l_\epsilon(y,\lambda)}{\varepsilon_\lambda\varepsilon_\lambda^\sigma(1-s_\epsilon(y))}, \frac{2\,l_\epsilon(y,\lambda)}{1-s_\epsilon(y)}\right).$$

Here we have
$$\frac{2\varepsilon_y\varepsilon_y^\sigma\,t_\epsilon(y,\lambda)}{\varepsilon_\lambda\varepsilon_\lambda^\sigma(1-s_\epsilon(y))} = -\frac{2\,s_\epsilon(y)}{1-s_\epsilon(y)} \equiv -\frac{2}{1-s_\epsilon(y)} \pmod{\mathcal{O}_F}.$$

Thus
$$\mathcal{Q}(\epsilon, y, \lambda) = |1 - s_\epsilon(y)|^{-1} \cdot \mathcal{K}\ell\left(\frac{2}{1-s_\epsilon(y)}, -\frac{2\,t_\epsilon(y,\lambda)}{1-s_\epsilon(y)}\right).$$

6.3.2. Evaluation of $\mathcal{Q}(\epsilon, y, \lambda)$ **when** $t_\epsilon(y, \lambda) \in F \setminus \mathcal{O}_F$ **and** $s_\epsilon(y) \in \mathcal{O}_F$. For simplicity let us write $s = s_\epsilon(y)$ and $t = t_\epsilon(y, \lambda)$. First we note that by (6.56), we have

$$(6.58) \qquad -2\mathrm{ord}(\lambda) - \mathrm{ord}(\epsilon) = 2n + \mathrm{ord}(1-s).$$

We remark that when $1 - s \in \mathcal{O}_F^\times$, we have $\epsilon = 1$ and $\mathrm{ord}(\lambda) = -n$ from (6.58). Also we note that $\mathrm{ord}(\epsilon)$ and $\mathrm{ord}(1-s)$ have the same parity by (6.58).

Since $\mathrm{ord}(\epsilon \lambda y^{-1}) = -m - n$ from (6.39), we have

$$(6.59) \qquad T_i = \begin{pmatrix} \varpi^{i-k} & \varpi^{m+n}\varepsilon_y \\ \varpi^{m+n}\varepsilon_y^\sigma & \varpi^{-i+2n+k}\varepsilon_y\varepsilon_y^\sigma \varepsilon_s^{-1} \end{pmatrix}$$

where $k = \mathrm{ord}(\lambda) + \mathrm{ord}(\epsilon)$. Thus $T_i \in \mathrm{Mat}_{2\times 2}(\mathcal{O}_E)$ if and only if $k \leq i \leq k + 2n$. Hence

$$(6.60) \qquad \mathcal{Q}(\epsilon, y, \lambda) = \Omega(\varpi)^{m+n} \sum_{i=k}^{k+2n} (-1)^{i+k-\mathrm{ord}(\epsilon)} \mathcal{H}(T_i, \varepsilon_\lambda).$$

Suppose that $m \geq n$. Then since $1 - s \in \mathcal{O}_F^\times$, we have $\epsilon = 1$ and $k = -n$. Thus (6.60) becomes

$$(6.61) \qquad \mathcal{Q}(1, y, \lambda) = \Omega(\varpi)^{m+n} \sum_{i=-n}^{n} (-1)^{i-n} \mathcal{H}(T_i, \varepsilon_\lambda)$$

where

$$\mathcal{H}(T_i, \varepsilon) = q^{2n} \left\{ \mathcal{K}\ell\left(\frac{2\varpi^{i-n}\varepsilon_t}{\varepsilon_\lambda \varepsilon_\lambda^\sigma (1-s)}, \frac{2\varpi^{-i-n}\varepsilon_t \varepsilon_y \varepsilon_y^\sigma \varepsilon_s^{-1}}{1-s}\right) + 1 + q^{-1} \right\}.$$

by Theorem 3.16. Here we note that for $i \neq 0$, we have $\min\{i-n, -i-n\} \leq -2$ and $i - n \neq -i - n$. Thus by Proposition 2.7, we have

$$\mathcal{Q}(1, y, \lambda) = \Omega(\varpi)^{m+n} q^{2n} (1 + q^{-1})$$
$$+ \Omega(\varpi)^{m+n} q^{2n} (-1)^n \mathcal{K}\ell\left(2\varpi^{-n}\varepsilon_t(\varepsilon_\lambda \varepsilon_\lambda^\sigma)^{-1}, 2\varpi^{-n}\varepsilon_t \varepsilon_y \varepsilon_y^\sigma \varepsilon_s^{-1}\right),$$

since $1 - s \equiv 1 \pmod{\varpi^n}$. Since $\varepsilon_t (\varepsilon_\lambda \varepsilon_\lambda^\sigma)^{-1} \varepsilon_y \varepsilon_y^\sigma \varepsilon_s^{-1} = -1$, we have

$$\mathcal{K}\ell\left(2\varpi^{-n}\varepsilon_t(\varepsilon_\lambda \varepsilon_\lambda^\sigma)^{-1}, 2\varpi^{-n}\varepsilon_t \varepsilon_y \varepsilon_y^\sigma \varepsilon_s^{-1}\right) = \mathcal{K}\ell\left(2\varpi^{-n}, -2\varpi^{-n}\varepsilon_t\right).$$

Thus

$$\mathcal{Q}(1, y, \lambda) = \Omega(\varpi)^{m+n} q^{2n} \left\{ (-1)^n \cdot \mathcal{K}\ell\left(2\varpi^{-n}, 2\varpi^{-n}\varepsilon_t\right) + 1 + q^{-1} \right\}.$$

Suppose that $0 \leq m < n$. Then we have

$$\mathcal{H}(T_i, \varepsilon_\lambda) = (-1)^{\mathrm{ord}(1-s)} q^{2n} |1-s|^{-1} \cdot \mathcal{K}\ell\left(\frac{2\varpi^{i-k-2n}\varepsilon_t}{\varepsilon_\lambda \varepsilon_\lambda^\sigma (1-s)}, \frac{2\varpi^{-i+k}\varepsilon_t \varepsilon_y \varepsilon_y^\sigma \varepsilon_s^{-1}}{1-s}\right)$$
$$+ q^{2n} |1-s|^{-1} (-1)^{m+n+\mathrm{ord}(1-s)} \cdot \mathcal{K}\ell\left(\frac{2\varpi^{m-n}\varepsilon_t}{\varepsilon_\lambda \varepsilon_\lambda^\sigma (1-s)}, \frac{2\varpi^{m-n}\varepsilon_y \varepsilon_y^\sigma \varepsilon_t}{1-s}\right)$$

by Theorem 3.16. By a similar argument as above, we obtain

$$\mathcal{Q}(\epsilon, y, \lambda) = \Omega(\varpi)^{m+n} q^{2n} |1-s|^{-1} (-1)^n$$
$$\cdot \left\{ \mathcal{K}\ell\left(\frac{2\varpi^{-n}}{1-s}, -\frac{2\varpi^{-n}\varepsilon_t}{1-s}\right) + (-1)^m \cdot \mathcal{K}\ell\left(\frac{2\varpi^{m-n}}{1-s}, -\frac{2\varpi^{m-n}\varepsilon_s\varepsilon_t}{1-s}\right) \right\}.$$

6.3.3. Evaluation of $\mathcal{Q}(1, y, \lambda)$ when $s_1(y) \in F \setminus \mathcal{O}_F$. First we note that

$$\operatorname{ord}(s_1(y)) = \operatorname{ord}\left(\frac{-yy^\sigma}{1-yy^\sigma}\right) = -2m < 0$$

implies that $\operatorname{ord}(y) = 0$ and $yy^\sigma \in 1 + \varpi^{2m}\mathcal{O}_F^\times$. Thus $\operatorname{ord}(\lambda) = -n$ and hence

(6.62)
$$\mathcal{Q}(1, y, \lambda) = \Omega(\varpi)^n (-1)^n \sum_{i=-n}^{n+2m} (-1)^i \mathcal{H}(T_i, \varepsilon_\lambda)$$

where

$$T_i = \varpi^n \begin{pmatrix} \varpi^i & y \\ y^\sigma & \varpi^{-i} yy^\sigma s_1(y)^{-1} \end{pmatrix}.$$

Suppose that $n = 0$. Then for $0 \leq i \leq 2m$, we have $T_i \in \operatorname{GL}_2(\mathcal{O}_E)$ and hence $\mathcal{H}(T_i, \varepsilon_\lambda) = 1$. Thus

$$\mathcal{Q}(1, y, \lambda) = \sum_{i=0}^{2m} (-1)^i = 1.$$

Suppose that $n > 0$. First we recall that (6.52) asserts

$$\mathcal{H}(T_i, \varepsilon_\lambda) = \mathcal{H}(T_{2m-i}, \varepsilon_\lambda).$$

By Theorem 3.16, each summand $\mathcal{H}(T_i, \varepsilon_\lambda)$ in (6.62) is given as follows.

1. When $i = -n$,
$$\mathcal{H}(T_i, \varepsilon_\lambda) = q^{2n} (-1)^n \cdot \mathcal{K}\ell\left(2\varpi^{-n}, 2\varpi^{-n}\varepsilon_\lambda\varepsilon_\lambda^\sigma yy^\sigma\right).$$

2. When $-n < i \leq 0$,
$$\mathcal{H}(T_i, \varepsilon_\lambda) = q^{2n} \mathcal{K}\ell\left(2\varpi^{i-n}, 2\varpi^{2m-n-i}\varepsilon_\lambda\varepsilon_\lambda^\sigma yy^\sigma \varepsilon_s^{-1}\right)$$
$$+ q^{2n} (-1)^n \mathcal{K}\ell\left(2\varpi^{-n}, 2\varpi^{-n}\varepsilon_\lambda\varepsilon_\lambda^\sigma yy^\sigma\right).$$

3. When $0 < i \leq \min\{m, n-1\}$,
$$\mathcal{H}(T_i, \varepsilon_\lambda) = q^{2n} (-1)^n \cdot \mathcal{K}\ell\left(2\varpi^{-n}, 2\varpi^{-n}\varepsilon_\lambda\varepsilon_\lambda^\sigma yy^\sigma\right)$$
$$+ q^{2n} \cdot \mathcal{K}\ell\left(2\varpi^{i-n}, 2\varpi^{2m-n-i}\varepsilon_\lambda\varepsilon_\lambda^\sigma yy^\sigma \varepsilon_s^{-1}\right).$$

Here we note that $yy^\sigma \in 1 + \varpi^{2m}\mathcal{O}_F^\times$ implies $\operatorname{sgn}(yy^\sigma) = 1$.

4. When $m \geq n$ and $n \leq i \leq m$,
$$\mathcal{Q}(T_i, \varepsilon_\lambda) = q^{2n} (-1)^n \mathcal{K}\ell\left(2\varpi^{-n}, 2\varpi^{-n}\varepsilon_\lambda\varepsilon_\lambda^\sigma yy^\sigma\right) + q^{2n} (1 - q^{-1}).$$

Thus by Proposition 2.7, we have

$$\mathcal{Q}(1, y, \lambda) = \Omega(\varpi)^n q^{2n} (-1)^n \cdot \mathcal{K}\ell\left(2\varpi^{-n}, 2\varpi^{-n}\varepsilon_\lambda\varepsilon_\lambda^\sigma yy^\sigma\right)$$
$$+ \begin{cases} \Omega(\varpi)^n q^{2n} (1 + q^{-1}), & \text{when } m \geq n \\ \Omega(\varpi)^n q^{2n} (-1)^{n+m} \cdot \mathcal{K}\ell\left(2\varpi^{n-m}, 2\varpi^{n-m}\varepsilon_\lambda\varepsilon_\lambda^\sigma yy^\sigma \varepsilon_s^{-1}\right), & \text{when } n > m. \end{cases}$$

Bibliography

[BEW] B. C. Berndt, R. J. Evans and K. S. Williams, *Gauss and Jacobi Sums*, Canadian Mathematical Society Series of Monographs and Advanced Texts, A Wiley-Interscience Publication, John Wiley & Sons, Inc., New York, 1998, xii+583pp.

[B] S. Böcherer, *Bemerkungen über die Drichletreihen von Koecher und Maaß*, Preprint Math. Gottingensis Heft 68, 1986.

[BSP] S. Böcherer and R. Schulze-Pillot, *The Dirichlet series of Koecher and Maaß and modular forms of weight 3/2*, Math. Z. **209** (1992), 273–287.

[Bu] D. Bump, *The Rankin-Selberg method: A survey*, Number Theory, trace formulas and discrete groups (Oslo, 1987) (K. E. Aubert, E. Bombieri and D. Goldfeld, eds.), Academic Press, Boston, MA, 1989, pp. 49–109.

[C] N. Chen, *Positivity of central values of twisted L-functions*, Ph.D. Thesis, Columbia University, New York, 2000.

[CJ] N. Chen and H. Jacquet, *Positivity of quadratic base change L-functions*, Bull. Soc. Math. France **129** (2001), 33–90.

[Ch] U. Christian, *Über Hilbert-Siegelsche Modulformen und Poincaréshcen Reihen*, Math. Ann. **148** (1962), 257–307.

[FS1] M. Furusawa and J. A. Shalika, *The fundamental lemma for the Bessel and Novodvorsky subgroup of* $GSp(4)$, C. R. Acad. Sci. Paris, t. **328** (1999), Série I, 105–110.

[FS2] M. Furusawa and J. A. Shalika, *The fundamental lemma for the Bessel and Novodvorsky subgroup of* $GSp(4)$ *II*, C. R. Acad. Sci. Paris, t. **331** (2000), Série I, 593–598.

[GP1] B. H. Gross and D. Prasad, *On the decomposition of a representation of* SO_n *when restricted to* SO_{n-1}, Canad. J. Math. **44** (1992), 974–1002.

[GP2] B. H. Gross and D. Prasad, *On irreducible representations of* $SO_{2n+1} \times SO_{2m}$, Canad. J. Math. **46** (1994), 930–950.

[Gu1] J. Guo, *On the positivity of the central critical values of automorphic L-functions for* $GL(2)$, Duke Math. J. **83** (1996), 157–190.

[Gu2] J. Guo, *On a generalization of a result of Waldspurger*, Canadian J. Math. **48** (1996), 104–142.

[Gur] N. Gurevich, *The relative trace formula for groups with involution*, Israel J. Math. **121** (2001), 125–141.

[I] H. Iwaniec, *Fourier coefficients of modular forms of half-integral weight*, Invent. math. **87** (1987), 385–401.

[Ja] N. Jacobson, *Basic Algebra* II, W. H. Freeman and Co., San Francisco, CA, 1980.

[J1] H. Jacquet, *Sur un résultat de Waldspurger*, Ann. scient. Éc. Norm. Sup., 4e série **19** (1986), 185–229.

[J2] H. Jacquet, *Sur un résultat de Waldspurger. II*, Compositio Mathematica **63** (1987), 315–389.

[J3] H. Jacquet, *On the nonvanishing of some L-functions*, Proc. Indian Acad. Sci. (Math. Sci.) **97** (1987), 117–155.

[JLR] H. Jacquet, E. Lapid and J. Rogawski, *Periods of automorphic forms*, J. Amer. Math. Soc. **12** (1999), 173–240.

[K] Y. Kitaoka, *Fourier coefficients of Siegel cusp forms of degree 2*, Nagoya Math. J. **93** (1984), 149–171.

[KK] W. Kohnen and M. Kuss, *Some numerical computations concerning the spinor zeta functions in genus 2 at the central point*, Math. Comp. **71** (2002), 1597–1607.

[LN] R. Lidl and H. Niederreiter, *Finite fields*, Second Edition, Encyclopedia of Mathematics and its Applications; v. 20, Cambridge University Press, Cambridge, 1997, xiv+755pp.

[N] M. E. Novodvorsky, *Automorphic L-functions for the symplectic group* $GSp(4)$ *(Proc. Sympos. Pure Math., Oregon State Univ., Corvallis, Ore., 1977)*, Part 2, pp. 87–95, Proc. Sympos. Pure Math., XXXIII, Amer. Math. Soc., Providence, RI, 1979.

[NP] M. E. Novodvorsky and I. I. Piatetski-Shapiro, *Generalized Bessel models for a symplectic group of rank* 2, Mat. Sbornik **90 (132)** (1973), 246–256; English transl. in Math. USSR Sbornik **19** (1973), 243–255.

[PS] I. I. Piatetski-Shapiro and D. Soudry, *On a correspondence of automorphic forms on orthogonal groups of order five*, J. Math. pures appl. **66** (1987), 407–436.

[S1] D. Soudry, *Rankin-Selberg convolutions for* $SO_{2\ell+1} \times GL_n$: *local theory*, Mem. Amer. Math. Soc. **105** (1993), no. 500, vi+100 pp.

[S2] D. Soudry, *On the Archimedean theory of Rankin-Selberg convolutions for* $SO_{2\ell+1} \times GL_n$, Ann. Sci. École Norm. Sup. (4) **28** (1995), no. 2, 161–224.

[T] R. Takloo-Bighash, *L-functions for the p-adic group* $GSp(4)$, Amer. J. Math. **122** (2000), 1085–1120.

[W1] J.-L. Waldspurger, *Sur les coefficients de Fourier des formes modulaires de poids demi-entier*, J. Math. Pures Appl. **60** (1981), 375–484.

[W2] J.-L. Waldspurger, *Sur les valeurs de certaines fonctions L automorphes en leur centre de symetrie*, Compositio Mathematica **54** (1985), 173–242.

[Wi] K. S. Williams, *Notes on Salié sums*, Proc. Am. Math. Soc. **30** (1971), 393–394.

[Y1] Y. Ye, *Kloosterman integrals and base change for* $GL(2)$, J. reine angew. Math. **400** (1989), 57–121.

[Y2] Y. Ye, *The lifting of Kloosterman sums*, J. Number Theory **51** (1995), 275–287.

[Z] D. Zagier, *Modular forms associated to real quadratic fields*, Invent. math. **30** (1975), 1–46.

Editorial Information

To be published in the *Memoirs*, a paper must be correct, new, nontrivial, and significant. Further, it must be well written and of interest to a substantial number of mathematicians. Piecemeal results, such as an inconclusive step toward an unproved major theorem or a minor variation on a known result, are in general not acceptable for publication. Papers appearing in *Memoirs* are generally longer than those appearing in *Transactions*, which shares the same editorial committee.

As of April 1, 2003, the backlog for this journal was approximately 4 volumes. This estimate is the result of dividing the number of manuscripts for this journal in the Providence office that have not yet gone to the printer on the above date by the average number of monographs per volume over the previous twelve months, reduced by the number of volumes published in four months (the time necessary for preparing a volume for the printer). (There are 6 volumes per year, each containing at least 4 numbers.)

A Consent to Publish and Copyright Agreement is required before a paper will be published in the *Memoirs*. After a paper is accepted for publication, the Providence office will send a Consent to Publish and Copyright Agreement to all authors of the paper. By submitting a paper to the *Memoirs*, authors certify that the results have not been submitted to nor are they under consideration for publication by another journal, conference proceedings, or similar publication.

Information for Authors

Memoirs are printed from camera copy fully prepared by the author. This means that the finished book will look exactly like the copy submitted.

The paper must contain a *descriptive title* and an *abstract* that summarizes the article in language suitable for workers in the general field (algebra, analysis, etc.). The *descriptive title* should be short, but informative; useless or vague phrases such as "some remarks about" or "concerning" should be avoided. The *abstract* should be at least one complete sentence, and at most 300 words. Included with the footnotes to the paper should be the 2000 *Mathematics Subject Classification* representing the primary and secondary subjects of the article. The classifications are accessible from www.ams.org/msc/. The list of classifications is also available in print starting with the 1999 annual index of *Mathematical Reviews*. The Mathematics Subject Classification footnote may be followed by a list of *key words and phrases* describing the subject matter of the article and taken from it. Journal abbreviations used in bibliographies are listed in the latest *Mathematical Reviews* annual index. The series abbreviations are also accessible from www.ams.org/publications/. To help in preparing and verifying references, the AMS offers MR Lookup, a Reference Tool for Linking, at www.ams.org/mrlookup/. When the manuscript is submitted, authors should supply the editor with electronic addresses if available. These will be printed after the postal address at the end of the article.

Electronically prepared manuscripts. The AMS encourages electronically prepared manuscripts, with a strong preference for $\mathcal{A}_{\mathcal{M}}\mathcal{S}$-LaTeX. To this end, the Society has prepared $\mathcal{A}_{\mathcal{M}}\mathcal{S}$-LaTeX author packages for each AMS publication. Author packages include instructions for preparing electronic manuscripts, the *AMS Author Handbook*, samples, and a style file that generates the particular design specifications of that publication series. Though $\mathcal{A}_{\mathcal{M}}\mathcal{S}$-LaTeX is the highly preferred format of TeX, author packages are also available in $\mathcal{A}_{\mathcal{M}}\mathcal{S}$-TeX.

Authors may retrieve an author package from e-MATH starting from `www.ams.org/tex/` or via FTP to `ftp.ams.org` (login as `anonymous`, enter username as password, and type `cd pub/author-info`). The *AMS Author Handbook* and the *Instruction Manual* are available in PDF format following the author packages link from `www.ams.org/tex/`. The author package can be obtained free of charge by sending email to `pub@ams.org` (Internet) or from the Publication Division, American Mathematical Society, 201 Charles St., Providence, RI 02904, USA. When requesting an author package, please specify $\mathcal{AMS}$-LaTeX or $\mathcal{AMS}$-TeX, Macintosh or IBM (3.5) format, and the publication in which your paper will appear. Please be sure to include your complete mailing address.

Sending electronic files. After acceptance, the source file(s) should be sent to the Providence office (this includes any TeX source file, any graphics files, and the DVI or PostScript file).

Before sending the source file, be sure you have proofread your paper carefully. The files you send must be the EXACT files used to generate the proof copy that was accepted for publication. For all publications, authors are required to send a printed copy of their paper, which exactly matches the copy approved for publication, along with any graphics that will appear in the paper.

TeX files may be submitted by email, FTP, or on diskette. The DVI file(s) and PostScript files should be submitted only by FTP or on diskette unless they are encoded properly to submit through email. (DVI files are binary and PostScript files tend to be very large.)

Electronically prepared manuscripts can be sent via email to `pub-submit@ams.org` (Internet). The subject line of the message should include the publication code to identify it as a Memoir. TeX source files, DVI files, and PostScript files can be transferred over the Internet by FTP to the Internet node `e-math.ams.org` (130.44.1.100).

Electronic graphics. Comprehensive instructions on preparing graphics are available at `www.ams.org/jourhtml/graphics.html`. A few of the major requirements are given here.

Submit files for graphics as EPS (Encapsulated PostScript) files. This includes graphics originated via a graphics application as well as scanned photographs or other computer-generated images. If this is not possible, TIFF files are acceptable as long as they can be opened in Adobe Photoshop or Illustrator. No matter what method was used to produce the graphic, it is necessary to provide a paper copy to the AMS.

Authors using graphics packages for the creation of electronic art should also avoid the use of any lines thinner than 0.5 points in width. Many graphics packages allow the user to specify a "hairline" for a very thin line. Hairlines often look acceptable when proofed on a typical laser printer. However, when produced on a high-resolution laser imagesetter, hairlines become nearly invisible and will be lost entirely in the final printing process.

Screens should be set to values between 15% and 85%. Screens which fall outside of this range are too light or too dark to print correctly. Variations of screens within a graphic should be no less than 10%.

Inquiries. Any inquiries concerning a paper that has been accepted for publication should be sent directly to the Electronic Prepress Department, American Mathematical Society, 201 Charles St., Providence, RI 02904, USA.

Editors

This journal is designed particularly for long research papers, normally at least 80 pages in length, and groups of cognate papers in pure and applied mathematics. Papers intended for publication in the *Memoirs* should be addressed to one of the following editors. In principle the Memoirs welcomes electronic submissions, and some of the editors, those whose names appear below with an asterisk (*), have indicated that they prefer them. However, editors reserve the right to request hard copies after papers have been submitted electronically. Authors are advised to make preliminary email inquiries to editors about whether they are likely to be able to handle submissions in a particular electronic form.

Algebra to KAREN E. SMITH, Department of Mathematics, University of Michigan, 525 University, Suite 2832, Ann Arbor, MI 48109-1109; email: `kesmith@lsa.umich.edu`

Algebraic geometry and commutative algebra to LAWRENCE EIN, Department of Mathematics, University of Illinois, 851 S. Morgan (M/C 249), Chicago, IL 60607-7045; email: `ein@uic.edu`

Algebraic topology and cohomology of groups to STEWART PRIDDY, Department of Mathematics, Northwestern University, 2033 Sheridan Road, Evanston, IL 60208-2730; email: `priddy@math.nwu.edu`

Combinatorics and Lie theory to SERGEY FOMIN, Department of Mathematics, University of Michigan, Ann Arbor, Michigan 48109-1109; email: `fomin@umich.edu`

Complex analysis and complex geometry to DUONG H. PHONG, Department of Mathematics, Columbia University, 2990 Broadway, New York, NY 10027-0029; email: `phong@math.columbia.edu`

*****Differential geometry and global analysis** to LISA C. JEFFREY, Department of Mathematics, University of Toronto, 100 St. George St., Toronto, ON Canada M5S 3G3; email: `jeffrey@math.toronto.edu`

Dynamical systems and ergodic theory to ROBERT F. WILLIAMS, Department of Mathematics, University of Texas, Austin, Texas 78712-1082; email: `bob@math.utexas.edu`

Functional analysis and operator algebras to DAN VOICULESCU, Department of Mathematics, University of California, Berkeley, 970 Evans Hall, Floor 9, Berkeley, CA 94720-0001; email: `dvv@math.berkeley.edu`

Geometric topology, knot theory and hyperbolic geometry to ABIGAIL A. THOMPSON, Department of Mathematics, University of California, Davis, Davis, CA 95616-5224; email: `thompson@math.ucdavis.edu`

Harmonic analysis to ALEXANDER NAGEL, Department of Mathematics, University of Wisconsin, 480 Lincoln Drive, Madison, WI 53706-1313; email: `nagel@math.wisc.edu`

Harmonic analysis, representation theory, and Lie theory to ROBERT J. STANTON, Department of Mathematics, The Ohio State University, 231 West 18th Avenue, Columbus, OH 43210-1174; email: `stanton@math.ohio-state.edu`

*****Logic** to THEODORE SLAMAN, Department of Mathematics, University of California, Berkeley, CA 94720-3840; email: `slaman@math.berkeley.edu`

Number theory to HAROLD G. DIAMOND, Department of Mathematics, University of Illinois, 1409 W. Green St., Urbana, IL 61801-2917; email: `diamond@math.uiuc.edu`

*****Ordinary differential equations, and applied mathematics** to PETER W. BATES, Department of Mathematics, Michigan State University, East Lansing, MI 48824-1027; email: `peter@math.msu.edu`

*****Partial differential equations** to PATRICIA E. BAUMAN, Department of Mathematics, Purdue University, West Lafayette, IN 47907-1395' email: `bauman@math.purdue.edu`

*****Probability and statistics** to KRZYSZTOF BURDZY, Department of Mathematics, University of Washington, Box 354350, Seattle, Washington 98195-4350; email: `burdzy@math.washington.edu`

*****Real analysis and partial differential equations** to DANIEL TATARU, Department of Mathematics, University of California, Berkeley, Berkeley, CA 94720; email: `tataru@math.berkeley.edu`

All other communications to the editors should be addressed to the Managing Editor, WILLIAM BECKNER, Department of Mathematics, University of Texas, Austin, TX 78712-1082; email: `beckner@math.utexas.edu`.

Titles in This Series

783 **Ethan Akin, Mike Hurley, and Judy A. Kennedy,** Dynamics of topologically generic homeomorphisms, 2003

782 **Masaaki Furusawa and Joseph A. Shalika,** On central critical values of the degree four L-functions for GSp(4): The Fundamental Lemma, 2003

781 **Marcin Bownik,** Anisotropic Hardy spaces and wavelets, 2003

780 **S. Marmi and D. Sauzin,** Quasianalytic monogenic solutions of a cohomological equation, 2003

779 **Hansjörg Geiges,** h-principles and flexibility in geometry, 2003

778 **David B. Massey,** Numerical control over complex analytic singularities, 2003

777 **Robert Lauter,** Pseudodifferential analysis on conformally compact spaces, 2003

776 **U. Haagerup, H. P. Rosenthal, and F. A. Sukochev,** Banach embedding properties of non-commutative L^p-spaces, 2003

775 **P. Lochak, J.-P. Marco, and D. Sauzin,** On the splitting of invariant manifolds in multidimensional near-integrable Hamiltonian systems, 2003

774 **Kai A. Behrend,** Derived ℓ-adic categories for algebraic stacks, 2003

773 **Robert M. Guralnick, Peter Müller, and Jan Saxl,** The rational function analogue of a question of Schur and exceptionality of permutation representations, 2003

772 **Katrina Barron,** The moduli space of $N = 1$ superspheres with tubes and the sewing operation, 2003

771 **Shigenori Matsumoto,** Affine flows on 3-manifolds, 2003

770 **W. N. Everitt and L. Markus,** Elliptic partial differential operators and symplectic algebra, 2003

769 **Jie Wu,** Homotopy theory of the suspensions of the projective plane, 2003

768 **R. Höpfner and E. Löcherbach,** Limit theorems for null recurrent Markov processes, 2003

767 **Po Hu,** S-modules in the category of schemes, 2003

766 **Su Gao and Alexander S. Kechris,** On the classification of Polish metric spaces up to isometry, 2003

765 **Robert Bieri and Ross Geoghegan,** Connectivity properties of group actions on non-positively curved spaces, 2003

764 **J. Spandaw,** Noether-Lefschetz problems for degeneracy loci, 2003

763 **Yasuyuki Kachi and Eiichi Sato,** Segre's reflexivity and an inductive characterization os hyperquadrics, 2002

762 **Leiba Rodman, Ilya M. Spitkovsky, and Hugo Woerdeman,** Abstract band method via factorization, positive and band extensions of multivariable almost periodic matrix functions, and spectral estimation, 2002

761 **Oliver Druet and Emmanuel Hebey,** The AB program in geometric analysis : Sharp Sobolev inequalities and related problems, 2002

760 **Markus Banagl,** Extending intersection homology type invarients to non-Witt spaces, 2002

759 **Donald M. Davis,** From representation theory to homotopy groups, 2002

758 **Alan Forrest, John Hunton, and Johannes Kellendonk,** Topological invariants for projection method patterns, 2002

757 **Douglas Bowman,** q-difference operators, orthogonal polynomials, and symmetric expansions, 2002

756 **José Ignacio Cogolludo-Agustín,** Topological invariants of the complement to arrangements of rational plane curves, 2002

755 **M. A. Mandell and J. P. May,** Equivariant orthogonal spectra and S-modules, 2002

TITLES IN THIS SERIES

754 **Edward L. Green, Idun Reiten, and Øyvind Solberg,** Dualities on generalized Koszul algebras, 2002

753 **Daniel Panazzolo,** Desingularization of nilpotent singularities in families of planar vector fields, 2002

752 **Linus Kramer,** Homogeneous spaces, Tits buildings, and isoparametric hypersurfaces, 2002

751 **Bruce Allison, Georgia Benkart, and Yun Gao,** Lie algebras graded by the root systems BC_r, $r \geq 2$, 2002

750 **Masaki Izumi and Hideki Kosaki,** Kac algebras arising from composition of subfactors: General theory and classification, 2002

749 **Nanhua Xi,** The based ring of two-sided cells of affine Weyl groups of type $\widetilde{A}_{n-1}$, 2002

748 **Jürgen Ritter and Alfred Weiss,** The lifted root number conjecture and Iwasawa theory, 2002

747 **Armand Borel, Robert Friedman, and John W. Morgan,** Almost commuting elements in compact Lie groups, 2002

746 **Peter Niemann,** Some generalized Kac-Moody algebras with known root multiplicities, 2002

745 **Mikhail A. Lifshits and Werner Linde,** Approximation and entropy numbers of Volterra operators with application to Brownian motion, 2002

744 **Roger Chalkley,** Basic global relative invariants for homogeneous linear differential equations, 2002

743 **Heng Sun,** Spectral decomposition of a covering of $GL(r)$: the Borel case, 2002

742 **J. E. Gilbert, Y. S. Han, J. A. Hogan, J. D. Lakey, D. Weiland, and G. Weiss,** Smooth molecular functions and singular integral operators, 2002

741 **Francisco Santos,** Triangulations of oriented matroids, 2002

740 **Rick Durrett,** Mutual invadability implies coexistence in spatial models, 2002

739 **Georgios K. Alexopoulos,** Sub-Laplacians with drift on Lie groups of polynomial volume growth, 2002

738 **Yasuro Gon,** Generalized Whittaker functions on $SU(2,2)$ with respect to the Siegel parabolic subgroup, 2002

737 **Arjen Doelman, Robert A. Gardner, and Tasso J. Kaper,** A stability index analysis of 1-D patterns of the Gray-Scott model, 2002

736 **Wojciech Chachólski and Jérôme Scherer,** Homotopy theory of diagrams, 2002

735 **Martina Brück, Xi Du, Joonsang Park, and Chuu-Lian Terng,** The submanifold geometries associated to Grassmannian systems, 2002

734 **Michel Van den Bergh,** Blowing up of non-commutative smooth surfaces, 2001

733 **Milé Krajčevski,** Tilings of the plane, hyperbolic groups and small cancellation conditions, 2001

732 **Jan O. Kleppe, Juan C. Migliore, Rosa Miró-Roig, Uwe Nagel, and Chris Peterson,** Gorenstein liaison, complete intersection liaison invariants and unobstructedness, 2001

731 **Jesús Bastero, Mario Milman, and Francisco J. Ruiz,** On the connection between weighted norm inequalities, commutators and real interpolation, 2001

730 **Suhyoung Choi,** The decomposition and classification of radiant affine 3-manifolds, 2001

729 **Michael Grosser, Eva Farkas, Michael Kunzinger, and Roland Steinbauer,** On the foundations of nonlinear generalized functions I and II, 2001

For a complete list of titles in this series, visit the AMS Bookstore at **www.ams.org/bookstore/**.